Enhancement of Industrial Energy Efficiency and Sustainability

Enhancement of Industrial Energy Efficiency and Sustainability

Editor

Andrea Trianni

MDPI • Basel • Beijing • Wuhan • Barcelona • Belgrade • Manchester • Tokyo • Cluj • Tianjin

Editor
Andrea Trianni
Faculty of Engineering and IT,
University of Technology Sydney
Australia

Editorial Office
MDPI
St. Alban-Anlage 66
4052 Basel, Switzerland

This is a reprint of articles from the Special Issue published online in the open access journal *Energies* (ISSN 1996-1073) (available at: https://www.mdpi.com/journal/energies/special_issues/ Energy_Efficiency_Sustainability).

For citation purposes, cite each article independently as indicated on the article page online and as indicated below:

LastName, A.A.; LastName, B.B.; LastName, C.C. Article Title. *Journal Name* **Year**, *Volume Number*, Page Range.

ISBN 978-3-0365-1565-6 (Hbk)
ISBN 978-3-0365-1566-3 (PDF)

Contents

About the Editor

Andrea Trianni is a Mechanical and Industrial engineer. His current research activities are focused on improved industrial energy efficiency and sustainability through the investigation of the energy efficiency productivity benefits within industrial activities, the evaluation of barriers and driving forces for the promotion of energy efficiency solutions, and the development of methodologies for energy audit and benchmarking, particularly for small and medium-sized enterprises. Andrea has promoted, led, and managed various energy efficiency projects with major industrial players as well as domestic and international policy-makers. Further, his research and consultancy activities in Europe and Australia revolve around industrial sustainability issues and business models for industrial energy productivity. Andrea has authored more than 70 international publications, and he is a member of several scientific committees as well as editorial boards of international peer-reviewed journals on industrial energy efficiency and sustainability.

Preface to "Enhancement of Industrial Energy Efficiency and Sustainability"

Industrial energy efficiency has been recognized as a major contributor, in the broader set of industrial resources, to improved sustainability and circular economy. Nevertheless, the uptake of energy efficiency measures and practices is still quite low, due to the existence of several barriers. Research has broadly discussed them, together with their drivers.

More recently, many researchers have highlighted the existence of several benefits, beyond mere energy savings, stemming from the adoption of such measures, for several stakeholders involved in the value chain of energy efficiency solutions. Nevertheless, a deep understanding of the relationships between the use of the energy resource and other resources in industry, together with the most important factors for the uptake of such measures—also in light of the implications on the industrial operations—is still lacking. However, such understanding could further stimulate the adoption of solutions for improved industrial energy efficiency and sustainability.

Andrea Trianni
Editor

Article

Energy-Saving Strategies and their Energy Analysis and Exergy Analysis for In Situ Thermal Remediation System of Polluted-Soil

Tian-Tian Li [1], Yun-Ze Li [1,2,3,*], Zhuang-Zhuang Zhai [4], En-Hui Li [1] and Tong Li [5]

[1] School of Aeronautic Science and Engineering, Beihang University, Beijing 100191, China; litiantian@buaa.edu.cn (T.-T.L.); lienhui@buaa.edu.cn (E.-H.L.)

[2] Institute of Engineering Thermophysics, North China University of Water Resources and Electric Power, Henan 450045, China

[3] Advanced Research Center of Thermal and New Energy Technologies, Xingtai Polytechnic College, Hebei 054035, China

[4] School of Automation Science and Electrical Engineering, Beihang University, Beijing 100191, China; zhaizz@buaa.edu.cn

[5] Chengyi Academy of PKUHS, Peking University, Beijing 100080, China; litong@i.pkuschool.edu.cn

* Correspondence: liyunze@buaa.edu.cn; Tel.: +86-10-82338778; Fax: +86-10-82315350

Received: 8 October 2019; Accepted: 20 October 2019; Published: 22 October 2019

Abstract: The environmental safety of soil has become a severe problem in China with the boost of industrialization. Polluted-soil thermal remediation is a kind of suitable remediation technology for large-scale heavily contaminated industrial soil, with the advantages of being usable in off-grid areas and with a high fuel to energy conversion rate. Research on energy-saving strategies is beneficial for resource utilization. Focused on energy saving and efficiency promotion of polluted-soil in situ thermal remediation system, this paper presents three energy-saving strategies: Variable-condition mode (VCM), heat-returning mode (HRM) and air-preheating mode (APM). The energy analysis based on the first law of thermodynamics and exergy analysis based on the second law of thermodynamics are completed. By comparing the results, the most effective part of the energy-saving strategy for variable-condition mode is that high savings in the amount of natural gas (NG) used can be achieved, from 0.1124 to 0.0299 kg·s^{-1} in the first stage. Energy-saving strategies for heat-returning mode and air-preheating mode have higher utilization ratios than the basic method (BM) for the reason they make full use of waste heat. As a whole, a combination of energy-saving strategies can improve the fuel savings and energy efficiency at the same time.

Keywords: contaminated soil; polluted soil; thermal desorption; thermal remediation; energy analysis and exergy analysis; energy saving

---

1. Introduction

Soil is the basic environmental element constituting the ecosystem, and the important material basis of human survival and development. The environmental safety of soil has become a severe problem in China with the boost of industrialization and urbanization. It was calculated that the amount of contaminated soil reached about 150 million mu up to 2012 [1]. Recent estimates indicate that 500,000 sites in Europe require cleanup, while nearly 3.5 million sites are potentially polluted [2]. Including heavy metals, soil contamination caused by so many contaminants is an urgent problem. It can be seen from the bulletin on Chinese domestic environmental conditions for the year 2000 that the heavy metals in 36,000 hectares of soil were out of limits in the surveyed 0.3 million hectares of soil and the over standard rate reached 12.1% of the total [3]. The prevention of contaminated soil is not only needed to control the sources such as heavy metals, but also enhance the remediation of contaminated soil [4]. In the last 30 years since 2013, more than 80,000 sites have been cleaned up in the European countries where data on remediation are available [5].

In situ thermal remediation is a kind of suitable remediation technology for heavily contaminated soil [6,7]. Thermal desorption removes pollutants from soil and other materials by using heat to change the chemicals into gases and speed up the cleanup of many pollutants from the ground [8–10]. All the soil contamination remediation mechanisms have their advantages and limitations. Moreover, they are contaminant specific and heavily dependent on the subsurface environmental conditions of the site [11]. In situ thermal remediation remedies contaminated soil on the contaminated site without excavation. Compared with ex situ thermal desorption (ESTD), it has the advantages of low investment and little impact on the surrounding environment, so it is a hotspot of soil remediation research [12–14]. In situ thermal remediation is a soil remediation process in which heat and vacuum are applied simultaneously to subsurface soils [15]. Volatile and semi-volatile organics are removed from contaminated soil in thermal desorbers at 100 to 300 °C for low temperature thermal desorption, or at 300 to 550 °C for high-temperature thermal desorption [16]. In the past decade, it has been applied at a number of sites and it has been used in various modes including surface heating with blankets, subsurface heating with an array of vertical heater/vacuum wells, and ex situ blankets [15]. During the remediation process, gases at high temperature (700–800 °C), coming from the combustion chamber, circulate within the heating elements, resulting in the heating of the soil and the evaporation of volatile pollutants (boiling point < 550 °C) contained in the soil [17]. Laboratory treatability studies and field project experience have confirmed that the combination of high temperature and long-time results in extremely high overall removal efficiency, even for high boiling point contaminants. Both thermal wells and thermal blankets have been demonstrated to be highly effective in removing a wide variety of low and high boiling point hydrocarbons, PCBs, pesticides, and chlorinated solvents from soils [15]. Shallow soil contamination (less than three feet deep) may be treated by thermal blankets or horizontal wells [11,18]. For soil contamination at depths greater than 3 feet, heating with surface blankets is ineffective and thermal wells are needed to attain high temperatures in the soil [15].

Figure 1 presents a general description of a traditional in situ thermal remediation system, that is a polluted-soil thermal remediation system including burner, pipe, well and soil. As Figure 1a shows, natural gas (NG) and air enter the burner through different inlets and an air-NG mixture is delivered to the burner, in which chemical energy of natural gas (NG) is converted to thermal energy in the exhaust gas by burning. The high temperature exhaust gas produced by the burner flows through the pipe into the heating well inserted vertically in the soil. The well is the heat transfer component of the whole system, in which the high temperature exhaust gas flows transferring heat to the soil to raise the soil temperature through the walls of the well. The volatile pollutants contained in the soil will then evaporate. As shown in Figure 1b, the gas flows directly in the system and is eventually discharged into the environment without recovery or recycling. In such a flow, the energy in the flowing gas is used only once to heat the soil. From the point of view of energy utilization, this is undoubtedly a huge waste.

At present, there are many studies on soil contamination, mainly about remediation methods, such as thermal desorption, chemical oxidation, phytoremediation etc. [19–25], assessment of contaminated soil [26], the process of soil contamination [27], areas for contaminated soil remediation, etc. [28]. However, few studies have focused on the energy saving and efficiency promotion of thermal desorption using natural gas (NG). Thus, it is very significant to analyze the energy loss and energy utilization ratio of the polluted-soil thermal remediation system. The 2008 gas flaring estimate of 139 billion cubic meters represents 21% of the natural gas consumption of the USA with a potential retail market value of $68 billion and the 2008 flaring added more than 278 million metric tons of carbon dioxide equivalents (CO_2e) into the atmosphere. That is to say, improved utilization of the gas is key to reducing global carbon emissions to the atmosphere [29].

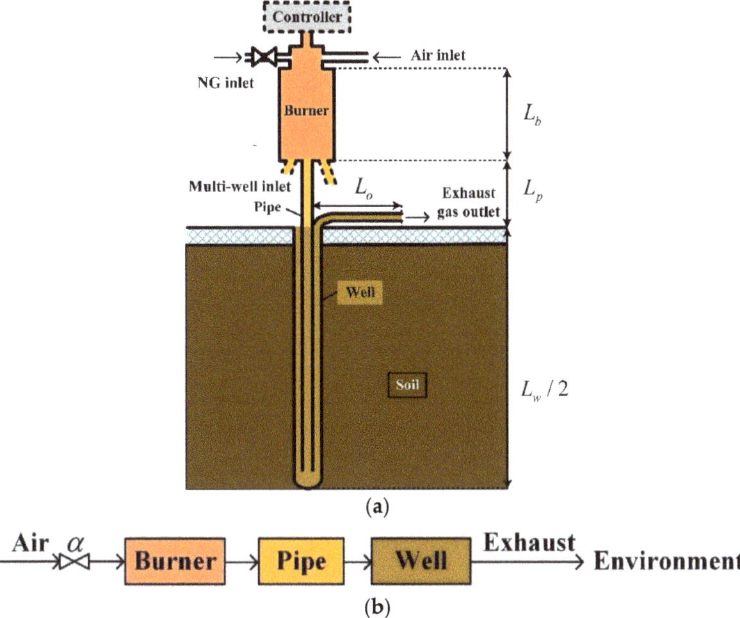

Figure 1. System diagram of polluted-soil thermal remediation system: (**a**) Structure diagram of polluted-soil thermal remediation system including burner, pipe, well and soil; (**b**) flowchart of air distribution in polluted-soil thermal remediation system.

Energy plays an important role in the history of human development [30]. In recent decades economic growth and increased human wellbeing around the globe have come at the cost of fast growing natural resource use (including materials and energy) and carbon emissions, leading to converging pressures of declining resource security, rising and increasingly volatile natural resource prices, and climate change [31]. Emissions of carbon dioxide from the combustion of fossil fuels, which may contribute to long-term climate change [32]. In recent decades, China has encountered serious environmental problem [33]. Some heavy industries and manufacturing enterprises are still characterized by extensive growth, facing enormous environmental challenges due to global climate change, rapid exhaustion of various non-renewable resources, and must improve their energy-save and emission-abate technology to favor the sustainable development [34–36]. Policies should aim to increase the efficiency of energy use [37].

In the energy system, energy analysis based on the first law of thermodynamics and exergy analysis based on the second law of thermodynamics are commonly used. The energy analysis is focused on the quantity of energy and the exergy analysis is focused on the quality of energy. Numerous studies have used these methods, such as the novel combined cooling, heating, and power (CCHP) system [38,39], ground source heat pumps [40], and exhaust waste heat recovery systems [41] and so on.

In the traditional polluted-soil thermal remediation system, the constant high temperature of exhaust is used to heat the soil with changing temperature and the exhaust is discharged directly into the atmosphere, which is disadvantageous for saving energy. Therefore, this paper is aimed at improving the existing problems in the traditional system, and so three energy-saving strategies were researched.

This paper proposes three energy saving strategies of polluted-soil thermal remediation system—variable-condition mode (VCM), heat-returning mode and air-preheating mode—and their thermal performance and efficiency are discussed by energy analysis and exergy analysis.

The mathematic models of a polluted-soil thermal remediation system including burner, pipe, well and soil for energy and exergy analysis are built based on thermodynamics, heat transfer and fluid mechanics. Keeping the energy (exergy) at the inlet to the system constant, and various energy (exergy) losses and energy (exergy) utilization ratios at different stages are calculated. The results are graphically formed to compare the energy-saving strategies with the basic method (BM) and to find where the specific embodiment of energy savings is.

2. Idea of Energy-Saving Strategies of Polluted-Soil Thermal Remediation System

The three energy-saving strategies are presented to improve on traditional systems as shown in Figure 1, and the environment is the same in the research except for the system. The area of soil researched in the paper is 3 meters long, 3 meters wide and 6 meters deep. The following Sections 2.1–2.3 introduce the three energy-saving strategies, respectively.

2.1. Description of Energy-Saving Strategy for Variable-Condition Mode

Energy-saving strategy for variable-condition mode (VCM) involves different exhaust gas temperatures used at different stages. The process of polluted-soil thermal remediation is divided into three stages lasting for 15, 20 and 10 days, respectively, in the study. In the first stage, the soil temperature rises from the initial temperature to the boiling point of water, and the soil moisture content is the initial moisture content. The second stage is the evaporation stage of water in the soil, and the soil keeps the temperature of boiling point of water unchanged. The third stage is to heat dry soil without water to increase the soil temperature to the final temperature. Therefore, the soil temperature is different as well as the soil heating requirements in the three stages, but in the basic method (BM) in use, the exhaust gas temperature at each stage of heating the soil is constant, that is, as shown in Figure 2, the constant high temperature of exhaust used to heat the soil with changing temperature, which is disadvantageous for saving energy. To solve the problem, variable-condition mode (VCM) is necessary, that is, different exhaust gas temperatures are used at different stages. In modeling and analysis, the most direct reflection is that the temperature inside the burner to the temperature outside the heating well are all different at three stages. The contrastive temperature configurations of variable-condition mode (VCM) and the basis method (BM) are presented in Table 1. In the variable-condition mode (VCM), the airflow circulation in polluted-soil thermal remediation system is the same as that in the basis method (BM), as shown in Figure 1b.

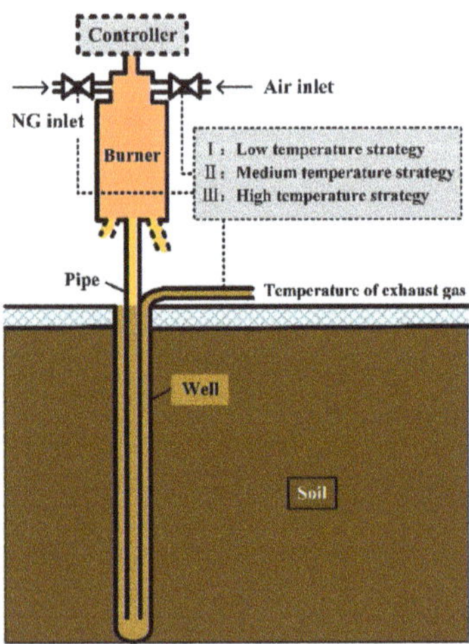

Figure 2. Structure diagram of polluted-soil thermal remediation system using the energy-saving strategy for variable-condition mode (VCM).

Table 1. Temperature configurations of VCM (variable-condition mode) and BM (basis method).

Strategy	Stage	t_b (°C)	$t_{b,out}$ (°C)	$t_{w,in}$ (°C)	t_{wout} (°C)	t_s (°C)	$t_{s,e}$ (°C)
BM	I	950	700	600	450	50	30
	II	950	700	600	450	100	80
	III	950	700	600	450	250	200
VCM	I	750	500	450	200	50	30
	II	800	550	500	300	100	80
	III	1050	800	750	600	250	200

Based on data from engineering practice and a preliminary estimate of the combustion process, the temperature of soil and the temperature in burner in different stage are set in Table 1. The outlet gas temperature of the heating well $t_{w,out}$ is 450 °C in BM, which is also a temperature often used in engineering practice. In VCM $t_{w,out}$ is the main way to achieve variable conditions to save energy, and it is set by the authors for the case.

2.2. Description of Energy-Saving Strategy for Heat-Returning Mode

The energy-saving strategy for heat-returning mode is returning the heat contained in the exhaust to the polluted-soil thermal remediation system again. In the basic method (BM), the exhaust containing a considerable amount of heat is discharged directly into the atmosphere and that is a great waste. To solve the problem, heat-returning mode is necessary, that is, the exhaust from the outlet of the heating well directly discharged to the environment is returned to the burner as the air in a certain proportion, and three schemes are made according to the different proportion of return gas. The rate of return gas is the rate of heat return β. The return air enters the burner from air inlet 2, and the amount of air required for combustion to remove this part is the amount of normal air required from air inlet. The airflow circulation of energy-saving strategy for heat-returning mode in polluted-soil thermal remediation system is different from that in the basis method (BM), as shown in Figure 3b.

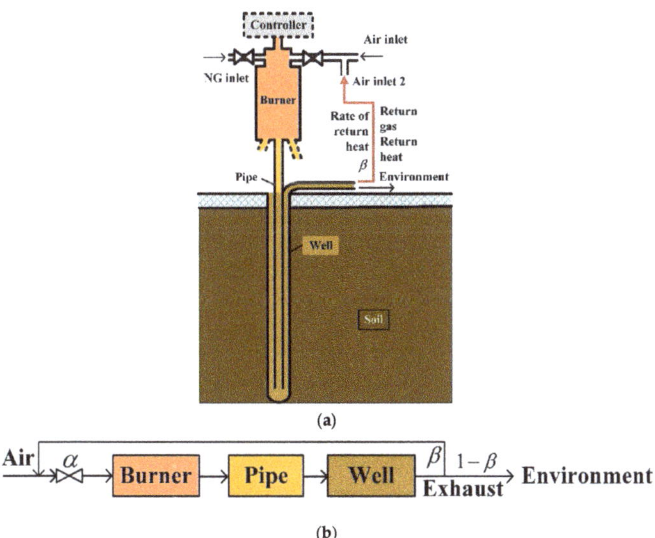

Figure 3. System diagram of polluted-soil thermal remediation system using the energy-saving strategy for heat-returning mode: (**a**) Structure diagram of polluted-soil thermal remediation system using the energy-saving strategy for heat-returning mode; (**b**) flowchart of air distribution in polluted-soil thermal remediation system using the energy-saving strategy for heat-returning mode.

2.3. Description of Energy-Saving Strategy for Air-Preheating Mode

The energy-saving strategy for air-preheating mode is to use the residual heat of the system to –preheat the air entering the burner for combustion. In the basic method (BM), heat from high-temperature parts directly exposed to the environment in the system is wasted and the residual heat can be used up. To solve the problem, preheaters for air-preheating mode are set. As shown in Figure 4, the air to be introduced into the burner is divided into three parts: The first part passes through preheater 1 between the burner and the inlet of heating well, the second part passes through preheater 2 at the outlet pipe of the heating well, and the third part enters the burner directly. Three schemes are designed according to different preheating ratio to different preheaters. The preheating ratio of air through preheater 1 is α_1, preheating ratio of air through preheater 2 is α_2 and the ratio of air that does not pass through the preheater directly into the burner is α_3. The airflow circulation of energy-saving strategy for air-preheating mode in polluted-soil thermal remediation system is different from that in the basis method (BM), as shown in Figure 4b.

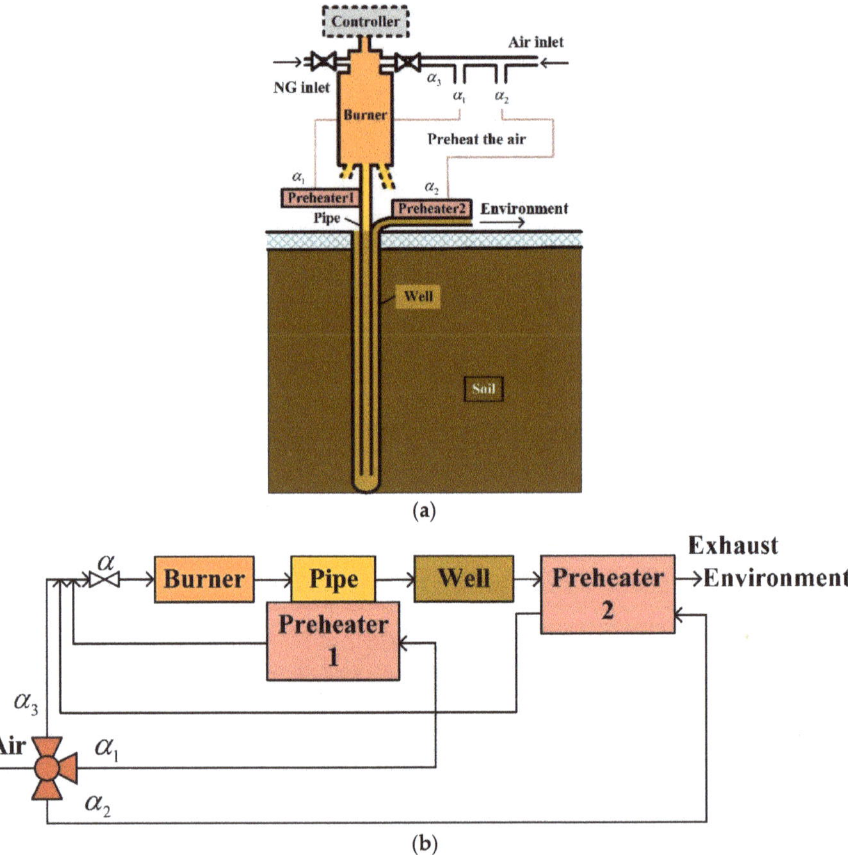

Figure 4. System diagram of polluted-soil thermal remediation system using the energy-saving strategy for air-preheating mode: (**a**) Structure diagram of polluted-soil thermal remediation system using the energy-saving strategy for air-preheating mode; (**b**) flowchart of air distribution in polluted-soil thermal remediation system using the energy-saving strategy for air-preheating mode.

3. Mathematic Models and Parameters Calculation Process

Mathematical models of the polluted-soil thermal remediation system established in this section are used to support the thermal performance analysis of energy-saving strategies. The thermal performance analysis includes an energy analysis based on the first law of thermodynamics and an exergy analysis based on the second law of thermodynamics, so the models are divided into two parts: Section 3.2 presents the energy analysis model and Section 3.3 the exergy analysis model. Energy utilization ratio and exergy utilization ratio, as the key parameters to evaluate the energy-saving strategies, are calculated at the end of the models in Sections 3.2.5 and 3.3.5. Before the specific model, the balance equation is indispensable. The following assumptions are made in the energy and exergy analysis:

(a) The soil is homogeneous and values of physical parameters of the soil remain unchanged in the heat transfer process at the same stage;
(b) The flow of fluid in porous media is called seepage, and the influence of seepage in soil, that is, water migration, was ignored;
(c) The influence of surface temperature fluctuation and depth of buried pipe on soil temperature was ignored, and the soil temperature was considered uniform in the initial stage.

The basic mathematical models in the energy-saving strategies are the same as the basic method (BM), except that the energy and exergy of the air entering the burner are different. In the calculation, paying attention to these parameters is the crucial key of the research. The process of parameters calculation is in the Section 3.4. The value of physical parameters used in the models is shown in Table A1.

3.1. Balance Models

The balance models are based on energy loss and exergy loss of each component in the process of energy flow and exergy flow. Figure 5 shows the energy loss of each component of the polluted-soil thermal remediation system. At the beginning of the energy flow throughout the system, the natural gas (NG) and air carry energy through their respective pipes into the burner. When the gas flows through the pipeline, there are throttling and friction process in the flow, which cause an energy loss. Throttling is a local flow loss, while friction is a path loss of flow. In reality, as long as there is flow in the pipeline, there will be flow loss, and as long as there is a pipe with fluid exposed to the environment, there will be heat leakage loss.

In the burner, incomplete combustion caused by inadequate mixes of fuel and air or the low temperature in the combustor cause energy losses. There are also heat leakage, air leakage and flow loss in the burner. After the energy loss is removed, the remaining energy flows out of the burner and through the pipe into the heating well. There are heat leakage and flow loss in the pipe. Local flow loss exists in the heating well because of the bent pipe. Part of the energy flowing to the heating well is transferred to the soil, heating it. The remaining energy is discharged directly to the environment by the outlet of the heating well through high-temperature exhaust gas, resulting in the maximal energy loss of the whole system. In addition to heating up the soil, the energy in the soil will also lose heat to the surrounding non-heating soil zone and to the air through the surface insulation layer.

Figure 6 shows the exergy loss of each component of the polluted-soil thermal remediation system. Energy loss is accompanied by exergy loss, so all of the energy loss described above has the consequent loss of exergy, including incomplete combustion, heat leakage, flow leakage and so on. Besides, Irreversible combustion, heat transfer, non-isothermal heat release and non-isothermal heat absorption also cause the exergy loss. Consequently, the energy and exergy balance of each component are modeled as shown in Sections 3.1.1 and 3.1.2, respectively.

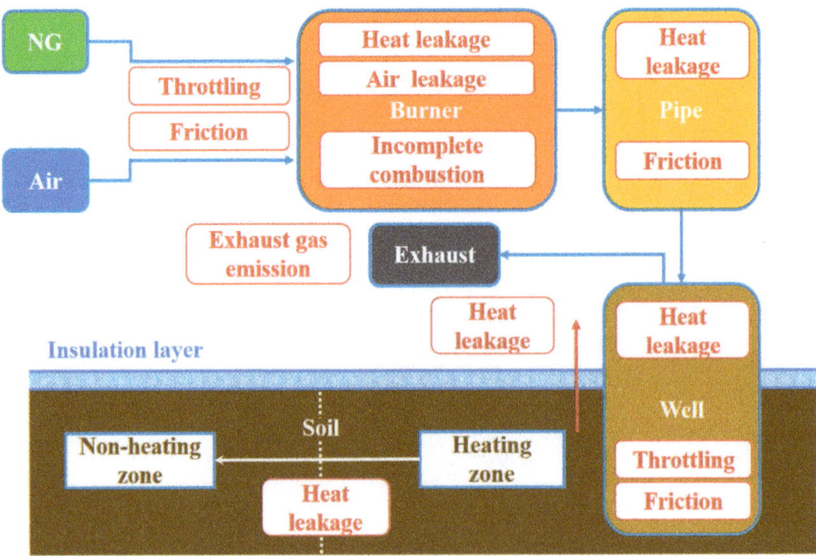

Figure 5. The locations of energy loss of components of polluted-soil thermal remediation system.

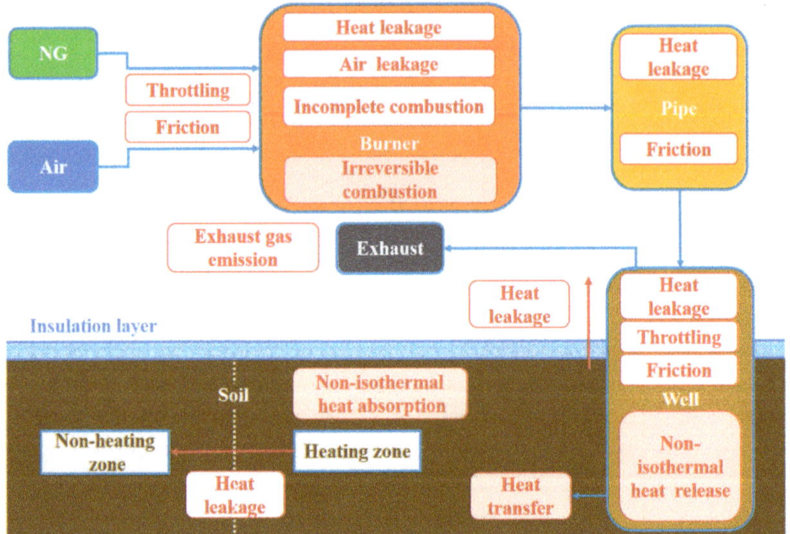

Figure 6. The locations of exergy loss of components of polluted-soil thermal remediation system.

3.1.1. Energy Balance Models

Based on the balance of energy principle and energy loss of each component described in Figure 5, the energy balance equation is established as follows. Equations (1)–(4) are the energy balance models of the burner, pipe, well and soil separately:

$$Q_{ar,net} + Q_{air} = Q_{b,to,p} + Q_{b,inc} + Q_{b,l} + Q_{b,f} \tag{1}$$

$$Q_{p,in} = Q_{b,to,p} = Q_{p,to,w} + Q_{p,l} + Q_{p,f} \tag{2}$$

$$Q_{w,in} = Q_{p,to,w} = Q_{w,to,s} + Q_{w,out} + Q_{w,l} + Q_{w,f} \tag{3}$$

$$Q_{s,in} = Q_{w,to,s} = Q_{s,a} + Q_{s,l} \tag{4}$$

3.1.2. Exergy Balance Models

The exergy balance models is similar to the energy balance model, based on the balance of exergy principle and exergy loss of each component described in Figure 6. Equations (5)–(8) are the exergy balance models of the burner, pipe, well and soil, respectively:

$$E_r + E_{air} = E_{b,to,p} + E_{b,irr} + E_{b,inc} + E_{b,l} + E_{b,f} \tag{5}$$

$$E_{p,in} = E_{b,to,p} = E_{p,to,w} + E_{p,l} + E_{p,f} \tag{6}$$

$$E_{w,in} = E_{p,to,w} = E_{w,to,s} + E_{x,Q} + E_{x,Q_1} + E_{w,out} + E_{w,l} + E_{w,f} \tag{7}$$

$$E_{s,in} = E_{w,to,s} = E_{s,a} + E_{x,Q_2} + E_{s,l} \tag{8}$$

3.2. Energy Analysis Model

One kilogram of natural gas (NG) is the total energy source of the system in the research and the study about the energy flow and the energy loss is started with the energy of one kilogram of natural gas (NG). The mass flow rates used in modeling is calculated in Appendix B (a). The convective heat transfer coefficient used in heat leakage modeling is calculated in Appendix B (b). The Reynolds number used in coefficient of path energy loss modeling is calculated in Appendix B (b) as well. The length of each component used in path energy loss modeling is shown in Figure 1 and the value of them is presented in Table A1. The specific energy analysis models of four components are as follows.

3.2.1. Burner

(a) $Q_{ar,net}$ is the lower calorific value of natural gas (NG), according to the value of the Utility Boiler Manual [42]:

$$Q_{ar,net} = 50200kJ \tag{9}$$

(b) Q_{air} is the energy of air and the value is approximately zero:

$$Q_{air} = 0 \tag{10}$$

(c) Energy loss of incomplete combustion is the product of incomplete combustion coefficient ε and the lower calorific value of natural gas (NG) $Q_{ar,net}$. In the calculation, the value of incomplete combustion coefficient ε is 0.3:

$$Q_{b,inc} = \varepsilon Q_{ar,net} \tag{11}$$

(d) The calculation of energy loss of heat leakage of burner is abstracted as a mathematical model of the heat transfer process of a cylinder tube with gas flowing in air, so are the energy loss of heat leakage of pipe and the extended part of well, as shown in Figure 5. The calculation of heat leakage energy is based on Fourier's Law and Newton's Law of Cooling of heat transfer theory:

$$Q_{b,l} = \frac{2\pi(t_{f,b} - t_0)L_b}{\frac{2}{h_{b,1}d_{b,1}} + \frac{1}{\lambda_b}\ln\frac{d_{b,2}}{d_{b,1}} + \frac{2}{h_{b,2}d_{b,2}}} \times \frac{1}{G_{NG}} \times 10^{-3} \tag{12}$$

$$Q_{p,l} = \frac{2\pi(t_{f,p} - t_0)L_p}{\frac{2}{h_{p,1}d_{p,1}} + \frac{1}{\lambda_p}\ln\frac{d_{p,2}}{d_{p,1}} + \frac{2}{h_{p,2}d_{p,2}}} \times \frac{1}{G_{NG}} \times 10^{-3} \tag{13}$$

$$Q_{w,l} = \frac{2\pi(t_{f,w} - t_0)L_o}{\frac{2}{h_{w,1}d_{w,1}} + \frac{1}{\lambda_w}\ln\frac{d_{w,2}}{d_{w,1}} + \frac{2}{h_{w,2}d_{w,2}}} \times \frac{1}{G_{NG}} \times 10^{-3} \tag{14}$$

The influence of air leaks is ignored. The convective heat transfer coefficient is different when there's wind and there's no wind. So the calculation of the convective heat transfer coefficient outside the tube is divided into forced convection and natural convection.

(e) Energy loss of flow $Q_{b,f}$ concludes path energy loss and local energy loss, the calculation is based on the algorithm in fluid mechanics:

$$Q_{b,f} = g(\varsigma_b\frac{L_b}{d_{b,1}}\frac{u_b^2}{2g} + 2\xi_b\frac{u_b^2}{2g}) \times \frac{G_e}{G_{NG}} \times 10^{-3} \tag{15}$$

The flow velocity u_b is calculated by mass flow rate and pipe diameter.

The flow in the tube is in the turbulent smooth zone in reality, so coefficient of path energy loss of the burner is determined by Equation (16), the flow in pipe and well as well. Therefore, the calculation equation of the coefficient of local energy loss of pipe and well will not be repeated below:

$$\varsigma_b = \frac{0.3164}{\text{Re}_b^{0.25}} \tag{16}$$

A right angle loss and a valve loss are considered for coefficient of local energy loss of the burner. The coefficient of local energy loss of the limit value of a pipe section expansion is 1 and a valve opening of 50% is 1.8. And the value of local loss coefficient is different for different components as show in Figure 1:

$$\zeta_b = 1 + 1.8 \tag{17}$$

(f) Energy out of the burner $Q_{b,to,p}$ is solved by the conservation of energy equation. The calculation of the energy out of the pipe, well and soil is similar, so the equation will not be repeated below:

$$Q_{b,to,p} = Q_{ar,net} + Q_{air} - Q_{b,inc} - Q_{b,l} - Q_{b,f} \tag{18}$$

3.2.2. Pipe

Energy loss of flow $Q_{p,f}$ includes path energy loss only:

$$Q_{p,f} = g\varsigma_p\frac{L_p}{d_{p,2}}\frac{u_p^2}{2g} \times \frac{G_e}{G_{NG}} \times 10^{-3} \tag{19}$$

3.2.3. Well

(a) Energy of exhaust gas that flows from the outlet of the well to the environment:

$$Q_{w,out} = \frac{1}{G_{NG}} \times G_e C_{p,e} t_{w,out} \tag{20}$$

(b) Energy loss of flow $Q_{w,f}$ includes path energy loss and local energy loss:

$$Q_{w,f} = g(\varsigma_w\frac{L_w}{d_{w,1}}\frac{u_w^2}{2g} + 2\zeta_w\frac{u_w^2}{2g}) \times \frac{G_e}{G_{NG}} \times 10^{-3} \tag{21}$$

Coefficient of local energy loss of the well:

$$\xi_w = 2.993 + 0.985 \tag{22}$$

3.2.4. Soil

(a) Energy loss of heat leakage in soil $Q_{s,l}$ includes the energy loss to the soil $Q_{s,l,s}$ and to the air $Q_{s,l,a}$. The leakage of heat from the soil to the air is conducted through an insulating layer covering the ground and the heat leakage to the surrounding soil is abstracted as heat transfer between two layers of cylindrical surfaces of a hollow cylinder:

$$\begin{cases} Q_{s,l} = Q_{s,l,s} + Q_{s,l,a} \\ Q_{s,l,s} = \dfrac{2\pi\lambda_s L_s (t_s - t_{s,e})}{\ln \frac{r_{s,e}}{r_s}} \times \dfrac{1}{G_{NG}} \times 10^{-3} \\ Q_{s,l,a} = \dfrac{\lambda_{il} A_{il}(t_s - t_0)}{\Delta il} \times \dfrac{1}{G_{NG}} \times 10^{-3} \end{cases} \tag{23}$$

(b) Energy that the soil eventually use to heat up $Q_{s,a}$:

$$Q_{s,a} = Q_{s,in} - Q_{s,l} \tag{24}$$

3.2.5. Energy Utilization Ratio

Energy utilization ratio as performance indicators of the energy analysis were calculated by the energy that the soil ultimately uses and low calorific value of 1 km of NG:

$$\eta_{en} = \dfrac{Q_{s,a}}{Q_{ar,net}} \tag{25}$$

3.3. Exergy Analysis Model

One kilogram of natural gas (NG) is the total exergy source of the system and the study about the exergy flow and the exergy loss is started with the exergy of one kilogram of natural gas (NG). The temperature used in modeling is shown in Table 1 or calculated in Appendix B (c).

3.3.1. Burner

(a) Reactant (natural gas) exergy E_r is calculated as follows: [43]

$$E_r = 0.95 Q_{ga,v,ad} \tag{26}$$

(b) E_{air} is the exergy of air:

$$E_{air} = 0 \tag{27}$$

(c) Exergy loss of irreversible combustion in the burner calculated by reactant exergy E_r and resultant exergy E_{rs} [43]:

$$\begin{cases} E_{b,irr} = E_r - E_{rs} \\ E_{rs} = (Q_{ar,net} - Q_{ar,net}\varepsilon)\left(1 - \dfrac{T_0}{T_b - T_0}\ln\dfrac{T_b}{T_0}\right) \end{cases} \tag{28}$$

(d) Exergy loss of incomplete combustion is energy loss of incomplete combustion according to the definition of exergy:

$$E_{b,inc} = Q_{b,inc} \tag{29}$$

(e) Exergy loss of heat leakage is calculated by heat leakage energy and heat leakage temperature, and the calculation of burner, pipe, well and soil is similar. The tube surface temperature $T_{b,w,o}$, $T_{p,w,o}$, $T_{w,w,o}$ are regarded as leakage temperature calculated in Appendix B (c). The heat leakage temperature of soil is the soil temperature itself:

$$E_{b,l} = Q_{b,l}\left(1 - \dfrac{T_0}{T_{b,w,o}}\right) \tag{30}$$

$$E_{p,l} = Q_{p,l}\left(1 - \dfrac{T_0}{T_{p,w,o}}\right) \tag{31}$$

$$E_{w,l} = Q_{w,l}(1 - \frac{T_0}{T_{w,w,o}})$$

(32)

$$E_{s,l} = Q_{s,l}(1 - \frac{T_0}{T_s})$$

(33)

(f) Exergy loss of flow is energy loss of flow on account of the definition of exergy and the pipe as well as well is calculated in the same way as the burner:

$$E_{b,f} = Q_{b,f}$$

(34)

$$E_{p,f} = Q_{p,f}$$

(35)

$$E_{w,f} = Q_{w,f}$$

(36)

(g) Exergy out of the burner $E_{b,to,p}$, is solved by the conservation of exergy equation. The calculation of the exergy out of the pipe, well and soil is similar, so the equation will not be repeated below:

$$E_{b,to,p} = E_r + E_{air} - E_{b,irr} - E_{b,inc} - E_{b,l} - E_{b,f}$$

(37)

3.3.2. Pipe

This part of the calculation has been mentioned in Section 3.3.1.

3.3.3. Well

(a) Exergy loss due to heat transfer process $E_{x,Q}$:

$$\begin{cases} E_{x,Q} = E_{x,Q_H} - E_{x,Q_L} \\ E_{x,Q_H} = (1 - \frac{T_0}{T_H})Q_1 \\ E_{x,Q_L} = (1 - \frac{T_0}{T_L})Q_2 \end{cases}$$

(38)

E_{x,Q_H} is the calorific exergy of Q_1 at the temperature $\overline{T_H}$ and E_{x,Q_L} is the calorific exergy of Q_2 at the temperature of $\overline{T_L}$. Q_1 and Q_2 are considered equal calculated in Appendix B(a), while the amount of heat transferred in the heat transfer process varies in each stage. And the calculation of $\overline{T_H}$ and $\overline{T_L}$ are in Appendix B (d).

(b) Exergy loss of non-isothermal heat release is caused by temperature change of exhaust gas when flowing in the well:

$$E_{x,Q_1} = T_0 G_e c_{p,w} ln \frac{T_{w,in}}{T_{w,out}}$$

(39)

(c) Exergy of exhaust gas that flows from the outlet of the well to the environment is connected to the energy and temperature of exhaust gas:

$$E_{w,out} = Q_{w,out}(1 - \frac{T_0}{T_{w,out} - T_0} ln \frac{T_{w,out}}{T_0})$$

(40)

3.3.4. Soil

(a).Exergy loss of non-isothermal heat absorption E_{x,Q_2} is caused by temperature change of soil at three stages: In the first stage, the soil temperature rises from the initial temperature (environment temperature) T_0 to the boiling point of water 373K. The soil keeps the temperature of 373K unchanged in the second stage. The third stage is to heat soil to increase the soil temperature to final temperature T_s:

$$\begin{cases} E_{x,Q_2,I} = T_0 m_s c_{p,s} ln \frac{373}{T_0} \\ E_{x,Q_2,II} = 0 \\ E_{x,Q_2,III} = T_0 m_{ps} c_{p,ps} ln \frac{T_s}{373} \end{cases}$$

(41)

(b) $E_{s,a}$ is the exergy that the soil eventually uses to heat up:

$$E_{s,a} = E_{s,in} - E_{x,Q_2} - E_{s,l} \tag{42}$$

3.3.5. Exergy Utilization Ratio

Exergy utilization ratio as performance indicators of the exergy analysis were calculated by the exergy that the soil ultimately uses and exergy value of 1 km of NG:

$$\eta_{ex} = \frac{E_{s,a}}{Er} \tag{43}$$

3.4. Process of Parameters Calculation in the Models

As Figure 7 shows, the calculation process of parameters of energy analysis starts at the thermal requirements and ends at the energy. The calculations of excess air coefficient α and mass flow rates G_e and G_{NG} are given in Appendix B (a). These three parameters are used to solve the time needed of flowing 1 km natural gas (NG). The thermal flux has been modeled in Section 3.2. The calculation process of parameters of exergy analysis is similar to that of energy analysis.

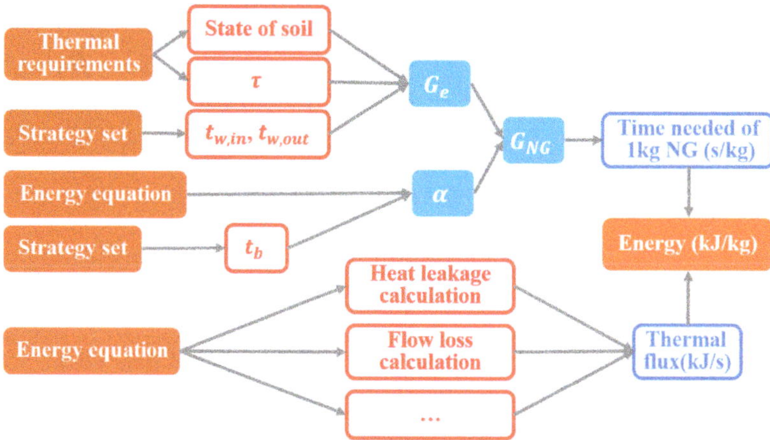

Figure 7. The flowchart of the parameters calculation in energy analysis.

4. Results and Discussions

In order to compare the effects of energy-saving strategies, the cases shown in Table 2 are designed. The traditional polluted-soil thermal remediation system, also named basic method (BM), is the fundamental case, Case BM. Case VCM applied energy-saving strategy for variable-condition mode with different exhaust gas temperatures at different stages. Energy-saving strategy for heat-returning mode is divided into 4 cases, among Case 3.1, Case 3.2 and Case 3.3, the difference is the rate of heat return. Case 3.4 combines variable-condition mode and heat-returning mode two energy-saving strategies. Energy-saving strategy for air-preheating mode is also divided into 4 cases, the difference is preheating ratio of air in Case 4.1, Case 4.2 and Case 4.3. As a comprehensive strategy for energy saving like Case 3.4, Case 4.4 combines variable-condition mode and air-preheating mode.

Table 2. Cases set of basic method (BM) and energy-saving strategies.

Strategy		Case		Stage	Rate of Heat Return β	Preheating Ratio of Air $\alpha_1, \alpha_2, \alpha_3$
1.Basic method (BM)		Case BM		I II III	0	1, 0, 0
	2.Variable-condition mode (VCM)	Case VCM		I	0	1, 0, 0
				II	0	1, 0, 0
				III	0	1, 0, 0
Energy-saving strategies	3.Heat-returning mode		Case 3.1	I II III	0.1	1, 0, 0
			Case 3.2	I II III	0.2	1, 0, 0
			Case 3.3	I II III	0.3	1, 0, 0
		With VCM	Case 3.4	I	0.3	1, 0, 0
				II	0.3	1, 0, 0
				III	0.3	1, 0, 0
	4.Air-preheating mode		Case 4.1	I II III	0	0.1, 0, 0.9
			Case 4.2	I II III	0	0.3, 0, 0.7
			Case 4.3	I II III	0	0.1, 0.1, 0.8
		With VCM	Case 4.4	I	0	0, 0.3, 0.7
				II	0	0, 0.3, 0.7
				III	0	0, 0.3, 0.7

Next, the effect analysis of three energy-saving strategies is established in Sections 4.1–4.3, respectively. In Section 4.4, a comprehensive analysis of the energy-saving strategies will be presented.

4.1. Energy Analysis and Exergy Anlysis of Variable-Condition Mode

The most effective part of the variable-condition mode is that under the premise of the same heating demand and heating time, high amounts of natural gas (NG) can be saved. Table 3 lists the results of the mass flow rates of exhaust gas and NG, as well as the calculated excess air coefficient. The number of mass flow rates in the first stage of variable-condition mode (VCM) is 0.0299 km per second, much smaller than 0.1124 km per second of basis method (BM). The differences of mass flow rates and excess air coefficient of basic method (BM) and variable-condition mode (VCM) result from different temperatures of exhaust out of the well.

Table 3. Results of mass flow rates and excess air coefficient.

Strategy	Stage	G_e (kg/s)	G_{NG} (kg/s)	α
BM	I II III	0.1124	0.0025	1.5510
VCM	I	0.0299	0.000533	2.2470
	II	0.1251	0.0024	1.9664
	III	0.0837	0.0022	1.1549

Besides savings in the amount of natural gas (NG) usage, this paper is mainly focused on improving the energetic and exergetic performance of the polluted-soil thermal remediation system depending on energy-saving strategies. Energy utilization ratios and exergy utilization ratios, two of the performance indicators of the analysis, were calculated as the important results of the mathematical model. The energy utilization ratios and exergy utilization ratios of the BM and VCM varies from different stages as well as different modes of heat convection. Detailed results of energy utilization ratios and exergy utilization ratios are shown in Tables 4 and 5. Modes of heat convection affects energy performance and exergy performance obviously. It can be observed that the energy utilization ratio of forced convection each stage is 2.6% lower than that of free convection, and exergy utilization ratio is 0.9% lower as well. It is because that the forced convection causes more loss of thermal leakage. While the two exergy utilization ratios of VCM is identical. It is because that the thermal leakage of forced convection is bigger, but the temperature of the outer wall of burner is lower. The larger quantity of thermal leakage and the lower temperature of thermal leakage lead to the same exergy loss.

Table 4. Energy utilization ratios of Case BM and Case VCM.

η_{en}	Forced Convection			Free Convection		
	I (%)	II (%)	III (%)	I (%)	II (%)	III (%)
BM	44.8	44.7	40.2	47.1	47	42.5
VCM	52.8	55.8	33.4	55.5	57.6	35.3

Table 5. Exergy utilization ratios of Case BM and Case VCM.

η_{ex}	Forced Convection			Free Convection		
	I (%)	II (%)	III (%)	I (%)	II (%)	III (%)
BM	26	19.9	22.8	26.9	20.8	23.8
VCM	12.1	23.2	9.6	12.1	23.9	10.4

Next, we assessed the energy and exergy efficiency using two curves more intuitively, as indicated in Figure 8. The utilization ratios' values of forced convection and free convection are different from the tables above, but the trend is the same, so the following analysis takes forced convection as an example. The energy utilization ratios of forced convection of the three stages are plotted in Figure 8a, while exergy utilization ratios of forced convection of the three stages are plotted in Figure 8b. We combine the two curves of BM and VCM together to make our analysis simpler to understand.

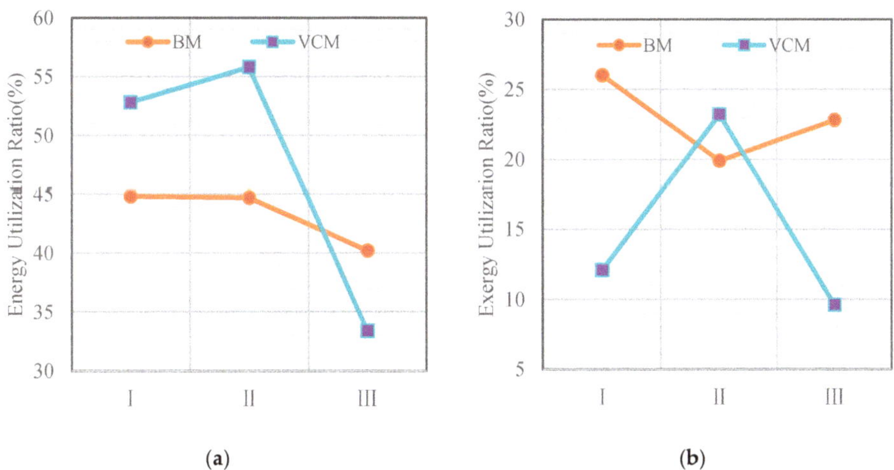

Figure 8. Energy utilization ratio and exergy utilization ratio comparisons of variable-condition mode (VCM) and basic method (BM): (**a**) Energy utilization ratio comparisons of variable-condition mode (VCM) and basic method (BM); (**b**) Exergy utilization ratio comparisons of variable-condition mode (VCM) and basic method (BM).

Figure 8a shows that energy utilization ratios of stage I in BM is not so very different from that of stage II. The lower temperature soil is heated by the higher temperature exhaust gas, which brings higher energy utilization ratios, but in stage III, the energy utilization ratio decreases significantly for maintaining the same temperature of exhaust gas, while the temperature of soil becomes higher and it becomes difficult to heat the soil. In VCM, stage II is the stage guaranteed the best performance of the whole heating process, as it maintains the highest values of the energy utilization ratio compared to other stages. In VCM, the energy utilization ratios of stage I and stage II are better than that of the same stage in BM, but stage III is worse because the temperature of exhaust gas in VCM is higher than that in BM bringing more heat loss due to exhaust gas and lower energy utilization ratio.

Figure 8b shows that the exergy utilization ratios of stage II in BM is smaller than that in stage I and stage III because the thermal requirement in stage II is larger and the loss of irreversible combustion is larger as well. Comparing BM and VCM, we can find it that the exergy utilization ratios of VCM in stage I and stage III are lower for the reason of small mass flow rates. That is because the small mass flow rates resulting in the bigger flow time. Thermal flux calculated by formulas multiplied by time is the eventual thermal leakage energy. In stage II, when the mass flow rates of BM and VCM is similar, the exergy utilization ratios of VCM is larger.

The utilization ratio curves express intuitively the energy saving situation, but where the specific embodiment of energy savings is to be analyzed from the diagrams of energy flow and exergy flow. From the calculation results obtained, the data of energy loss and exergy loss of each component are used to draw energy flow diagrams and exergy flow diagrams of basic method (BM) and variable-condition mode (VCM) representing the flow of energy and exergy visually. The following analysis is concentrated on forced convection. The thickness of the arrows represents the size of the value. Regarding energy of 1 km natural gas (NG) as 100%, the energy and exergy distribution fraction of various losses in each component of the system is presented in the Figure 9, so that comparing the losses of the two strategies is not difficult. The meanings of the parameters in all flow diagrams, including Figure 9, list in Table A2.

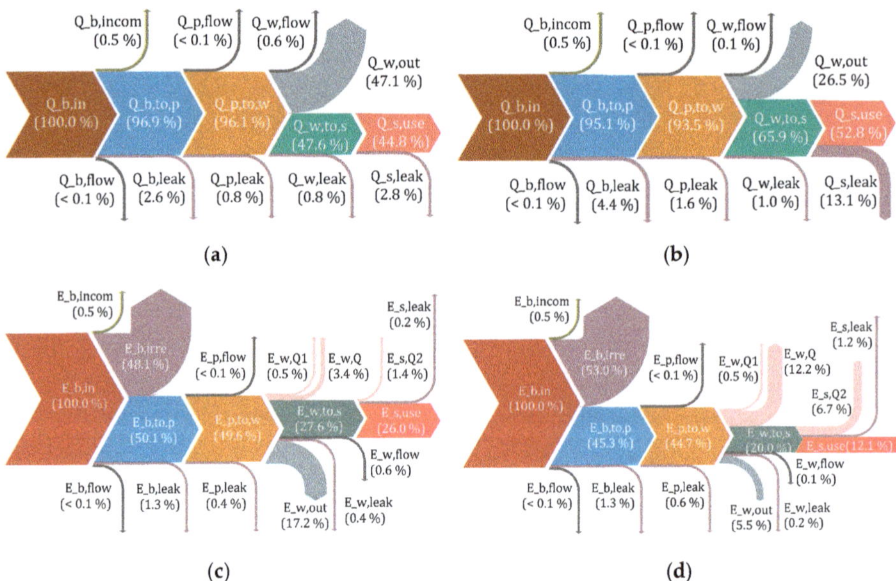

Figure 9. Flow diagrams of forced convection in stage I of of BM and VCM: (**a**) Energy flow diagram of forced convection in stage I of BM; (**b**) Energy flow diagram of forced convection in stage I of VCM; (**c**) Exergy flow diagram of forced convection in stage I of BM; (**d**) Exergy flow diagram of forced convection in stage I of VCM.

Comparing Figure 9a,b, it is observed that the energy in the exhaust gas is smaller in VCM compared with BM. The reason is that the temperature of exhaust gas in VCM is 200 °C, lower than that in BM with the value of 450 °C and this is the key to saving energy for VCM. It can be found by comparing Figure 9c,d that the exergy loss of irreversible combustion in BM is smaller than that in VCM. The reason is that exergy loss of irreversible combustion is associated with the adiabatic combustion temperature. The higher the adiabatic combustion temperature, the smaller the exergy loss of irreversible combustion. The adiabatic combustion temperature in VCM is lower in stage I, so that the exergy loss of irreversible combustion is bigger. Relative to energy, the energy saving strategy reduces more exergy loss of exhaust gas than energy loss. That is because the low temperature and the low energy of exhaust gas bring double effect of low exergy of exhaust gas. The increase of exergy loss impacted by mass flow rates is reflected in heat transfer process, non-isothermal heat release of exhaust gas and non-isothermal absorption of heat of soil. For example, when computing the exergy loss of non-isothermal heat release of exhaust gas of 1 km of natural gas (NG), the quantity of heat of non-isothermal heat release is an important factor. While the total quantity of heat of non-isothermal heat release in the first stage is settled, which is decided by the thermal requirements, in other words, the state of the soil. In the case of the same heat requirements and the same heating time, changing the temperature of exhaust gas from 450 °C to 200 °C in the first stage due to energy saving purpose results in the small amount of natural gas (NG). The total quantity of heat release maintained invariant, so that the exergy loss of one kilogram natural gas (NG) on average is bigger.

4.2. Energy Analysis and Exergy Anlysis of Heat-Returning Mode

The energy-saving strategy for heat-returning mode is returning the exhaust used to discharge to the atmospheric environment directly to the burner as the air in a certain proportion, Case 3.1 is with the rate of heat return of 0.1, Case 3.2 with the rate of 0.2 and Case 3.3, the rate 0.3. Using curves to assess the energy and exergy efficiency is the more intuitive way, as shown in Figure 10. And the specific distribution of energy and exergy loss as well as energy and exergy flow comparing with basic method (BM) are shown in Figure 11.

In Figure 10, it can be seen that all three case have higher utilization ratios than the basic method (BM), and the utilization ratios increase with increasing rate of heat return. The Case 3.3 with the largest rate of heat-returning has the best energy utilization ratio and exergy utilization ratio no matter what stage, which means the most significant energy-saving effect. However, the rate of heat return cannot always be increased without limit due to equipment and practical conditions. Compared with utilization ratios of Case 3.2 for Case 3.1, the Case 3.3 for Case 3.2 is more significant.

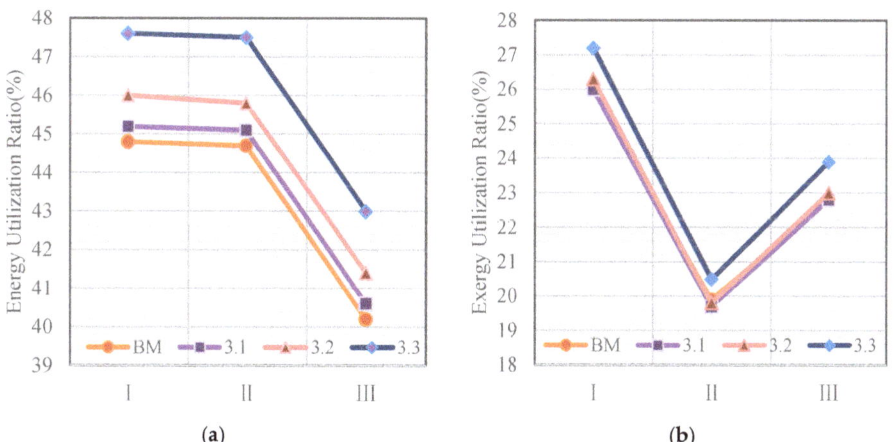

Figure 10. Energy utilization ratio and exergy utilization ratio comparisons of Case 3.1, Case 3.2 and Case 3.3 of heat-returning mode and basic method (BM): (**a**) Energy utilization ratio comparisons of Case 3.1, Case 3.2 and Case 3.3 of heat-returning mode and basic method (BM); (**b**) Exergy utilization ratio comparisons of Case 3.1, Case 3.2 and Case 3.3 of heat-returning mode and basic method (BM).

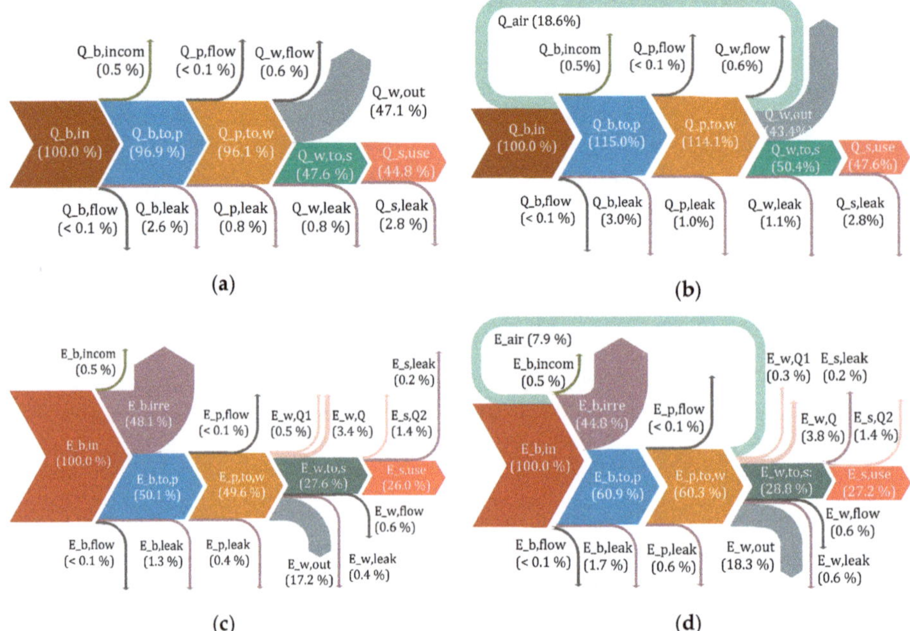

Figure 11. Flow diagrams of forced convection in stage I of BM and Case 3.3 of heat-returning mode: (a) Energy flow diagram of forced convection in stage I of BM; (b) Energy flow diagram of forced convection in stage I of Case 3.3 of heat-returning mode; (c) Exergy flow diagram of forced convection in stage I of BM; (d) Exergy flow diagram of forced convection in stage I of Case 3.3 of heat-returning mode.

The Case 3.3 with the best energy saving effect of the three cases of energy-saving strategy for heat-returning mode is selected to draw the energy flow diagram. From Figure 11a,b, there is a backflow of energy to the burner that is most obvious in Case 3.3, and it represents the heat-returning mode of exhaust. A Sankey diagram is a good way to show the flow of energy and the thickness of the arrows represents the size of the value in the diagram. With an initial energy of 1 km of natural gas, it is using the regenerative energy that to make more energy go into the system initially and it also results in more energy being used to heat the soil ultimately. While the energy loss of heat leakage of each component in Case 3.3 is bigger than that in BM for the reason of the higher temperature caused by more initial energy in the polluted-soil thermal remediation system. By comparing Figure 11c,d that the exergy loss of irreversible combustion in Case 3.3 is smaller than that in BM as a result of higher adiabatic combustion temperature. And the analysis of exergy loss of heat leakage is the same as the energy analysis above, the higher temperature, the more energy loss, and the more exergy loss.

4.3. Energy Analysis and Exergy Anlysis of Air-Preheating Mode

The energy-saving strategy for air-preheating mode is setting preheaters to air-preheating mode using residual heat of the system. The cases in this section selected two places with high temperature and enough space to set the preheaters. In Case 4.1, the ratio of air through preheater 1 to be preheated is 0.1, the ratio of air through preheater 2 to be preheated is 0, and the ratio of air that does not pass through the preheater directly into the burner is 0.9. In Case 4.2, the ratio of air through preheater 1 to be preheated is 0.3, the ratio of air through preheater 2 to be preheated is 0, and the ratio of air that does not pass through the preheater directly into the burner is 0.7. In Case 4.3, the ratio of air through preheater 1 to be preheated is 0.1, the ratio of air through preheater 2 to be preheated is 0.1 as well, and the ratio of air that does not pass through the preheater directly into the burner is 0.8. Curves are used to assess the energy and exergy efficiency as shown in Figure 12. And the specific distribution of energy and exergy loss as well as energy and exergy flow comparing with basic method (BM) are shown in Figure 13.

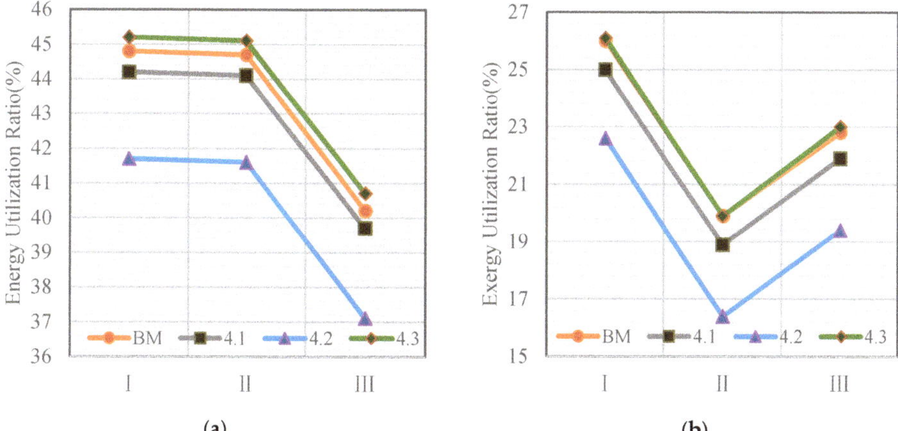

(**a**)   (**b**)

Figure 12. Energy utilization ratio and exergy utilization ratio comparisons of Case 4.1, Case 4.2 and Case 4.3 of heat-returning mode and basic method (BM): (**a**) Energy utilization ratio comparisons of Case 4.1, Case 4.2 and Case 4.3 of heat-returning mode and basic method (BM); (**b**) Exergy utilization ratio comparisons of Case 4.1, Case 4.2 and Case 4.3 of heat-returning mode and basic method (BM).

In Figure 12, it can be seen that the energy and exergy utilization ratio of Case 4.1 is smaller than that of BM, and it is proved that the air preheating through the preheater 1 is not conducive to the improvement of utilization ratio and energy saving. The energy and exergy utilization ratio of Case 4.2 is even smaller than that of Case 4.1, that is to say, the effect of the preheater 1 wasting energy increases with the proportion of air passing through it. While the preheater 2 performs better, the energy utilization ratio of Case 4.3 is bigger than that of BM and the exergy utilization ratio is similar to that of BM under the bad interference of preheater 1. The underlying reason is that preheater 1 uses the energy to flow to the next component, while preheater 2 uses the waste heat to be drained into the air, so making full use of waste heat is the wonderful way to save energy, so in the Section 4.4 comprehensive energy-saving strategies, in Case 4.4, the ratio of air through preheater 1 to be preheated is 0, the ratio of air through preheater 2 to be preheated is 0.3, and the ratio of air that does not pass through the preheater directly into the burner is 0.7.

Case 4.3 with the best energy saving effect of the three cases of energy-saving strategy for air-preheating mode is selected to draw the energy flow diagram. From Figure 13a,b, there are two backflows of energy to the burner in Case 4.3, and they represent preheated air with energy. Although the proportion of air through the preheater is not high, not much heat is brought back. The amount of energy used eventually increases a little with the increase in heat leakage accompanied by an increase in temperature.

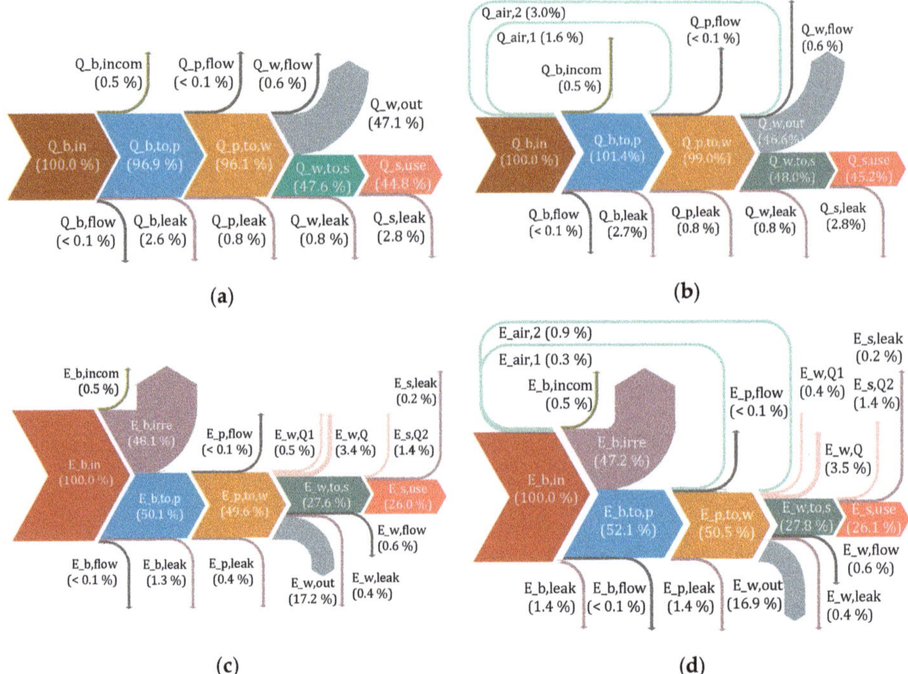

Figure 13. Flow diagrams of forced convection in stage I of BM and Case 4.3 of air-preheating mode: (**a**) Energy flow diagram of forced convection in stage I of BM; (**b**) Energy flow diagram of forced convection in stage I of Case 4.3 of air-preheating mode; (**c**) Exergy flow diagram of forced convection in stage I of BM; (**d**) Exergy flow diagram of forced convection in stage I of Case 4.3 of air-preheating mode.

4.4. Energy Analysis and Exergy Anlysis of Comprehensive Energy-Saving Strategies

The comprehensive energy-saving strategies are mainly to compare air-preheating mode combined with variable-condition mode (VCM) and heat-returning mode combined with variable-condition mode (VCM), that is Case 3.4 and Case 4.4. The rate of heat return in Case 3.4 is 0.3 and in Case 4.4 the ratio of air through preheater 2 to be preheated is 0.3 as well as the ratio of air that does not pass through the preheater directly into the burner is 0.7. Because the variable-condition mode (VCM) were applied in both cases, so draw the energy utilization ratio curves for the three cases on one graph, Figure 14a, and draw the exergy utilization ratio curves on the other graph, Figure 14b, for easy comparison.

By comparing Figure 14a,b, in the three cases with variable-condition mode (VCM), the trend of energy utilization ratio and exergy utilization ratio curve is the same in the three stages. The second stage has the highest utilization ratio, then the first stage, then the third stage. The results in Figure 14a indicate that the Case 3.4 has the best energy utilization ratio in all three stages by combining the advantages of variable-condition mode (VCM) and heat-returning mode, that is to say, using two energy-saving strategies can bring the improvement of fuel saving and energy efficiency at the same time. The results of Case 4.4 have the same implications that Case 4.4 has better energy utilization ratio and exergy utilization ratio than Case VCM by combining the advantages of variable-condition mode (VCM) and air-preheating mode. If all three energy-saving strategies are combined, the structure of the system will become too complex and uncontrollable. Therefore, the combination of the two energy-saving strategies is recommendable.

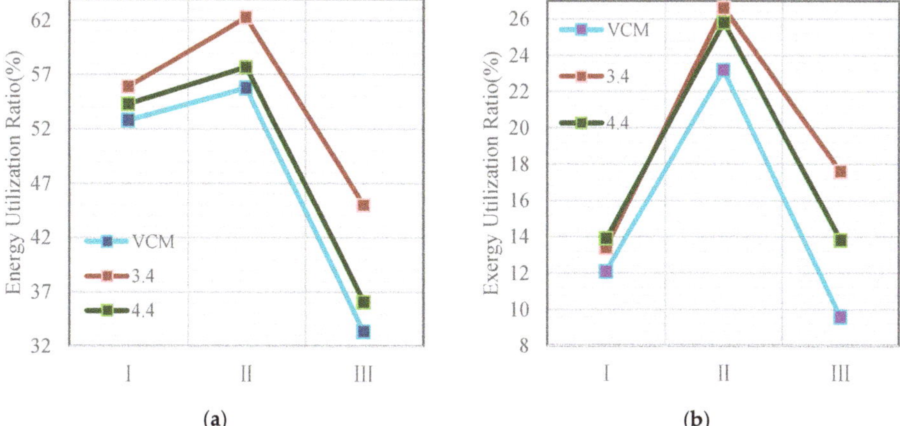

Figure 14. Energy utilization ratio and exergy utilization ratio comparisons of Case 3.4, Case 4.4 and variable-condition mode (VCM): (**a**) Energy utilization ratio comparisons of Case 3.4, Case 4.4 and variable-condition mode (VCM); (**b**) Exergy utilization ratio comparisons of Case 3.4, Case 4.4 and variable-condition mode (VCM).

5. Conclusions

This paper proposes three energy-saving strategies for a polluted-soil thermal remediation system—variable-condition mode (VCM), heat-returning mode and air-preheating mode—and their thermal performance and efficiency were discussed by energy analysis and exergy analysis. The mathematical models of a polluted-soil thermal remediation system including burner, pipe, well nd soil for energy and exergy analysis are built. The following main conclusions are reached:

- The most effective part of the energy-saving strategy for variable-condition mode (VCM) is that under the premise of the same heating demand and heating time, the usage amount of natural gas (NG) can be saved highly. The number of mass flow rates in the first stage of variable-condition mode (VCM) is 0.0299 km per second, much smaller than 0.1124 km per second of basis method (BM). It can be observed that the energy utilization ratio of forced convection each stage is 2.6% lower than that of free convection, and exergy utilization ratio is 0.9% lower as well. In VCM, the energy utilization ratio of stage I and stage II is better than that of the same stage in BM.

- All three energy-saving strategy cases for heat-returning mode have utilization ratios of 3% on average higher than the basic method (BM), and the utilization ratios increase with increasing rate of heat return. Case 3.3 with the largest rate of heat return has the best energy utilization ratio and exergy utilization ratio no matter what stage, which means the most significant energy-saving effect. That is because it is using the regenerative energy that to make more energy go into the system initially and it also results in more energy being used to heat the soil ultimately with an initial energy of 1 km of natural gas.

- In the analysis of energy-saving strategies for air-preheating mode, the air flowing through the preheater 1 to be preheated is not conducive to the improvement of utilization ratio and energy saving and the effect of the preheater 1 wasting energy increases with the proportion of air passing through it. While the preheater 2 performs better, the energy utilization ratio of Case 4.3 is bigger than that of BM and the exergy utilization ratio is similar to that of BM under the bad interference of preheater 1. The underlying reason is that preheater 1 uses the energy to flow to the next component, while preheater 2 uses the waste heat to be drained into the air. So making full use of waste heat is the wonderful way to save energy.

- The comprehensive energy-saving strategies are mainly to compare air-preheating mode combined with variable-condition mode (VCM) and heat-returning mode combined with variable-condition mode (VCM). The results indicate that the Case 3.4 has the best energy utilization ratio in all three stages by combining the advantages of variable-condition mode (VCM) and heat-returning mode, and the results of Case 4.4 have the same implications. That is to say, combination of two energy-saving strategies can bring the improvement of fuel saving and energy efficiency at the same time and it is recommendable.

Research on methods and effects of energy-saving strategies is very beneficial for the full use of energy, for the reason of reduced natural gas consumption and higher energy utilization. Energy analysis and exergy analysis are also intuitive for the presentation of results. Further to say, resource problem can be solved better through the research of efficient use of energy.

Author Contributions: Conceptualization, Y.-Z.L.; methodology, T.-T.L.; software, T.-T.L.; validation, Y.-Z.L.; formal analysis, Z.-Z.Z.; investigation, E.-H.L.; writing—original draft preparation, T.-T.L.; writing—review and editing, T.L.; visualization, E.-H.L.; supervision, Y.-Z.L.; project administration, Y.-Z.L.

Funding: The project was funded by the Open Research Fund of Key Laboratory of Space Utilization, Chinese Academy of Sciences through grant no.LSU-JCJS-2017-1.

Conflicts of Interest: The authors declare no conflict of interest.

Nomenclature

A	Area (m^2)	λ	Thermal conductivity (W/(m*K))
C	Constant, depending on the Reynolds number	ν	Kinematic viscosity (m^2/s)
d	Diameter (m)	ξ	Coefficient of local energy loss
E	Exergy (kJ/kg)	τ	Total time of the stage (days)
E_r	Reactant exergy (kJ/kg)	Subscript	
E_{rs}	Resultant exergy (kJ/kg)	*air*	Air
$E_{x,Q}$	Exergy loss due to heat transfer process(kJ/kg)	b	Burner
E_{x,Q_1}	Exergy loss due to non-isothermal heat release(kJ/kg)	b,f	Flow loss in the burner
E_{x,Q_2}	Exergy loss due to non-isothermal absorption of heat(kJ/kg)	b,inc	Incomplete combustion in the burner
E_{x,Q_H}	Heat exergy of Q_1 at $\overline{T_H}$(kJ/kg)	b,irr	Irreversible combustion in the burner
E_{x,Q_L}	Heat exergy of Q_2 at $\overline{T_L}$(kJ/kg)	b,l	Thermal leakage in the burner
g	Gravitational acceleration (m/s^2)	b,to,p	From burner to pipe
G	Mass flow rate (kg/s)	b,w,o	The outer wall of burner
Gr	Grashof number	$b,1$	Interior burner
h	Convective heat transfer coefficient (W/($m^2 \cdot {}^\circ C$)	$b,2$	External burner
l	characteristic length (m)	e	Exhaust
L	Length(m)	f,b	Fluid in the burner
m_{ps}	Quality of dry soil (kg)	f,p	Fluid in the pipe
m_s	Quality of soil (kg)	f,w	Fluid in the well
m_w	Quality of water in the soil (kg)	il	Insulating layer
n	Constant, depending on the Reynolds number	m	At qualitative temperature
Nu	Nusselt number	NG	Natural gas

Pr	Prandtl number	o	Extended well
Q	Energy (kJ/kg)	p	Pipe
$Q_{ar,net}$	Lower calorific value (kJ/kg)	p,f	Flow loss in the pipe
$Q_{ga,v,ad}$	Higher calorific value (kJ/kg)	p,in	Input of pipe
Re	Reynolds number	p,l	Thermal leakage in the pipe
t	Temperature (°C)	p,to,w	From pipe to well
t_b	Adiabatic combustion temperature (°C)	p,w,o	The outer wall of pipe
t_0	Ambient temperature (°C)	s	Soil
t_s	Soil temperature at the end of stage (°C)	s,a	Absorbed by the soil
T	Kelvin temperature (K)	s,e	Unheated soil
$\overline{T_H}$	Kelvin temperature of intermediate high temperature heat source (K)	s,in	Input of soil
$\overline{T_L}$	Kelvin temperature of intermediate low temperature heat source (K)	s,l	Thermal leakage in the soil
T_0	Ambient kelvin temperature (K)	s,l,a	Thermal leakage from soil to soil
u	Velocity (m/s)	s,l,s	Thermal leakage from soil to air
Greek symbols		w	Well
α	Excess air coefficient	w,f	Flow loss in the well
α_V	Cubic expansion coefficient	w,in	Input of well
γ	Latent heat of vaporization (kJ/kg)	w,l	Thermal leakage in the well
Δ	Thickness (m)	w,out	Output of well
Δt	Temperature difference (°C)	w,to,s	Well to soil
ε	Incomplete combustion coefficient	w,w,o	The outer wall of extended well
η_{en}	Energy utilization ratio	I	The first stage
η_{ex}	Exergy utilization ratio	II	The second stage
ς	Coefficient of path energy loss	III	The third stage

Appendix A

Table A1. The Value of Physical Parameters.

Parameters	Value
$c_{p,b}$ Specific heat of gas in burner	1.2390 kJ/(kg*K)
$c_{p,e}$ Specific heat of exhaust	1.1850 kJ/(kg*K)
$c_{p,ps}$ Specific heat of pure soil	0.84 kJ/(kg*K)
$c_{p,s}$ Specific heat of soil	1.8480 kJ/(kg*K)
$c_{p,w}$ Specific heat of well	4.2 kJ/(kg*K)
$d_{b,1}$ Inner diameter of burner	190 mm
$d_{b,2}$ External diameter of burner	200 mm
$d_{p,1}$ Inner diameter of pipe	89 mm
$d_{p,2}$ External diameter of pipe	97 mm
$d_{w,1}$ Inner diameter of well	89 mm
$d_{w,2}$ External diameter of well	97 mm
g Acceleration of gravity	9.8 m/s^2
L_b Length of burner	0.5 m
L_o Length of extension of the well	0.3 m
L_p Length of pipe	0.2 m
L_w Length of well	13 m
r_s Radius of heated soil	1.5 m
$r_{s,e}$ Radius of unheated soil	2 m
λ_b Thermal conductivity of burner	0.0311 W/(m*K)
λ_f Thermal conductivity of fluid	Differs from parts
λ_p Thermal conductivity of pipe	0.0231 W/(m*K)
λ_s Thermal conductivity of soil	1.2 W/(m*K)
λ_w Thermal conductivity of well	0.0209 W/(m*K)
v_{air} Kinematic viscosity of air	30×10^{-6} m^2/s

Table A2. The meanings of the parameters in flow diagrams.

Parameter	Meaning	Parameter	Meaning
Q_air	Energy of air to burner	E_air	Exergy of air to burner
Q_b, in	Input energy to burner of 1kg NG	E_b, in	Input exergy to burner of 1kg NG
Q_b, to, p	Energy from burner to pipe	E_b, to, p	Exergy from burner to pipe
$Q\_b, incomp$	Energy loss of incomplete combustion	$E\_b, irre$	Exergy loss of irreversible combustion
$Q\_b, flow$	Energy loss of flow in burner	$E\_b, incomp$	Exergy loss of incomplete combustion
$Q\_b, leak$	Energy loss of heat leakage in burner	$E\_b, flow$	Exergy loss of flow in burner
Q_p, to, w	Energy from pipe to well	$E\_b, leak$	Exergy loss of heat leakage in burner
$Q\_p, flow$	Energy loss of flow in pipe	E_p, to, w	Exergy from pipe to well
$Q\_p, leak$	Energy loss of heat leakage in pipe	$E\_p, flow$	Exergy loss of flow in pipe
Q_w, out	Output energy of exhaust from well	$E\_p, leak$	Exergy loss of heat leakage in pipe
Q_w, to, s	Energy from well to soil	E_w, out	Output exergy of exhaust from well
$Q\_w, flow$	Energy loss of flow in well	E_w, to, s	Exergy from well to soil
$Q\_w, leak$	Energy loss of heat leakage in well	$E\_w, flow$	Exergy loss of flow in well
Q_s, use	Energy that soil use	$E\_w, leak$	Exergy loss of heat leakage in well
$Q\_s, leak$	Energy loss of heat leakage in soil	$E\_w, Q1$	Exergy loss of non-isothermal heat release
$E\_s, Q2$	Exergy loss of non-isothermal heat absorption	E_w, Q	Exergy loss of heat transfer
$E\_s, leak$	Exergy loss of heat leakage in soil	E_s, use	Exergy that soil use

Appendix B Calculation of Intermediate Parameters

(a)   Excess Air Coefficient and Mass Flow Rates Calculation

Mass flow rate of exhaust gas is calculated by state of soil, duration of stages and the set temperature of the exhaust. However, in the basic method (BM) the required conditions for calculation in each stage are uniform, and in variable-condition mode (VCM) those are distinguishing because of different exhaust gas temperatures at different stages. Excess air coefficient is also different caused by varied adiabatic combustion temperature.

Thermal requirement is the quantity of heat in the heat transfer process from well to soil by exhaust gas. Q_I is the heat in the first stage, Q_{II} is in the second stage and Q_{III} is the third. $c_{p,s}$ is specific heat with the moisture content of 0.3:

$$\begin{cases} Q_I = m_s c_{p,s}(100 - t_0) \\ Q_{II} = m_w \gamma \\ Q_{III} = m_{ps} c_{p,ps}(t_s - 100) \end{cases} \tag{A1}$$

Mass flow rate of exhaust gas in BM of the three stages are calculated uniformly, as follows. And in VCM the mass flow rates of exhaust gas of the three stages are calculated respectively using separate thermal requirement. Because the calculation method is same as that in BM, it will not be repeated here.

Mass flow rate of exhaust gas in BM:

$$G_e = \frac{Q_I + Q_{II} + Q_{III}}{\tau c_{p,e}(t_{w.in} - t_{w,out})} = \frac{m_s c_{p,s}(100 - t_0) + m_w \gamma + m_{ps} c_{p,ps}(t_s - 100)}{\tau c_{p,e}(t_{w,in} - t_{w,out})} \tag{A2}$$

In order to calculate the excess air coefficient α, chemical equation: Equation (48) and energy equation: Equation (49) in burner is essential. The excess air coefficient α is the ratio of redundant air volume to the right air volume in a complete reaction:

$$CH_4 + 2(1+\alpha)O_2 + \frac{2\times78}{21}(1+\alpha)N_2 \to CO_2 + 2H_2O + \frac{2\times78}{21}(1+\alpha)N_2 + 2\alpha O_2 + Q \tag{A3}$$

$$G_{NG}q_{ar,net}(1-\varepsilon) + G_{NG}c_{p,NG}t_{NG} + G_{air}c_{p,air}t_{air} = G_e c_{p,b}t_b \tag{A4}$$

$c_{p,NG}$ is the specific heat of natural gas (NG) and $c_{p,air}$ is the specific heat of air. Because the temperature of natural gas (NG) and air are close to zero, the two parameters are not involved in the calculation.

Equation (50) is the excess air coefficient α:

$$\alpha = \frac{\left(\dfrac{16q_{ar,net}(1-\varepsilon)}{c_{p,b}t_b} - 288.012\right)}{272.012} \tag{A5}$$

Mass flow rate of natural gas (NG) is corresponding to mass flow rate of exhaust gas one by one in each stage:

$$G_{NG} = \frac{G_e \times 16}{272.012\alpha + 288.012} \tag{A6}$$

(b)   Convective Heat Transfer Coefficient Calculation

Convective heat transfer coefficient in the calculation of energy loss of heat leakage is divided to three categories:

(a) Convective heat transfer coefficient of forced convection in tube is as follows:

$$\begin{cases} Re = \frac{ud}{v} \\ Nu = 0.023Re^{0.8}Pr^{0.3} \\ h = \frac{\lambda_f}{d}Nu \end{cases} \tag{A7}$$

(b) Convective heat transfer coefficient of forced convection outside the tube is solved in Equation (53). The speed of the wind is 3 meters per second in the paper:

$$\begin{cases} Re_{air} = \frac{u_{air}d}{v_{air}} \\ Nu_{air} = CRe_{air}^{n}Pr_{air}^{1/3} \\ h_{air} = \frac{\lambda_{air}}{d}Nu_{air} \end{cases} \tag{A8}$$

(c) Equation (54) is the convective heat transfer coefficient of natural convection outside the tube:

$$\begin{cases} Gr_{air} = \frac{g\alpha_v\Delta t l^3}{v_{air}^2} \\ Nu_{air} = C(Gr_{air}Pr_{air})_m^n \\ h_{air} = \frac{\lambda_{air}}{d}Nu_{air} \end{cases} \tag{A9}$$

(c)   Tube Surface Temperature Calculation

$$\begin{cases} t_{b,w,o} = \frac{Q_{b,l}}{L_b h_{b,2}\pi d_{b,2}} + t_0 \\ T_{b,w,o} = t_{b,w,o} + 273.15 \end{cases} \tag{A10}$$

$$\begin{cases} t_{p,w,o} = \frac{Q_{p,l}}{L_p h_{p,2}\pi d_{p,2}} + t_0 \\ T_{p,w,o} = t_{p,w,o} + 273.15 \end{cases} \tag{A11}$$

$$\begin{cases} t_{w,w,o} = \frac{Q_{w,l}}{L_o h_{w,2}\pi d_{w,2}} + t_0 \\ T_{w,w,o} = t_{w,w,o} + 273.15 \end{cases} \tag{A12}$$

(d)   Temperature of Intermediate Heat Source Calculation

In the actual heat transfer process, the temperature of the heat source will change as Figure A1 shows, so the temperature of intermediate heat source is essential to calculate. The logarithmic mean temperature difference is used to calculate the temperature of intermediate high temperature heat source $\overline{T_H}$, so as the temperature of intermediate low temperature heat source $\overline{T_L}$, and the $\overline{T_L}$ is different in each stage:

$$\overline{T_H} = \frac{T_{w,in} - T_{w,out}}{\ln \frac{T_{w,in}}{T_{w,out}}} \tag{A13}$$

$$\begin{cases} \overline{T_{L,I}} = \frac{373-T_0}{\ln \frac{373}{T_0}} \\ \overline{T_{L,II}} = 373 \\ \overline{T_{L,III}} = \frac{T_s-373}{\ln \frac{T_s}{373}} \end{cases} \tag{A14}$$

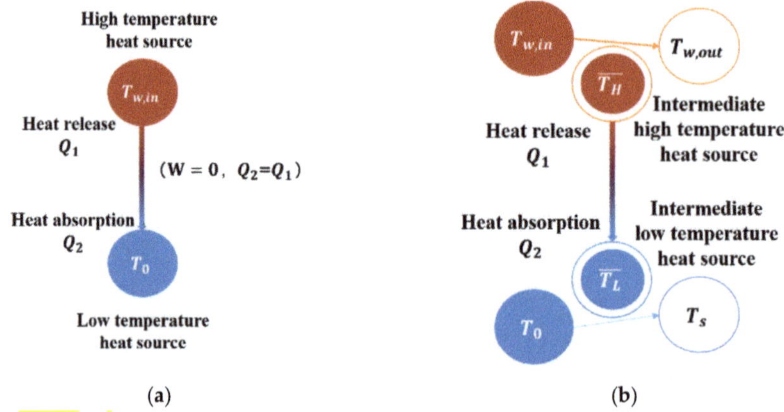

Figure A1. The heat transfer process: (**a**) The ideal heat transfer and (**b**) The actual heat transfer.

References

1. Yao, Z.T.; Li, J.H.; Xie, H.H.; Yu, C.H. Review on remediation technologies of soil contaminated by heavy metals. In Proceedings of the 7th International Conference on Waste Management and Technology, Beijing, China, 5–7 September 2012.

2. European Environment Agency. *The European Environment e State and Outlook 2010: Soil*; European Environment Agency: Copenhagen, Denmark, 2010; p. 44.

3. Liu, K.; Zhang, R.H.; Wang, S.J. Development and application of in situ thermal desorption for the remediation of contaminated sites. *China Chlor. Alkali* **2017**, *12*, 31–37.

4. Zhou, D.M.; Hao, X.Z.; Xue, Y. Advances in remediation technologies of contaminated soils. *Ecol. Environ. Sci.* **2004**, *13*, 234–242.

5. Cappuyns, V. Environmental impacts of soil remediation activities: Quantitative and qualitative tools applied on three case studies. *J. Clean. Prod.* **2013**, *52*, 145–154. [CrossRef]

6. Ministry of Environmental Protection of the People's Republic of China. *Bulletin on Chinese Domestic Environmental Conditions of 2000 Year*; Ministry of Environmental Protection of the People's Republic of China: Beijing, China, 2001.

7. Aresta, M.; DiBenedetto, A.; Fragale, C.; Giannoccaro, P.; Pastore, C.; Zammiello, D.; Ferragina, C. Thermal desorption of polychlorobiphenyls from contaminated soils and their hydrodechlorination using Pd- and Rh-supported catalysts. *Chemosphere* **2008**, *70*, 1052–1058. [CrossRef]

8. United States Environmental Protection Agency, Office of Solid Waste and Emergency Response. *A Citizen's Guide to Thermal Desorption*; 542-F-01-003; United States Environmental Protection Agency, Office of Solid Waste and Emergency Response: Washington, DC, USA, 2001; p. 2.

9. Luo, Y.M. Current Research and Development in Soil Remediation Technologies. *Prog. Chem.* **2009**, *21*, 558–565.

10. Araruna, J.T.J.; Portes, V.L.O.; Soares, A.P.L. Oil spills debris clean up by thermal desorption. *J. Hazard. Mater.* **2004**, *110*, 161–171. [CrossRef] [PubMed]

11. Yeung, A.T. Remediation technologies for contaminated sites. In *Advances in Environmental Geotechnics*; Zhejiang University Press: Hangzhou, China, 2010; pp. 328–369.

12. Feng, J.S.; Zhang, Q.C. A Review on the Study on Practice of Soil Remediation in situ. *Ecol. Environ. Sci.* **2014**, *23*, 1861–1867.

13. Fang, X.L.; Wang, Z.Z. Advances in in situ soil remediation technology. *Agric. Technol.* **2015**, *18*, 29–30.

14. Lu, L.Q.; Ye, C.M. Advances in in situ remediation of heavy metal contaminated soils. *Anhui Agric. Sci. Bull.* **2017**, *21*, 69–70.

15. George, L.S.; Harold, J.V. Thermal conduction heating for in situ thermal desorption of soils. *J. Hazard. Radioact. Waste Treat. Technol. Handb.* **2001**, 1–37.

16. Soesilo, J.A.; Stephanie, R.W. *Site Remediation: Planning and Management*; CRC Press: Boca Raton, FL, USA, 1997.

17. Saadaoui, H.; Jourdain, S.; Falcinelli, U.; Haemers, J. In situ thermal treatment in urban polluted areas: Application of thermopile. In Proceedings of the 10th Consoil Conference, Milan, Italy, 3–6 June 2008; pp. 1–10.

18. Julia, E.V.; Kyriacos, Z.; Caroline, A.M.; Gabriel, S.; Pedro, J.J.A. Masiello Thermal Treatment of Hydrocarbon-Impacted Soils: A Review of Technology Innovation for Sustainable Remediation. *J. Eng.* **2016**, *2*, 426–437.

19. Rada, E.C.; Andreottola, G.; Istrate, I.A.; Viotti, P.; Conti, F.; Magaril, E.R. Remediation of Soil Polluted by Organic Compounds Through Chemical Oxidation and Phytoremediation Combined with DCT. *Int. J. Environ. Res. Public Health* **2019**, *16*, 3179. [CrossRef] [PubMed]

20. Sun, Y.; Guan, F.; Yang, W.; Wang, F. Removal of Chromium from a Contaminated Soil Using Oxalic Acid, Citric Acid, and Hydrochloric Acid: Dynamics, Mechanisms, and Concomitant Removal of Non-Targeted Metals. *Int. J. Environ. Res. Public Health* **2019**, *16*, 2771. [CrossRef] [PubMed]

21. Geiger, E.M.; Sarkar, D.; Datta, R. Evaluation of Copper-Contaminated Marginal Land for the Cultivation of Vetiver Grass (Chrysopogon zizanioides) as a Lignocellulosic Feedstock and its Impact on Downstream Bioethanol Production. *Appl. Sci.* **2019**, *9*, 2685. [CrossRef]

22. Tian, J.; Huo, Z.; Ma, F.; Gao, X.; Wu, Y. Application and Selection of Remediation Technology for OCPs-Contaminated Sites by Decision-Making Methods. *Int. J. Environ. Res. Public Health* **2019**, *16*, 1888. [CrossRef] [PubMed]

23. Elyamine, A.M.; Moussa, M.G.; Afzal, J.; Rana, M.S.; Imran, M.; Zhao, X.; Hu, C.X. Modified Rice Straw Enhanced Cadmium (II) Immobilization in Soil and Promoted the Degradation of Phenanthrene in Co-Contaminated Soil. *Int. J. Mol. Sci.* **2019**, *20*, 2189. [CrossRef]

24. Park, K.; Jung, W.; Park, J. Decontamination of Uranium-Contaminated Soil Sand Using Supercritical CO_2 with a TBP–HNO_3 Complex. *Metals* **2015**, *5*, 1788–1798. [CrossRef]

25. Navarro, A.; Cañadas, I.; Rodríguez, J. Thermal Treatment of Mercury Mine Wastes Using a Rotary Solar Kiln. *Minerals* **2014**, *4*, 37–51. [CrossRef]

26. Chonokhuu, S.; Batbold, C.; Chuluunpurev, B.; Battsengel, E.; Dorjsuren, B.; Byambaa, B. Contamination and Health Risk Assessment of Heavy Metals in the Soil of Major Cities in Mongolia. *Int. J. Environ. Res. Public Health* **2019**, *16*, 2552. [CrossRef]

27. Huang, X.; Luo, D.; Chen, X.; Wei, L.; Liu, Y.; Wu, Q.; Xiao, T.; Mai, X.; Liu, G.; Liu, L. Insights into Heavy Metals Leakage in Chelator-Induced Phytoextraction of Pb- and Tl-Contaminated Soil. *Int. J. Environ. Res. Public Health* **2019**, *16*, 1328. [CrossRef] [PubMed]

28. Lin, W.-C.; Lin, Y.-P.; Anthony, J.; Ding, T.-S. Avian Conservation Areas as a Proxy for Contaminated Soil Remediation. *Int. J. Environ. Res. Public Health* **2015**, *12*, 8312–8331. [CrossRef] [PubMed]

29. Christopher, D.E.; Daniel, Z.; Kimberly, E.B.; Benjamin, T.T.; Tilottama, G.; Dee, W.P.; Edward, H.E.; Mikhail, Z. A fifteen year record of global natural gas flaring derived from satellite data. *Energies* **2009**, *2*, 595–622.

30. Chen, J.; Wang, Y.-H.; Lang, X.-M.; Fan, S.-S. Energy-efficient methods for production methane from natural gas hydrates. *J. Energy Chem.* **2015**, *24*, 552–558. [CrossRef]

31. Heinz, S.; Steve, H.-D.; Thomas, W.; Arne, G.; Yiyong, C.; James, W.; David, N.; Tim, B.; Manfred, L.; Anne, O. Decoupling global environmental pressure and economic growth: Scenarios for energy use, materials use and carbon emissions. *J. Clean. Prod.* **2016**, *132*, 45–56.

32. Schmalensee, R.; Stoker, T.M.; Judson, R.A. World Carbon Dioxide Emissions: 1950–2050. *Rev. Econ. Stat.* **1998**, *80*, 15–27.

33. Yu, H.; Zhang, Z.Y.; Hua, L.; Wei, Y.M. China's farewell to coal: A forecast of coal consumption through 2020. *Energy Policy* **2015**, *86*, 444–455.

34. Chen, S.Y. Energy Consumption, CO_2 Emission and Sustainable Development in Chinese Industry. *Econ. Res. J.* **2009**, *4*, 88–96.

35. Tang, D.; Dai, M. Energy-efficient Approach to Minimizing the Energy Consumption in an Extended Job-shop Scheduling Problem. *Chin. J. Mech. Eng.* **2015**, *28*, 1048–1055. [CrossRef]

36. Zou, C.; Zhao, Q.; Zhang, G.; Xiong, B. Energy revolution: From a fossil energy era to a new energy era. *Nat. Gas Ind. B* **2016**, *3*, 1–11. [CrossRef]

37. Carlos, G.M.; Alejandro, G.S.; Laurence, S.; Adisa, A. Life cycle environmental impacts of electricity from fossil fuels in Chile over a ten-year period. *J. Clean. Prod.* **2019**, *232*, 1499–1512.

38. Wu, J.; Wang, J.; Wu, J.; Ma, C. Exergy and Exergoeconomic Analysis of a Combined Cooling, Heating, and Power System Based on Solar Thermal Biomass Gasification. *Energies* **2019**, *12*, 2418. [CrossRef]
39. Rangel-Hernandez, V.H.; Torres, C.; Zaleta-Aguilar, A.; Gomez-Martinez, M.A. The Exergy Costs of Electrical Power, Cooling, and Waste Heat from a Hybrid System Based on a Solid Oxide Fuel Cell and an Absorption Refrigeration System. *Energies* **2019**, *12*, 3476. [CrossRef]
40. Bobbo, S.; Fedele, L.; Curcio, M.; Bet, A.; De Carli, M.; Emmi, G.; Poletto, F.; Tarabotti, A.; Mendrinos, D.; Mezzasalma, G.; et al. Energetic and Exergetic Analysis of Low Global Warming Potential Refrigerants as Substitutes for R410A in Ground Source Heat Pumps. *Energies* **2019**, *12*, 3538. [CrossRef]
41. Valencia, G.; Fontalvo, A.; Cárdenas, Y.; Duarte, J.; Isaza, C. Energy and Exergy Analysis of Different Exhaust Waste Heat Recovery Systems for Natural Gas Engine Based on ORC. *Energies* **2019**, *12*, 2378. [CrossRef]
42. Lin, Z.H.; Xu, T.M. *Utility Boiler Manual*; Chemical Industry Press: Beijing, China, 1999.
43. Exergy efficiency testing technical procedures for industrial boilers. *DB21-T 2567—2016*.

Article

Operability and Technical Implementation Issues Related to Heat Integration Measures—Interview Study at an Oil Refinery in Sweden

Sofie Marton [1,*], **Elin Svensson** [2] **and Simon Harvey** [1]

1   Division of Energy Technology, Department of Space, Earth and Environment, Chalmers University of Technology, 412 96 Gothenburg, Sweden; simon.harvey@chalmers.se
2   CIT Industriell Energi AB, 412 58 Gothenburg, Sweden; elin.svensson@chalmersindustriteknik.se
*   Correspondence: sofie.marton@chalmers.se

Received: 6 May 2020; Accepted: 3 July 2020; Published: 5 July 2020

Abstract: In many energy-intensive industrial process plants, significant improvements in energy efficiency can be achieved through increased heat recovery. However, retrofitting plants for heat integration purposes can affect process operability. The aim of this paper is to present a comprehensive overview of such issues by systematically relating different types of heat recovery retrofit measures to a range of technical barriers associated with process operability and practical implementation of the measures. The paper presents a new approach for this kind of study, which can be applied in the early-stage screening of heat integration retrofit measures. This approach accounts for the importance of a number of selected operability factors and their relative significance. The work was conducted in the form of a case study at a large oil refinery. Several conceptual heat exchanger network retrofit design proposals were prepared and discussed during semi-structured interviews with technical staff at the refinery. The results show that many operability and practical implementation factors, such as spatial limitations, pressure drops and non-energy benefits, influence the opportunities for implementation of different types of heat exchanger network retrofit measures. The results indicate that it is valuable to consider these factors at an early stage when designing candidate heat exchanger network retrofit measures. The interview-based approach developed in this work can be applied to other case studies for further confirmation of the results.

Keywords: heat integration; operability; retrofit; oil refinery; interviews

1. Introduction

There are currently many driving forces to incentivize increasing energy efficiency in industry. For example, the European Union's Energy Efficiency Directive [1] has resulted in national laws requiring large companies to perform energy audits to identify measures for energy efficiency. Environmental legislation, various incentives, and policy support programs for energy efficiency, as well as economic and environmental concerns from customers and business partners, all motivate a stronger focus on energy efficiency in process industry companies.

However, technical, economical, and organizational barriers often hinder implementation of energy efficiency measures. Fleiter et al. [2] stress the importance of distinguishing between different types of energy efficiency measures when discussing barriers for energy efficiency. For example, the risk of production disruption is one of the most important barriers when the energy efficiency measures can affect the core process. Dieperink et al. [3] discussed difficulties associated with implementing energy efficiency measures that affect the core process for selected industrial sites in the Netherlands. Rhodin et al. [4] presented an example from the Swedish foundry industry where technical risks such as production disruptions was the second largest barrier for implement energy efficiency measures. Thollander and Ottosson [5] discussed an example from the Swedish pulp and paper industry in which technical difficulties were also ranked as a large obstacle for implementing energy efficiency measures. Cagno and Trianni [6] also addressed the importance of considering barriers for specific energy efficiency measures, rather than assuming that barriers are the same for all types of measure. This implies that research is needed that addresses a variety of industrial sectors, as well as different types of energy efficiency measures, in order to thoroughly evaluate and investigate which factors affect the implementation potential. Much research has been conducted concerning drivers and barriers in small and medium-sized enterprises. Johansson and Thollander [7] discussed drivers and barriers and suggested recommendations for in-house energy management procedures for several industrial sectors in Sweden, including both small and medium-sized enterprises and energy-intensive industry. Cagno et al. [8] presented a framework for assessing non-energy benefits and non-energy losses that is targeted at industrial decision-makers and covers both technical and management perspectives. However, only a few studies have focused on large energy-intensive industrial plants, such as the Swedish pulp and paper industry [5], the Swedish iron and steel industry [9], and the German steel industry [10], and the lack of studies from the petrochemical process industry is noteworthy.

Changes to an industrial process can have major effects on process operability. It is therefore imperative that operational issues are considered when planning such changes. Process operability includes different operational aspects such as flexibility, controllability, reliability, availability, and start-up and shutdown of the process [11]. For example, if a process is not flexible it cannot adapt to different operating conditions, such as varying feedstock, change of product specifications and/or product mix, and varying ambient conditions. Equipment reliability/availability issues can cause expected and unexpected operational disruptions and controllability problems can lead to major safety issues and production disruptions. Therefore, it is important to investigate how energy efficiency can affect operability. Furthermore, energy efficiency measures can also improve operability, for example by reducing the load on a process capacity-limiting furnace, leading to valuable non-energy benefits for the process. Non-energy benefits refer to benefits other than the direct energy cost savings from the energy efficiency improvement, e.g., reduced carbon dioxide emissions, increased production, and better work environment [12].

Although there are several options to increase energy efficiency in industry, thermal energy is used in large quantities in chemical process plants and heat recovery is therefore one important option to improve energy efficiency in such plants. In previous research, many case studies have identified large techno-economic potentials for energy savings by heat integration in existing industrial plants. To evaluate feasibility of new processes and increased energy efficiency through increased heat recovery, a better estimation of the techno-economic potentials of process heat integration measures is necessary as well as a better understanding of the drivers and barriers affecting the implementation potential. Rebuilding an existing industrial plant to increase heat integration affects the process in several ways. In particular, the number of interdependencies between different parts of the process increases. In previous studies it has been repeatedly discussed that the risk for operability or control problems is strongly connected to the number of interdependencies and interconnections within a process. Subramanian and Georgakis [13] investigated plant-wide steady-state operability issues for an integrated process plant. Setiawan and Bao also discussed the connection between an integrated process and operability issues, both in a study considering interactions between process units [14] and in a study that investigates interaction effects connected to operability [15]. Such operability problems may lead to production disruptions, which must be avoided since they are extremely costly. This underlines the importance of considering operability of heat integration measures at an early stage when investigating retrofits of industrial energy systems.

Heat integration analysis can be based on mathematical programming or graphical insights (e.g., pinch analysis) (see e.g., [16]). A wide variety of case studies have shown a large potential for increased energy efficiency through retrofitting of heat exchanger networks (HENs) at different industrial process sites. There are many different methodologies to identify HEN retrofit designs that achieve high energy savings at low cost, each of which has their own benefits and drawbacks (see e.g., [17] for a review of HEN retrofit methodologies and applications). It is common that several HEN designs can be identified that achieve approximately the same energy saving at similar costs. However, such HEN designs can vary significantly regarding network complexity, placement of new heat exchangers, as well as utility heaters and coolers for target temperature control, etc. It is thus clear that technical and operational factors need to be considered together with investment cost and fuel cost savings when investigating HEN retrofit options.

In the existing literature, there are many studies presenting methods for accounting for specific practical considerations and associated costs in HEN retrofit studies. For example, Becker and Maréchal [18] presented a method to consider heat exchange restrictions using mixed integer linear programming, and Cerda and Westerburg [19] presented a study of HEN synthesis with restricted stream matches. Polley and Kumana [20] suggested dividing larger networks into a number of smaller networks to deal with large heat integration projects. Practical considerations and associated costs are especially important when considering integration at large sites or even across company boundaries, which is the case, for example, for piping and pressure drops. To include plant-specific factors such as piping costs, pressure drop, and heat losses, Bütün et al. [21] proposed a mixed integer linear programming framework. Hiete et al. [22] also included piping costs when considering energy integration between industrial plants with different owners. Jegla and Freisleben [23] also considered pressure drops in their practical method for energy retrofit, but in addition they also considered available heat exchanger space. Reddy et al. [24] presented an optimization method for retrofits of cooling water systems including pressure drops, cooling tower operation, and piping costs. Hackl and Harvey [25] developed a methodology for identifying cost-effective retrofit measures in a chemical cluster adopting a total site perspective. Nemet et al. [26] developed methods for included piping costs, pressure drops, and temperature drops in total site analyses.

Other methods have also been proposed to account for certain specific operability considerations such as flexibility and controllability in network design. Escobar et al. [27] suggested a method to include flexibility and controllability consideration in HEN synthesis. Another method for including operability and observability in HEN design was recently proposed by Leitod et al. [28]. Andiappan and Ng [29] presented a methodology to consider energy systems operability, feasibility and debottlenecking opportunities connected to retrofit design. Abu Bakar et al. [30] suggested including operability in addition to investment and utility cost savings in the choice of ΔT_{min} for HEN design. Several authors have used mixed-integer programming for multi-period optimization to consider flexibility in process integration problems, for example, for integration of utility systems [31], flexible HEN design [32], and integration of biomass and bioenergy supply networks [33]. Bütün et al. [34] presented an approach for including multiple investment periods for a longer time horizon for energy integration.

Although many studies have focused on specific individual aspects of process operability, there is, to the best of our knowledge, no scientific literature that provides a comprehensive overview of the wide variety of process operability issues and that systematically investigates their impact on decision processes related to implementation of HEN retrofit measures. Furthermore, operability issues are traditionally not considered at the early conceptual design stage for techno-economic ranking of alternative heat integration measures. One common approach in HEN retrofit studies is to identify pinch rules violations in the existing HEN, and thereafter attempt to remove or reduce such violations starting with the largest violation. At this early design stage, it is unusual to consider costs other than heat exchanger and utility costs. Operability issues are usually not included until the pre-feasibility or feasibility study phases of the decision-making process for heat integration projects, see Figure 1. However, since energy efficiency measures for increased heat integration are closely connected to the core process of industrial plants, technical aspects can be assumed to be important barriers for their implementation. By considering possible technical barriers and operability issues at an earlier stage of the screening process of energy efficiency options, it could be possible to avoid spending resources on detailed design and feasibility studies of projects that are highly unlikely to be implemented. This is crucial to enable a rapid and relevant screening process of energy efficiency measures and to be able to estimate accurate technical and economical potentials of heat integration. To enable more explicit consideration of operability issues earlier in the screening process, it is, however, important to know which operability factors are most important to consider in the techno-economic evaluation.

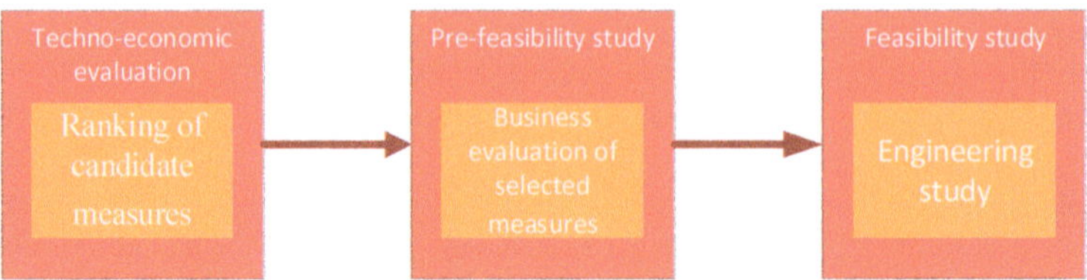

Figure 1. Decision-making process for process development projects.

The aim of this paper is to present a comprehensive overview of operability and technical implementation issues related to heat integration measures by mapping, discussing and clarifying how such measures relate to a comprehensive set of key operability factors. This approach differs from previous studies that have primarily investigated these issues individually. The work was conducted in the form of an interview study at a large oil refinery in Sweden, and as such suggests a new approach for inclusion of operability considerations at an early stage of screening of alternative heat integration options. The paper aims to present an in-depth discussion of the theoretical definitions of operability and the practical considerations of heat integration by investigating its relevance in a real industrial process plant. The case study contributes to expanding the knowledge base for operability and practical implementation issues related to heat integration retrofits.

2. Definition and Categorization of Operability

The following definitions were proposed in previous work by the authors, based on a review of the literature in the area of operability issues related to heat integration [10]. Operability is defined as

" … the ability to operate equipment, process units and total sites at different external conditions and operating conditions, without negatively affecting safety or product quality and quantity. This includes both steady-state and dynamic aspects of operation."

It was also proposed to distinguish between a number of operability aspects that can be sorted into the following sub-categories: flexibility, controllability, feasibility of start-up/shutdown transitions, reliability, availability, and other practical considerations. These sub-categories were based on the following considerations:
Flexibility:

"A flexible process has the ability to maintain feasible operation for different operating scenarios. For oil refining processes, flexibility includes, for example, being able to handle different crude recipes, product mixes and ambient conditions. Flexibility also includes the ability for the operation to handle long-term variations within the process, such as decreased reactivity in catalyst beds and decreased heat transfer due to fouling."

Controllability refers to

" … the ability to maintain a stable process, while handling disturbances and short-term variations to the process. According to our choice of definition, it also includes being able to maintain a stable process during transition from one operating scenario to another."

Feasibility of startup/shutdown transitions refers to

" … the ability to start up or shut down the process in a controlled and safe manner. Due to the special characteristics of startup/shutdown transitions, this is important to consider separately, although it is essentially included in the afore-mentioned definition of controllability."

Reliability refers to

" … the ability to operate a process without unexpected equipment failure."

Whereas availability on the other hand refers to

" … the expected operating time for equipment during a time period that also includes planned maintenance."

Practical considerations are not per se strictly related to operability only but are nevertheless included in the analysis, given their importance related to implementation of HEN retrofit measures [16]. Examples of practical implementation issues include space for new equipment, time availability for retrofitting during major process maintenance shut-down periods and accessibility for erecting new equipment.

3. Industrial Case Study Plant

To thoroughly discuss operability of heat integration measures, a single plant was considered in the case study, which gave the opportunity to design and evaluate retrofit proposals based on real process data. Large industrial plants include many interconnected process units and extensive utility systems. A comprehensive data collection and analysis is thus essential to obtain the details necessary to identify candidate HEN retrofit measures and related operability aspects. In this work, a single process plant was investigated in detail, which provided the opportunity to design HEN retrofits (see Section 4.2) that include many of the aforementioned operability aspects and discuss the proposed retrofit measures in detail with refinery staff. This level of detail would not have been possible if several plants were included in the study.

The case study was conducted at one of the most modern and energy efficient complex oil refineries in Europe, with a crude oil capacity of 11.5 million tons per year and total CO_2 emissions of 1.6 million tons in 2017 [35]. The main products are petrol, diesel, propane, propylene, butane, and bunker oil.

The heat demand of the refinery is satisfied mainly by direct fired process furnaces and by steam that is produced in steam boilers, flue-gas heat recovery boilers and process coolers. The process furnaces and steam boilers are fired by fuel gas that consists mainly of non-condensable gases from the refinery distillation columns. Liquefied Natural Gas (LNG) is used as make-up fuel when the non-condensable gas fuel stream is insufficient. An overview of the main material and energy flows is presented in Figure 2.

The refinery steam network consists of four main pressure headers, which are connected by let-down valves and turbines. The turbines are used in direct drive configuration to operate compressors and pumps, a number of which can be switched to electric motor drive. There is no electrical power generation on site. The refinery regularly has an excess of low-pressure steam. During 25% of the year, the excess of low-pressure steam is particularly high due to an excess of non-condensable gases from the refinery processes. Flaring is strictly regulated, and the refinery has no storage capacity for the non-condensable gases, thus excess gas is combusted in the steam boilers, leading to an excess of steam that is vented.

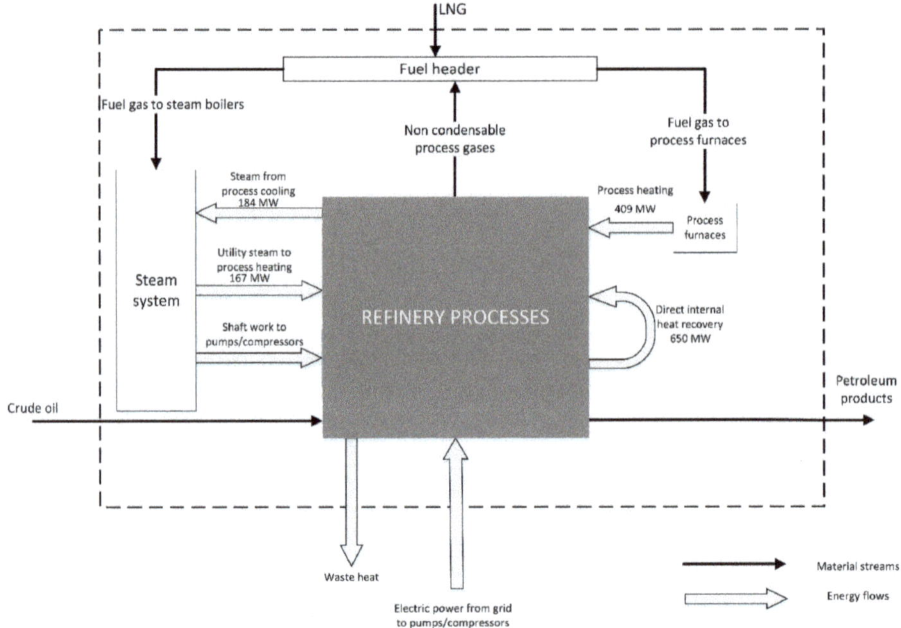

Figure 2. Major material and energy flows of the studied refinery. Data for shaft work from steam turbines and electricity to pumps/compressors was not collected at the same time as the process stream data. The refinery steam system is described in detail in reference [36].

In connection to earlier research projects, energy targeting [37] and retrofit studies [38] have been carried out for the case study refinery. In this work, stream temperature and heat load data were collected for the majority of the refinery heat exchangers. Process stream data was collected 23 April 2010 from production data control room screen shots and data logs. The date was chosen in collaboration with refinery engineers to represent stable full capacity operation. For these operating conditions, the process hot utility demand that was covered by process furnaces was determined to be 409 MW [37]. Minimum utility requirements were determined for the different process units using pinch analysis. Details about the results of this energy targeting are presented in Section 4.2.

4. Methodology

In this section, the framework for the interview study is described as well as the methods used for the interviews and HEN retrofit design.

Although scientific literature is scarce on the subject of operability related to heat integration measures, many experienced engineers and operators in industry possess a deep knowledge and understanding of their processes and the way they operate under various conditions. To be able to tap into this extensive knowledge base, a case study approach based on interviews was adopted. As discussed by Sovacool in references [39,40], this approach provides a broader and more detailed perspective of process operation compared to simulation using a computer model which includes only known parameters and variables. Since limited research is available on the operability aspects that are most important to consider in a HEN retrofit study, a mapping is needed which can thereafter provide guidance for future HEN retrofit evaluations.

An overview of the methodology used for the interview study is shown in Figure 3. As the figure shows, HEN retrofit proposals were designed specifically for the case study process (see Section 4.2), based on a literature review of operability issues related to heat integration measures [10]. The process data for the retrofit proposals were taken from a previous energy targeting study at the refinery [37]. The proposals were discussed with refinery experts in eleven interviews (see Section 4.3). The results were then summarized and presented to the refinery experts again for confirmation and further discussion at a validation seminar.

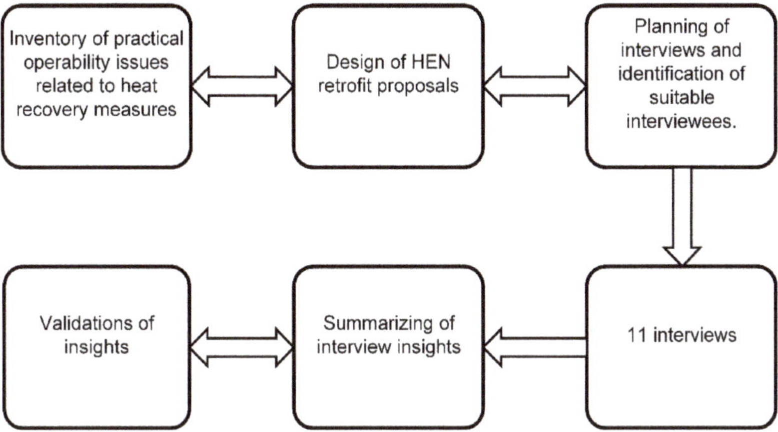

Figure 3. The work flow for the interview study.

4.1. Inventory of Possible Process Operability Implications Related to Heat Integration Measures

Implementation of new heat recovery measures involves many changes to the process equipment and operation, ranging from new and modified heat exchanger units to changes in pressures, steam balances and interactions between different parts of the process. In this paper, we refer to these changes as "process implications".

In order to discuss different aspects of operability during the interviews, a number of heat integration retrofit proposals were designed that cover different process implications related to operability. To ensure an exhaustive coverage of process implications and operability aspects, a list of potential process implications was compiled based on literature examples and experience from previous process integration projects. The implications included on the list were matched with the operability aspects that were most likely to be affected (see Table 1). After the first round of interviews with plant staff, the list of possible implications was extended if new process implications were identified. Table 1 was also used to design the list of questions to be addressed during the interviews, as described in Section 4.3.

Table 1. The assumed relations between the process implications of heat integration retrofit measures and the most affected operability factors and practical implementation issues (Marton et al. 2016).

Implications of Heat Integration Retrofit Measures	Operability Factors and Implementation Issues	Flexibility	Controllability	Startup/ Shutdown	Reliability/ Availability	Practical Considerations
1. De-bottlenecking		X				
2. Stream splitting			X			
3. HEN complexity		X	X			
4. Reduced load on a furnace		X	X			
5. Reduced load on an air cooler		X	X			
6. Increased pressure drop in heat exchangers		X	X			
7. Change in steam balance		X	X			
8. Shut down of furnace before reactor		X	X	X		
9. Heat exchange between process units		X	X	X		
10. New equipment installation					X	X
11. Rebuilding existing equipment					X	X
12. Pressure differences between streams or high pressures					X	X

4.2. Design of Retrofit Proposals

Nine HEN retrofit proposals were designed for discussion during the interviews. The proposals were designed to include specific process implications connected to operability and technical implementation aspects. The proposed HEN retrofits were designed within selected process units at the refinery, using a pinch technology approach [41] based on stream data representing normal operating conditions and information about the current placement of heat exchangers.

Previous studies at the refinery, also based on pinch analysis, showed that five process units account for 90% of the current hot utility use and also have the greatest potentials for utility savings [38]. One of these units has been rebuilt since the data was collected and pinch analysis targeting was conducted. Therefore, the remaining four units were chosen for this study. To be able to investigate operability aspects of heat integration between process units, two units located close to each other were grouped together. Current hot utility usage and theoretical minimum heat demand for the chosen units are listed in Table 2. The analysis and design were conducted for a single operating point which represents normal refinery operation. It should be noted, however, that process operation and ambient conditions vary over time. Although the study includes a single operating point, flexible operation for other process conditions was discussed in the interviews.

Table 2. Current and minimum heat demands for process units included in this paper. For the pinch analysis, the following minimum temperature differences contributions were considered; $\Delta T_{min}/2 = 10\,K$ for condensing/boiling hydrocarbons, $\Delta T_{min}/2 = 5\,K$ for water, $\Delta T_{min}/2 = 2.5\,K$ for boiling water, and $\Delta T_{min}/2 = 15\,K$ for other process streams.

Unit	Current Heat Demand (MW)	Min Heat Demand (MW)
A + B	125	104
C	26	10
D	46	9

The design of retrofit proposals was based on the list of implications presented in Section 4.1. Each retrofit proposal was designed to investigate the effect of some of the specific implications. The retrofit proposals were also designed so that all implications are covered, which can be seen in Table 3. All retrofit proposals are described in detail in Appendix A.

For Unit A + B, the main objective was to include heat exchange between two process units (Implication #9) in the retrofit proposals (Retrofit proposal 1A–C and 2). All proposals for Unit A + B include reduced load on the same furnace, but with different paths for stream pre-heating. The stream pre-heating configurations differ with respect to complexity, additional heat exchanger area requirements, and heat source (hot process streams or hot flue gases). Another aspect included in Retrofit proposal 1B is the replacement of steam heating in a distillation column reboiler by heating by internal heat exchange within the process units. For Unit C, three different ways of increasing the pre-heating before a process furnace were considered. The first retrofit proposal, 4A, involves heat recovery from other process streams currently cooled with overhead air fans. An excess of low-pressure (LP) steam is available at the refinery during most of the year. In Retrofit proposal 4B, excess LP steam is used for the pre-heating, decreasing the number of process interconnections. Retrofit proposal 4C also uses LP steam for pre-heating, but the proposal includes a stream split. For unit D, two retrofit proposals were designed. The retrofit proposals for unit D involve heat savings in two different process furnaces that also result in a reduction in high-pressure steam production. The furnace in Retrofit proposal 5 is placed prior to an exothermic reactor and is suggested to be taken out of operation. Both retrofit proposals in unit D also include process streams with high pressures and heat exchangers with large pressure differences between the process streams.

Table 3. Implications included in each retrofit proposal.

Implications of Retrofit Measures	Retrofit Proposal	Unit A + B				Unit C			Unit D	
		1A	1B	1C	2	4A	4B	4C	5	6
1. De-bottlenecking						X	X	X		X
2. Stream splitting								X		X
3. HEN complexity			X			X		X		
4. Reduced load on a furnace		X	X	X	X	X	X	X		X
5. Reduced load on an air cooler		X	X	X		X	X	X		
6. Increased pressure drop in heat exchangers		X	X	X		X	X			X
7. Change in steam balance			X		X		X	X		
8. Shut down of furnace before reactor								X		X
9. Heat exchange between process units		X	X	X		X	X	X		
10. New equipment installation		X	X	X	X	X	X			X
11. Rebuilding existing equipment		X	X	X		X			X	X
12. Pressure differences between streams or high pressures									X	X

4.3. Interview Procedure

All interviews in the study were semi-structured interviews that were conducted face-to-face. This enabled good communication as well as the possibility to discuss printouts of flow charts in detail. In addition, semi-structured interviews enable a good combination of structure as well as flexibility with respect to opportunities for follow-up questions and discussion during the interviews [42]. To our knowledge, this method has not been used before for discussing and investigating technical aspects related to implementation of heat integration retrofit measures. The interviews were conducted with technical staff with significant knowledge about operational and technical aspects of the refinery. Most interviews were about one hour long, but there was no time limit. The interviews were conducted in Swedish and all material was transcribed afterwards.

The interview procedure was the same for all technical staff responsible for the process units included in the study. HEN retrofit proposals were shared in advance to give the interviewees an opportunity to prepare for questions and check anything uncertain about the affected part of the process unit. The same set-up of open questions was used to discuss all retrofit proposals. Firstly, open questions were asked about the interviewee's thoughts about potential consequences of implementing the retrofit proposal. For all issues that were identified, solution suggestions were requested and discussed. Following the open questions, more specific questions were asked about operability aspects considered in the design phase of the retrofit proposal. To conclude, the interviewee was asked to list the top three obstacles and grade the retrofit proposals implementation potential from one (low) to four (high). The interviews with mechanical engineers, control engineers, and the process utility system engineer started with a general discussion about their expertise related to process integration. Retrofit proposals that were discussed were sent beforehand. This allowed the interviewees to collect necessary information about the processes affected by the proposals and less experienced engineers could discuss the proposals with more experienced engineers beforehand. The interviews with mechanical engineers, control engineers, and the process utility system engineer also provided an opportunity to verify anything unclear brought up in the previous interviews with operations and process engineers regarding equipment, utility systems, or control systems. Table 4 lists the content discussed in each interview.

Finally, results from the interviews were summarized and presented at a validation seminar which was attended by several of the interviewed engineers as well as managers responsible for process development. The results and main conclusions from the interviews were presented to the refinery experts involved in the study. The refinery experts confirmed and clarified the results. Consequently, a comprehensive and systematic in-depth coverage of the included topics was achieved.

Table 4. List of interviewees and content discussed in the interviews.

	Refinery Responsibilities	**Content Discussed**
1	Operations engineer, Unit A and B	Retrofit proposals 1A–C, 2
2	Process engineer, Unit A and B	Retrofit proposals 1A–C, 2
3	Operations engineer, Unit C	Retrofit proposals 4A–C
4	Process engineer, Unit C	Retrofit proposals 4A–C
5	Operations engineer, Unit D	Retrofit proposals 5, 6
6	Process engineer, Unit D	Retrofit proposals 5, 6
7	Control engineer	Process control system Retrofit proposals 1A, 4C, 5
8	Control engineer	Control of the steam utility system
9	Process engineer, energy systems	Steam utility system and fuel gas system Retrofit proposal 4A, 6
10	Mechanical engineer, heat exchangers and air coolers	Heat exchangers and air coolers Retrofit proposal 1A, 4A, 5
11	Mechanical engineer, boilers and process heaters	Fired heaters and boilers Retrofit proposal 1A, 2, 5

5. Observations and Insights

In this section, the main observations and insights from the interviews are presented and discussed. The discussion refers to the retrofit proposals that are presented in Appendix A.

5.1. Practical Considerations

In almost all interviews, practical considerations were stated as being important for implementing HEN retrofit proposals. The most commonly discussed practical consideration was spatial restrictions in the plant. Other issues that were stated repeatedly were limitation in time and space for making process modifications during expensive turn-around periods and the high cost of equipment operating at high pressure. Although most retrofit proposals involve practical difficulties, it was generally considered that these issues can usually be solved. However, the proposed solutions to the practical problems lead to higher costs that should be accounted for when designing HEN retrofit proposals. For example, retrofit proposal 1A includes doubling of the surface area of a large existing heat exchanger to achieve higher internal heat recovery. There is limited space available in the affected process unit, which makes doubling of the heat exchanger size difficult. However, if the existing shell-and-tube exchanger were to be replaced with new efficient plate heat exchangers, the increased heat recovery could be achieved in a smaller space than the space currently occupied by the existing heat exchanger. To enable cleaning of the new plate heat exchanger during operation, it would be necessary to have two plate exchangers in parallel, but they would still occupy less space than the original shell-and-tube heat exchanger. This was confirmed by a mechanical engineer during interview 10 as well as during the validation seminar:

> *"Yes, if they are replaced with plate exchangers that should not be an issue. It would require more piping but that would require less space, so I do not see any issues with that."*
>
> *Mechanical engineer, heat exchangers and air coolers*

Increased pressure drop caused by increased heat exchanger area was highlighted during several interviews as something that always needs to be taken into consideration. Both new and extended existing heat exchangers will lead to increased pressure drop. Pressure drop was mentioned both during the interviews and the seminar. In particular, several interviewees stated that if the pressure drop is too large for the current pump capacity, new pumps will be necessary, and the total investment cost will probably be too high for the retrofit proposal to be implemented.

5.2. Maintenance

During the interviews, increased maintenance was mentioned as a potential issue, particularly the need for both space and time to clean the heat exchangers. If the heat exchangers are not properly cleaned, pressure drops increase significantly, which can cause issues for operation of downstream process units. Heat exchangers that already experience problems with fouling are likely to be penalized by decreased reliability/availability if enlarged, due to the increased need for maintenance during operation. The decreased reliability/availability for tube-and-shell heat exchangers is caused by the need to lower the feed flowrate to the unit to enable cleaning on both tube and shell side of the heat exchanger. One solution stated by several interviewees for fouling issues is to remove existing shell-and-tube exchangers and replace them with parallel plate exchangers. In a number of interviews, it was stated that a simultaneous investment to improve current operability issues caused by fouling could increase the prospect of investing in an energy saving project. This is the case for the previously discussed retrofit proposal 1A. Combining an expansion of the heat exchanger with a replacement of the existing shell-and-tube heat exchanger would not only decrease utility usage but would also decrease fouling problems. This reasoning was confirmed in several of the interviews as well as at the validation seminar.

5.3. De-Bottlenecking

Several retrofit proposals turned out to provide opportunities for removing bottlenecks in the production process. De-bottlenecking increases the flexibility of the process. Retrofit proposals that result in load reduction in process furnaces that currently constitute production bottlenecks were ranked higher than other proposals. These debottlenecking implications were identified by the interviewees even though they were not intentionally included in the retrofit proposals during the design procedure. The possibility to increase production or yield of desirable products was declared as important and recurrently discussed in the interviews. For example, the operations engineer in interview 3 stated that

> *"Well, I am very interested in this, if we can reduce the energy consumption here … it will not only be an energy aspect, but I think we can increase the flow through the unit as well"*
>
> *Operations engineer Unit C*

Regarding the process furnace, HTR-C, included in retrofit proposals 4A–C.

5.4. Controllability and Flexibility

The effect on flexibility and controllability from increased interconnections and complexity was mentioned as a potential issue, but often needs further investigation to evaluate its significance. For retrofit proposal 1B (see Appendix A), it is suggested to heat a distillation column reboiler by internal heat exchanging instead of with utility steam. The retrofit proposal contains several new interconnections, both within the process unit (Unit B) and between Unit A and Unit B. It was considered a potential problem that the reboiler would become dependent on other parts of the unit. Whether the increased number of interdependencies would have a significant effect on reboiler operability needs to be further investigated. Similar issues were discussed regarding retrofit proposal 4C (see Appendix A) in which a stream split is included. Stream splits are not used to a great extent in the process units for which the interviewed process and operations engineers are responsible and they therefore had no clear opinion about possible impacts on operability. The control engineer, on the other hand, stated that the stream split is possible but new control valves and measurements are needed, as well as a more thorough analysis of the control structure. However, at the validation seminar it was acknowledged that almost all refineries have several well-functioning stream splits in the crude oil pre-heating unit. Both examples (the integration of the reboiler in 1B and the stream split in 4C) show that a large increase in interdependencies might cause operability issues, but to know whether this is the case, and how it then can be managed, a more thorough analysis is needed. Issues to be investigated include, for example, modeling and simulation and potentially more advanced control structure design.

Another important aspect discussed concerning flexibility and controllability is that temperatures in several process units change over time due to the deactivation of reactor catalysts. Because of the changing temperatures during the catalyst cycle, it is important to consider the HEN design for more than one operational point.

Large negative effects on flexibility or controllability caused by heat exchange between process units were not discussed to any great extent during the interviews. Units A + B are almost always operated simultaneously, but a back-up solution for heating/cooling needs to be available when one of the units is not in operation. During the validation seminar, the same aspects were discussed, and it was confirmed that it is possible to heat exchange between two process units without any significant decrease in flexibility or controllability. However, this is only the case for units with similar operation patterns and, as previously stated, if there is a back-up solution available when heat exchange is not possible. Heat exchange between two process units that are not operated according to similar schedules was considered very unlikely to be feasible.

5.5. Safety Aspects

Safety aspects were discussed in many interviews, especially regarding retrofit proposal 5 (see Appendix A) which involves taking a process furnace (HTR-D1) out of operation since it is not needed from an energy point of view. Increased internal heat exchange could easily replace the heat provided by the furnace. The process furnace is placed immediately upstream of an exothermic reactor which is very sensitive to inlet temperature. During the interviews it was clearly stated that the retrofit proposal would not create a controllability issue with respect to the stabilizing function of the reactor inlet temperature during normal operation. The temperature control would not be affected since the existing control is located upstream of the furnace. However, it was very clear during the interviews that a safety issue could occur. In all interviews regarding retrofit proposal 5, it was explained that it is necessary to be able to rapidly lower the reactor inlet temperature if a runaway reaction occurs. This is currently achieved by shutting down the furnace. If the furnace is to be taken out of operation, another solution would be necessary to stop potential runaway reactions. Possible solutions were discussed during the interviews, but the safety control for the retrofit proposal needs to be thoroughly investigated if the retrofit proposal is to be implemented.

5.6. Start-Up and Shutdown

In the interviews, operability aspects related to start-up and shut-down of the process were mostly discussed concerning heat exchange between different process units and taking furnace HTR-D1 out of operation in retrofit proposal 5 (see Appendix A). Regarding heat exchange between different process units, the interviewees stated that the process units are not usually started up and shut down simultaneously. Consequently, back-up solutions for heating/cooling during start-up/shutdown are required for heat integration designs in which heating is provided from a different process unit during normal operation. Regarding retrofit proposal 5, the process furnace is not needed during normal operation. However, HX-D1 in the proposal is dependent on the hot reactor effluent. To start the reaction, heat needs to be added to the process, making the process furnace necessary during start-up.

5.7. Non-Energy Benefits

Other than operability considerations and practical implementation issues, non-energy benefits were stated as important for several of the retrofit proposals during the interviews. Many of the non-energy benefits discussed were not considered in the original design of the retrofit proposals, but pointed out by the refinery experts during the interviews. Examples of non-energy benefits that were discussed are de-bottlenecking, reduced load on overloaded air coolers and improved product quality. If the retrofit proposal included a non-energy benefit, the interviewees claimed that this increased the incentive to implement the measure by simultaneously providing an opportunity to increase production or solve an operational issue. This was very clear during both the interviews and the validation seminar. Additionally, the retrofit proposals that included non-energy benefits were ranked higher in the interviews. For example, the process engineer in interview number 6 stated that

> *"If you only consider fuel savings, if we would implement this to save those 8 MW, I would give the retrofit score two. But if you consider that we could achieve a bigger revamp and also consider the effects on the tower it would be a three."*

Process Engineer, Unit D

Regarding retrofit proposal 6 (see Appendix A). In the validation seminar, it was confirmed that non-energy benefits are important to consider alongside energy efficiency measures and the refinery experts stated that non-energy benefits have been a major decisive factor for previously implemented energy saving projects. Energy projects without other process gains have usually been discarded when planning refinery turn-arounds.

6. Discussion

Many of the insights presented in Section 5 have been highlighted in previous research. However, previous studies have only investigated the related aspects individually and separately and have not presented a comprehensive overview of operability and practical implementation issues. Lundberg et al. [43] identified the positive effects of de-bottlenecking the recovery boiler at a Kraft pulp mill if heat integration measures are implemented simultaneously when rebuilding the plant. Dhole and Buckingham [44] proposed a methodology to simultaneously consider pinch analysis and column targeting (modification of column design to fit thermodynamic profiles obtained from pinch analysis) for a refinery, in order to achieve de-bottlenecking without increasing the existing furnace load. See also Li et al. [45] for a description of combining de-bottlenecking and pinch analysis in oil refining industry. These examples indicate that the importance of de-bottlenecking highlighted in the interview study is applicable for other cases than the selected oil refinery. Similar comparisons can be made for technical difficulties such as controllability or flexibility. Although it has previously been shown that non-energy benefits and technical and practical difficulties will have an impact on decision making process for selecting new projects to implement, the results in this paper present a wider perspective. The results show these aspects together rather than separately which enables a discussion of their relative importance. For example, the results indicate that non-energy benefits can outweigh the negative effects of technical and practical difficulties for heat integration retrofit proposals. Economic considerations are included in traditional pinch analysis based design but were not in focus in the retrofit proposals in this work. In traditional pinch design, the profitability of heat integration rebuilds is assumed to depend primarily on the energy cost savings and the investment cost for new heat exchangers. Operability considerations are likely to affect both the operating and investment costs for heat integration retrofit measures. Traditional pinch analysis is conducted for steady-state operation. In order to achieve good dynamic operability, additional equipment might be required, such as advanced control systems and/or over-capacity or back-up systems for flexible production. Additionally, if flexibility is considered, the heat savings can vary for different operating scenarios which change the expected heat savings, affecting the cash flows and the expected profitability. Non-energy benefits also affect the profitability of the heat integration retrofit proposals by increasing the revenues or decreasing the capital costs.

Practical issues discussed during the interviews indicate that spatial limitations should be considered earlier in the screening of alternative investment options. It is usually only the investment cost for a new heat exchanger or for extension of an existing heat exchanger that is included in the analysis. Since practical issues, especially spatial restrictions, were a major part of the issues discussed during the interviews, this is an important parameter to include in the early energy targeting analysis. One way to take this into account is to include spatial restrictions in the choice of ΔT_{min}. The optimal ΔT_{min} value is affected by investment and operating costs as well as operability considerations [30].

For retrofit studies where space is limited, spatial restrictions should be reflected by a higher value of ΔT_{min} in the screening phase.

Steam balances were discussed in several of the interviews. At large refineries, and other chemical industrial plants, utility systems are often large and complicated. The fuel balances at the studied refinery vary during the year, and during 25% of the year the refinery has an excess of fuel gas for the steam boilers and process furnaces. As a consequence, the studied refinery often has an excess of low-pressure steam, which is considered as free. For steam at higher pressure levels it is more difficult to know how the overall steam and fuel balances are affected by changes. Since it is not obvious how changes in steam production and consumption affect the overall steam and fuel balances at the refinery, a steam model for the refinery was developed after the interview study and applied to the retrofits that included the steam system (see [36]).

7. Conclusions

This paper presented the results of a comprehensive inventory of potential technical barriers and process operability constraints related to the implementation of heat recovery measures in industrial process plants. Unlike previous studies, which mainly focus on single implementation barriers or operability aspects of heat integration measures, we presented a complete inventory of such technical constraints including aspects related to flexibility, controllability, start-up/shutdown, reliability/availability, and other practical considerations related to the implementation of measures.

We also suggested a mapping of these potential technical issues to certain design characteristics that may appear in HEN retrofit designs. This mapping could be used to identify the issues that are most likely to be of importance when evaluating a certain heat recovery design. As such, it was proposed to support a new approach for investigating operability and implementation constraints in the early-stage screening of candidate heat recovery measures in industrial process plants. The approach is based on qualitative analysis by means of an interview study, in which operability and technical implementation issues related to specific heat recovery projects are identified, characterized and described with respect to their impact on implementation potential.

The results from interviews at a large oil refinery used as a case study indicated that technical and practical constraints in the plant have a major influence on how the potential for implementation of the heat integration retrofits are ranked. The process considerations that were highlighted most often in the interview study were spatial limitations and maintenance requirements. Another conclusion from the interviews is that non-energy benefits of an energy-saving project can be far more important than the energy savings as such and may also outweigh potential technical difficulties. An example of this is when the heat savings create an opportunity to remove a production bottleneck.

Based on the ranked importance of technical constraints and productivity benefits, it can be concluded that it would be valuable to take such process aspects into consideration at an earlier design stage than usual when constructing HEN retrofits for increased heat integration. If operability, non-energy benefits and practical implementation issues were considered in earlier-stage techno-economic assessments and screenings of heat recovery measures, several issues could be avoided, and large process benefits could be achieved. The inclusion of those factors would also lead to a better estimation of techno-economic potentials for heat integration measures and thereby a more accurate screening process for different energy efficiency and climate mitigation options.

Future work including additional case studies is needed to further confirm the impacts of operability and other technical constraints on the feasibility and cost-efficiency of HEN retrofit measures, prior to suggesting ways to include the effect of these at an earlier design stage and better quantify their values. It would also be of great value to study how these aspects have affected the evaluation of previously implemented and rejected heat integration measures.

Author Contributions: Conceptualization, S.M. and E.S.; Data curation, S.M.; Formal analysis, S.M.; Funding acquisition, E.S.; Investigation, S.M.; Methodology, S.M. and E.S.; Project administration, E.S. and S.H.; Resources, S.H.; Supervision, E.S. and S.H.; Validation, E.S. and S.H.; Visualization, S.M.; Writing—original draft, S.M.;

Writing—review and editing, E.S. and S.H. All authors have read and agreed to the published version of the manuscript.

Funding: This research was funded by Preem AB through Chalmers Energy Area of Advance.

Acknowledgments: A number of participants from Preem are gratefully acknowledged for their time, input data and participation in interviews.

Conflicts of Interest: The authors declare no conflict of interest. The funders had no role in the design of the study; in the collection, analyses, or interpretation of data; in the writing of the manuscript, or in the decision to publish the results.

Appendix A Heat Exchanger Network Retrofit Proposals

In this appendix all retrofit proposals included in the interview study are presented.

Appendix A.1 Unit A/B

All retrofit proposed for the Units A and B reduce the load on furnace HTR-AB1. For the existing configuration of the included process equipment of Unit A, see Figure A1.

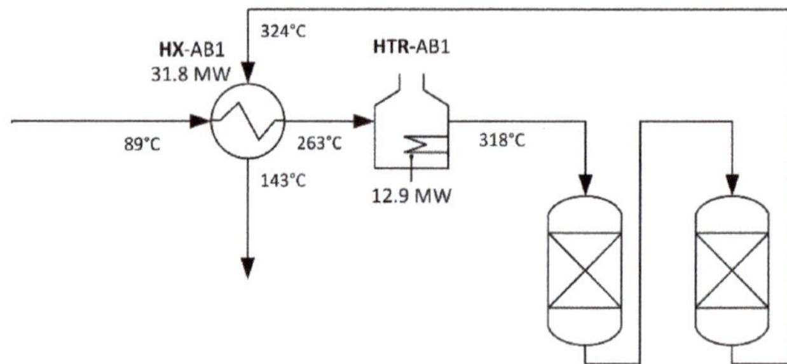

Figure A1. Existing process configuration of effected part of Unit A.

Appendix A.1.1 Retrofit Proposal 1A

In this retrofit proposal, the feed to furnace HTR-AB1 is pre-heated before entering heat exchanger HX-AB1. The feed is pre-heated by a process stream from Unit B, which thereby gets a lower cooling demand, reducing the load on air cooler CLR-AB. Since the heat recovery is very high in this retrofit proposal, the need for new heat exchanger area is large. The retrofit proposal is displayed in Figure A2.

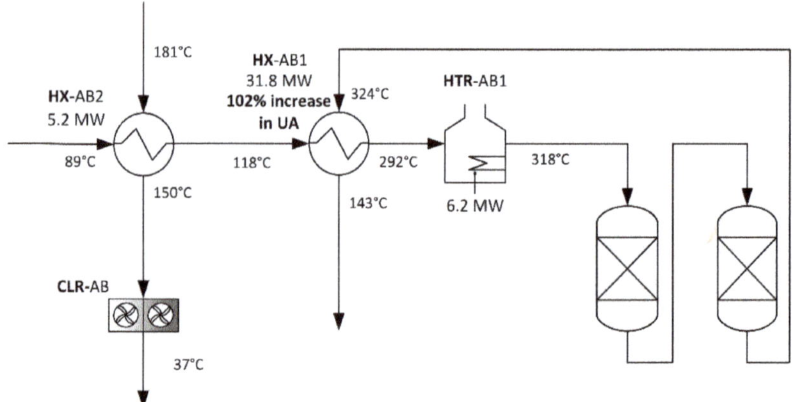

Figure A2. Retrofit proposal 1A.

Appendix A.1.2 Retrofit Proposal 1B

Retrofit proposal 1B involves increased heat recovery and network complexity compared to retrofit proposal 1A. The hot stream from Unit B is in this proposal used as a heat source for a distillation column's reboiler, HX-AB3, before preheating the reactor feed in HX-AB2. The proposal is shown in Figure A3.

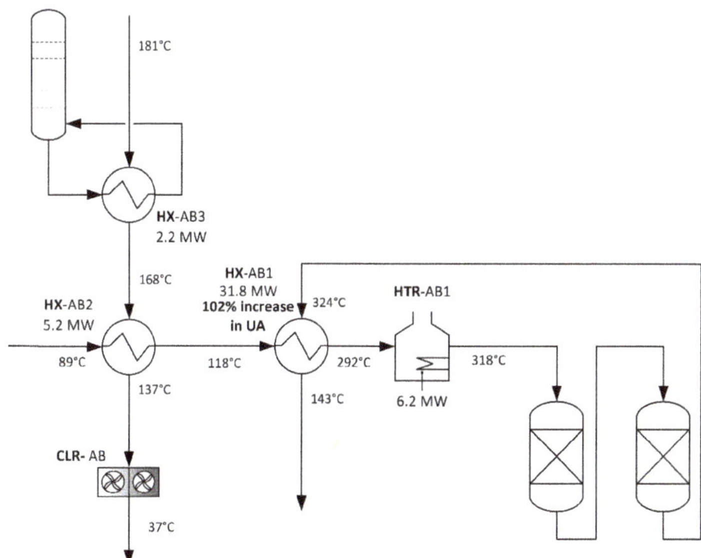

Figure A3. Retrofit proposal 1B.

Appendix A.1.3 Retrofit Proposal 1C

Retrofit proposal 1C is structurally the same as retrofit proposal 1A. The difference in this proposal is that HX-AB1 is not extended, thus changing the temperature of the hot process stream leaving HX-AB1 and reducing the fuel saving in the furnace compared to proposal 1A. The proposal is shown in Figure A4.

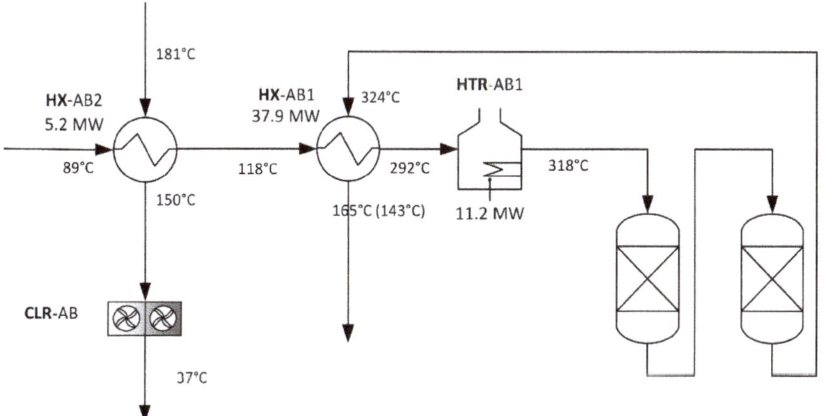

Figure A4. Retrofit proposal 1C.

Appendix A.1.4 Retrofit Proposal 2

Also in retrofit proposal 2, fuel gas is saved in HTR-AB1. In this case the increased pre-heating is achieved by an additional convection part in HTR-AB2, recovering more energy from the hot flue gases of that furnace. The retrofit proposal is shown in Figure A5.

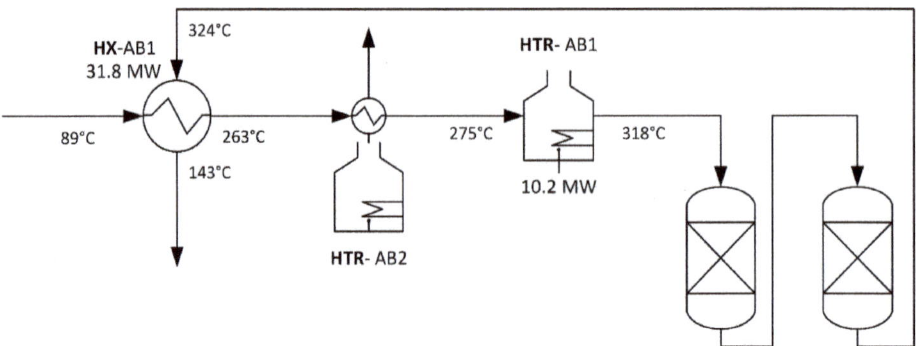

Figure A5. Retrofit proposal 2.

Appendix A.2 Unit C

All retrofit proposals for unit C reduces the load on HTR-C. The current process layout is shown in Figure A6.

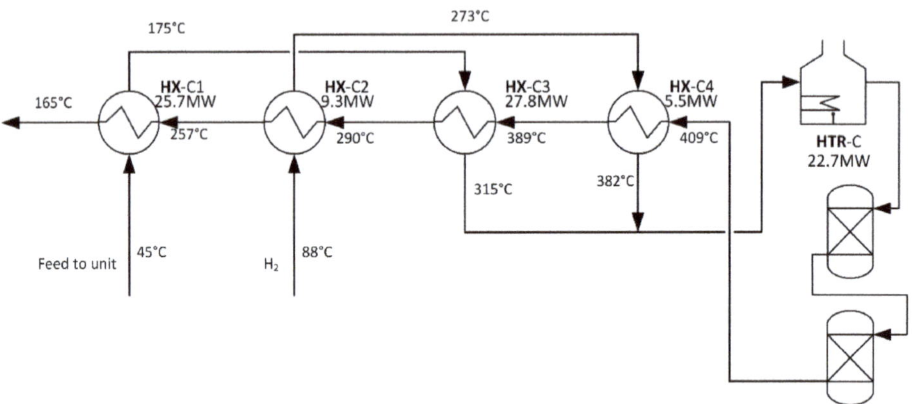

Figure A6. Current process scheme for selected part of Unit C.

Appendix A.2.1 Retrofit Proposal 4A

In retrofit proposal 4A, the feed is heated by internal heat exchange with other process streams in the same process unit. The pre-heating is smaller than in retrofit proposal 4B and 4C, which leads to less fuel savings but also less need for new heat exchanger area. The proposal is shown in Figure A7.

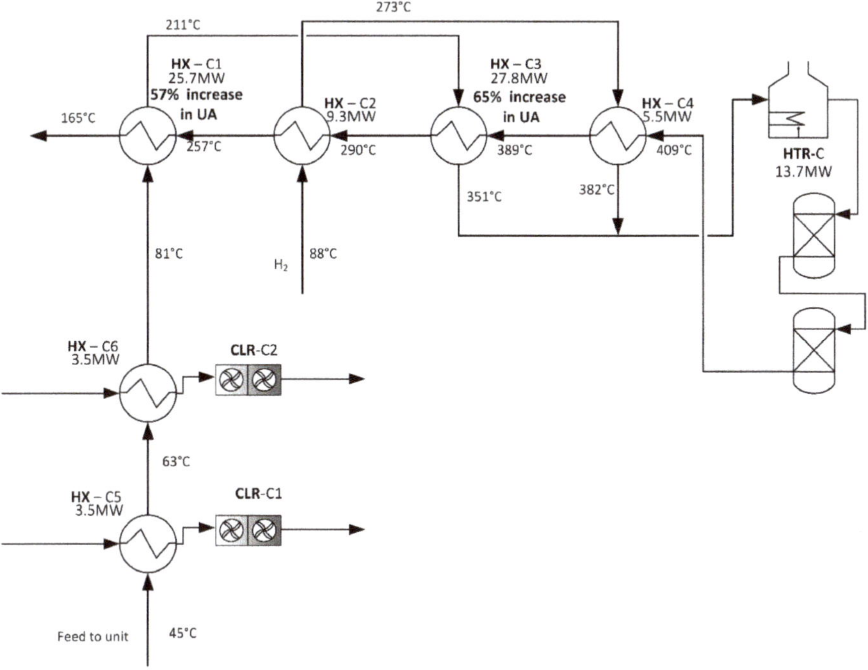

Figure A7. Retrofit proposal 4A.

Appendix A.2.2 Retrofit Proposal 4B

In retrofit proposal 4B, the feed is pre-heated with LP steam. A large part of the year, excess LP steam is vented to the atmosphere. In the proposal, some of this excess LP steam is used. The feed is pre-heated more than in retrofit proposal 4A, which gives larger fuel savings but bigger area increase in existing heat exchangers. Retrofit proposal 4B is shown in Figure A8.

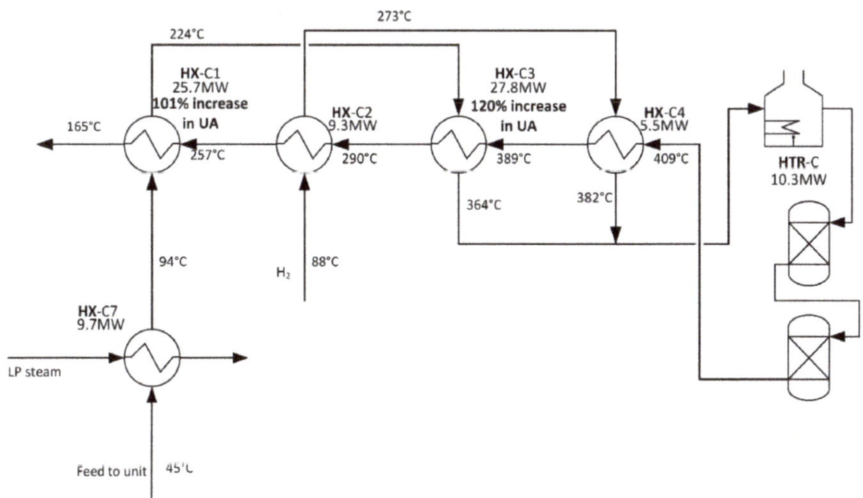

Figure A8. Retrofit proposal 4B.

Appendix A.2.3 Retrofit Proposal 4C

In retrofit proposal 4C, the feed is pre-heated with LP steam as in retrofit proposal 4B. The difference is that the piping is changed and the hot reactor effluent is split to two parallel streams. This gives a slightly smaller area increase compared to 4C, but requires more piping and likely a more complex control system. The proposal is shown in Figure A9.

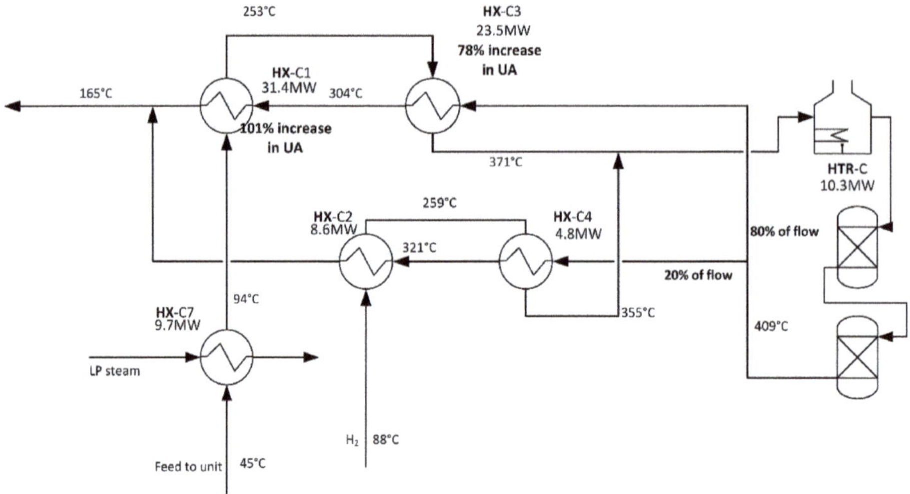

Figure A9. Retrofit proposal 4C.

Appendix A.3 Unit D

For Unit D, two retrofit proposals were designed.

Appendix A.3.1 Retrofit Proposal 5

In this retrofit proposal, the area in HX-D1 is increased to increase the heat recovery from the hot reactor effluent. When the pre-heating before furnace HTR-D1 is increased, the furnace will not be needed during normal operation. As the hot inlet temperature to SG-D is decreased, the steam generation is also decreased. In addition, the steam generation is further decreased as a consequence of that the pre-heating and superheating that occurs in furnace HTR-D1 is removed if the furnace is shut down during normal operation. The proposal is shown in Figure A10.

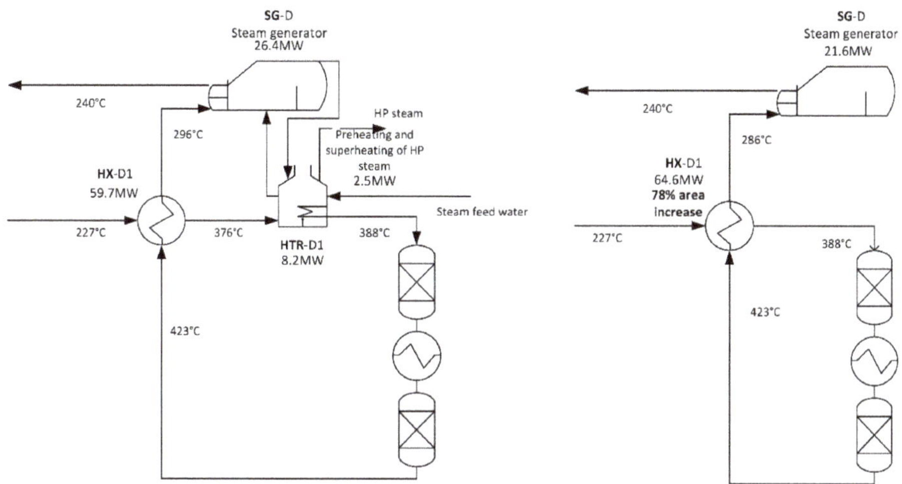

Figure A10. The process current configuration is shown to the left in the figure and retrofit proposal 5 is shown to the right.

Appendix A.3.2 Retrofit Proposal 6

In retrofit proposal 6, fuel is saved in furnace HTR-D2 by pre-heating the feed by another hot process stream. As in retrofit proposal 5, retrofit proposal 6 leads to lower steam production in steam generator SG-D. The proposal is shown in Figure A11.

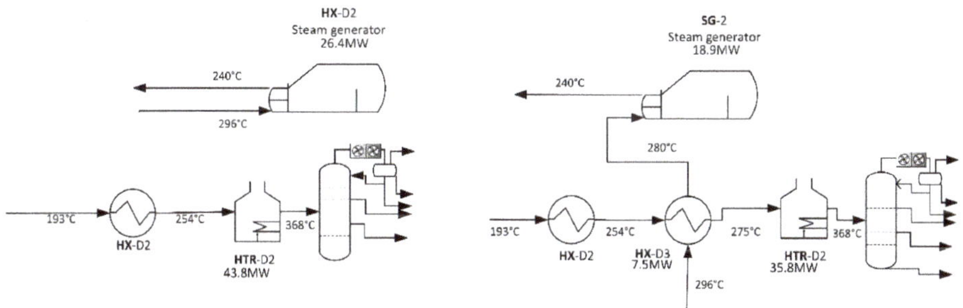

Figure A11. Left part of the figure shows current process configuration and right part of the figure shows the suggested retrofit proposal 6.

References

1. European Union. Directive 2012/27/EU of the European Parliament and of the Council of 25 October 2012 on Energy Efficiency, Amending Directives 2009/125/EC and 2010/30/EU and Repealing Directives 2004/8/EC and 2006/32/EC Text with EEA Relevance. 2012. Available online: http://eur-lex.europa.eu/eli/dir/2012/27/oj (accessed on 16 April 2019).
2. Fleiter, T.; Hirzel, S.; Worrell, E. The characteristics of energy-efficiency measures—A neglected dimension. *Energy Policy* **2005**, *51*, 502–513. [CrossRef]
3. Dieperink, C.; Brand, L.; Vermeulen, W. Diffusion of energy-saving innovations in industry and the built environment: Dutch studies as inputs for a more integrated analytical framework. *Energy Policy* **2004**, *32*, 773–784. [CrossRef]
4. Rohdin, P.; Thollander, P.; Solding, P. Barriers to and drivers for energy efficiency in the Swedish foundry industry. *Energy Policy* **2007**, *35*, 672–677. [CrossRef]

5. Thollander, P.; Ottosson, M. An energy efficient Swedish pulp and paper industry—Exploring barriers to and driving forces for cost-effective energy efficiency investments. *Energy Effic.* **2008**, *1*, 21–34. [CrossRef]

6. Cagno, E.; Trianni, A. Evaluating the barriers to specific industrial energy efficiency measures: An exploratory study in small and medium-sized enterprises. *J. Clean. Prod.* **2014**, *82*, 70–83. [CrossRef]

7. Johansson, M.T.; Thollander, P. A review of barriers to and driving forces for improved energy efficiency in Swedish industry-Recommendations for successful in-house energy management. *Renew. Sustain. Energy Rev.* **2018**, *82*, 618–628. [CrossRef]

8. Cagno, E.; Moschetta, D.; Trianni, A. Only non-energy benefits from the adoption of energy efficiency measures? A novel framework. *J. Clean. Prod.* **2019**, *212*, 1319–1333. [CrossRef]

9. Brunke, J.C.; Johansson, M.; Thollander, P. Empirical investigation of barriers and drivers to the adoption of energy conservation measures, energy management practices and energy services in the Swedish iron and steel industry. *J. Clean. Prod.* **2014**, *84*, 509–525. [CrossRef]

10. Arens, M.; Worrell, E.; Eichhammer, W. Drivers and barriers to the diffusion of energy-efficient technologies-a plant-level analysis of the German steel industry. *Energy Effic.* **2017**, *10*, 441–457. [CrossRef]

11. Marton, S.; Svensson, E.; Harvey, S. Investigating Operability Issues of Heat Integration for Implementation in the Oil Refining Industry. In Proceedings of the ECEEE Industrial Efficiency, Berlin, Germany, 12–14 September 2016.

12. IEA. *Capturing the Multiple Benefits of Energy Efficiency*; International Energy Agency: Paris, France, 2014.

13. Subramanian, S.; Georgakis, C. Methodology for the steady-state operability analysis of plantwide systems. *Ind. Eng. Chem. Res.* **2005**, *44*, 7770–7786. [CrossRef]

14. Setiawan, R.; Bao, J. Operability Analysis of Chemical Process Network based on Dissipativity. In Proceedings of the ASCC: 7th Asian Control Conference, Hong Kong, China, 27–29 August 2009; Volume 1–3, pp. 847–852.

15. Setiawan, R.; Bao, J. Analysis of Interaction Effects on Plantwide Operability. *Ind. Eng. Chem. Res.* **2011**, *50*, 8585–8602. [CrossRef]

16. Klemeš, J. *Sustainability in the Process Industry: Integration and Optimization*; McGraw-Hill Professional: New York, NY, USA, 2011.

17. Sreepathi, B.K.; Rangaiah, G.P. Review of heat exchanger network retrofitting methodologies and their applications. *Ind. Eng. Chem. Res.* **2014**, *53*, 11205–11220. [CrossRef]

18. Becker, H.; Maréchal, F. Energy integration of industrial sites with heat exchange restrictions. *Comput. Chem. Eng.* **2012**, *37*, 104–118. [CrossRef]

19. Cerda, J.; Westerburg, A.W. Synthesizing heat-exchanger networks having restricted stream stream matches using transportation problem formulations. *Chem. Eng. Sci.* **1983**, *38*, 1723–1740. [CrossRef]

20. Polley, G.T.; Kumana, J.D. Practical Process Integration Retrofit: Part 4. Dealing with large projects. In Proceedings of the 2012 AIChE Spring Meeting and 8th Global Congress on Process Safety, Houston, TX, USA, 1–5 April 2012.

21. Bütün, H.; Kantor, I.; Maréchal, F. Incorporating Location Aspects in Process Integration Methodology. *Energies* **2019**, *12*, 3338. [CrossRef]

22. Hiete, M.; Ludwig, J.; Schultmann, F. Intercompany Energy Integration Adaptation of Thermal Pinch Analysis and Allocation of Savings. *J. Ind. Ecol.* **2012**, *16*, 689–698. [CrossRef]

23. Jegla, Z.; Freisleben, V. Practical Energy Retrofit of Heat Exchanger Network Not Containing Utility Path. *Energies* **2020**, *13*, 2711. [CrossRef]

24. Reddy, C.C.S.; Rangaiah, G.; Long, L.W.; Naidu, S.V. Holistic Approach for Retrofit Design of Cooling Water Networks. *Ind. Eng. Chem. Res.* **2013**, *52*, 13059–13078. [CrossRef]

25. Hackl, R.; Harvey, S. From heat integration targets toward implementation—A TSA (total site analysis)-based design approach for heat recovery systems in industrial clusters. *Energy* **2015**, *90*, 163–172. [CrossRef]

26. Nemet, A.; Klemeš, J.J.; Kravanja, Z. Designing a Total Site for an entire lifetime under fluctuating utility prices. *Comput. Chem. Eng.* **2015**, *72*, 159–182. [CrossRef]

27. Escobar, M.; Trierweiler, J.O.; Grossmann, I.E. Simultaneous synthesis of heat exchanger networks with operability considerations: Flexibility and controllability. *Comput. Chem. Eng.* **2013**, *55*, 158–180. [CrossRef]

28. Leitold, D.; Vathy-Fogarassy, A.; Abonyi, J. Evaluation of the Complexity, Controllability and Observability of Heat Exchanger Networks Based on Structural Analysis of Network Representations. *Energies* **2019**, *12*, 513. [CrossRef]

29. Andiappan, V.; Ng, D.K.S. Systematic analysis for operability and retrofit of energy systems. In *Green Energy and Technology*; Springer: Singapore, 2018; pp. 147–166. [CrossRef]
30. Abu Bakar, S.H.; Abd Hamid, M.K.; Alwi, S.R.W.; Manan, Z.A. Selection of minimum temperature difference (Delta T-min) for heat exchanger network synthesis based on trade-off plot. *Appl. Energy* **2016**, *162*, 1259–1271. [CrossRef]
31. Maréchal, F.; Kalitventzeff, B. Targeting the integration of multi-period utility systems for site scale process integration. *Appl. Therm. Eng.* **2003**, *23*, 1763–1784. [CrossRef]
32. Verheyen, W.; Zhang, N. Design of flexible heat exchanger network for multi-period operation. *Chem. Eng. Sci.* **2006**, *61*, 7730–7753.
33. Čuček, L.; Martin, M.; Grossmann, I.E.; Kravanja, Z. Multi-period synthesis of optimally integrated biomass and bioenergy supply network. *Comput. Chem. Eng.* **2014**, *66*, 57–70. [CrossRef]
34. Bütün, H.; Kantor, I.; Maréchal, F. An Optimisation Approach for Long-Term Industrial Investment Planning. *Energies* **2019**, *12*, 4076. [CrossRef]
35. Naturvårdsverket | Swedish Pollutant Release and Transfer Register. Available online: https://utslappisiffror.naturvardsverket.se (accessed on 31 March 2018).
36. Marton, S.; Svensson, E.; Subiaco, R.; Bengtsson, F.; Harvey, S. A Steam Utility Network Model for the Evaluation of Heat Integration Retrofits—A Case Study of an Oil Refinery. *JSDEWES* **2017**, *4*, 560–578. [CrossRef]
37. Andersson, E.; Franck, P.-Å.; Åsblad, A.; Berntsson, T. *Pinch Analysis at Preem LYR*; Chalmers University of Technology: Gothenburg, Sweden, 2013.
38. Åsblad, A.; Andersson, E.; Eriksson, K.; Franck, P.-Å.; Svensson, E.; Harvey, S. *Pinch Analysis at Preem LYR II—Modifications*; Chalmers University of Technology: Gothenburg, Sweden, 2014.
39. Sovacool, B.K. Energy studies need social science. *Nature* **2014**, *511*, 529–530. [CrossRef]
40. Sovacool, B.K. What are we doing here? Analyzing fifteen years of energy scholarship and proposing a social science research agenda. *Energy Res. Soc. Sci.* **2014**, *1*, 1–29. [CrossRef]
41. Kemp, I.C. *Pinch Analysis and Process Integration: A User Guide on Process Integration for the Efficient Use of Energy*, 2nd ed.; Butterworth-Heinemann: Oxford, UK, 2007.
42. Kvale, S. *Doing Interviews*; Sage Publications: Thousand Oaks, CA, USA, 2007.
43. Lundberg, V.; Axelsson, E.; Mahmoudkhani, M.; Berntsson, T. Converting a kraft pulp mill into a multi-product biorefinery—Part 1: Energy aspects. *Nord. Pulp. Pap. Res. J.* **2013**, *28*, 480–488. [CrossRef]
44. Dhole, V.R.; Buckingham, P.R. Refinery Column Integration for De-bottlenecking and Energy-saving. In Proceedings of the ESCAPE-4, the fourth European Symposium on Computer Aided Process, Dublin, Ireland, 30 July 1994; pp. 309–316.
45. Li, S.Y.; Zhang, H.B.; Meng, S.; Lu, H. Improving Energy Efficiency for Refinery through Bottleneck Elimination. *Chem. Eng. Trans.* **2015**, *45*, 1159–1164.

Article

Heat Transfer Characteristics of High-Temperature Dusty Flue Gas from Industrial Furnaces in a Granular Bed with Buried Tubes

Shaowu Yin [1,2,*], Feiyang Xue [1], Xu Wang [1], Lige Tong [1,2], Li Wang [1,2] and Yulong Ding [3]

[1] School of Energy and Environmental Engineering, University of Science and Technology Beijing, Beijing 100083, China; xuefeiyang@ustb.edu.cn (F.X.); g20178154@xs.ustb.edu.cn (X.W.); tonglige@me.ustb.edu.cn (L.T.); liwang@me.ustb.edu.cn (L.W.)

[2] Beijing Key Laboratory of Energy Saving and Emission Reduction in Metallurgical Industry, University of Science and Technology Beijing, Beijing 100083, China

[3] College of Chemical Engineering, University of Birmingham, Birmingham B15 2TT, UK; Y.Ding@bham.ac.uk

* Correspondence: yinsw@ustb.edu.cn

Received: 1 June 2020; Accepted: 9 July 2020; Published: 11 July 2020

Abstract: Experimental heat transfer equipment with a buried tube granular bed was set up for waste heat recovery of flue gas. The effects of flue gas inlet temperature (1096.65–1286.45 K) and cooling water flow rate (2.6–5.1 m^3/h) were studied through experiment and computational fluid dynamics' (CFD) method. On the basis of logarithmic mean temperature difference method, the total heat transfer coefficient of the granular bed was used to characterize its heat transfer performance. Experimental results showed that the waste heat recovery rate of the equipment exceeded 72%. An increase in the cooling water flow rate and inlet gas temperature was beneficial to recovering waste heat. The cooling water flow rate increases from 2.6 m^3/h to 5.1 m^3/h and the recovery rate of waste heat increases by 1.9%. Moreover, the heat transfer coefficient of the granular bed increased by 4.4% and the inlet gas temperature increased from 1096.65 K to 1286.45 K. The recovery rate of waste heat increased by 1.7% and the heat transfer coefficient of the granular bed rose by 26.6%. Therefore, experimental correlations between the total heat transfer coefficient of a granular bed and the cooling water flow rate and inlet temperature of dusty gas were proposed. The CFD method was used to simulate the heat transfer in the granular bed, and the effect of gas temperature on the heat transfer coefficient of granular bed was studied. Results showed that the relative error was less than 2%.

Keywords: heat transfer; waste heat recovery; dusty flue gas; granular bed; buried tubes

---

1. Introduction

Granular beds are extensively used in the metallurgical industry, environmental protection, and other fields given their simple structure, convenient operation, and strong environmental adaptability. The reaction heat inside the bed can be moved out effectively by using buried tubes. Heat transfer process is complex and affects the normal operation of granular beds importantly. Numerous studies have been conducted on the heat transfer process in granular beds.

Nasr et al. [1] used air as the working medium to study the influence of filling particle diameter and heat transfer coefficient of the heat transfer process in granular beds. Their results showed that small particles indicate improved heat transfer performance of a granular bed. Pivem et al. [2] used a granular layer as a porous medium to establish a model and studied the influence of porosity, Reynolds number, and other factors on the heat transfer process by using a two-energy equation model. In engineering practice, the random accumulation of particles explains the difference in bed porosity in various locations. Zumbrunnen et al. [3] designed an equipment to measure thermal conductance for several packed beds over a wide temperature range, and the thermal conductance of packed beds increased with the temperature difference across the bed thickness. Ram et al. [4] developed a simple numerical method to determine the interparticle radiation heat transfer in granular bed, which can handle large numbers of surfaces without involving matrix inversion and independent of coordinate system. Shen et al. [5] studied the heat transfer performance of a parallel flow heat exchanger. Their results showed that the heat transfer efficiency of a parallel flow heat exchanger is between 95% and 98% and is affected by a pulsation phenomenon caused by a small tube diameter. Thus, further research is required to determine the appropriate heat transfer tube diameter to eliminate the influence of the pulsation phenomenon. Zhang et al. [6] studied the influence of vertical buried tubes on heat transfer in a large-particle fluidized bed. Their results showed that the average heat transfer coefficient in the circumferential direction of the vertical buried tube remains stable after the fluidization speed reaches the bubble speed, and the heat transfer coefficient in the lateral direction of the horizontal buried tube with the same diameter is approximately 20% higher than that under the same condition. Royston [7] conducted experiments to investigate the heat transfer of gas–solid two-phase mixtures flowing through a column granular bed vertically under the adiabatic wall conditions. The experimental results showed a significant enhancement of heat transfer in comparison with single gas phase conditions. Doherty et al. [8] conducted experiments to investigate the heat transfer coefficient of a horizontal smooth tube immersed in a gas–liquid bed and the results showed that the heat transfer coefficient of the gas and liquid phase decreases at first as the outer diameter of tube is increased but increases as the diameter is further increased. Zhang and Wang [9] studied the heat transfer for a fluidized granular bed air receiver experimentally and numerically with a non-uniform energy flux and the fluidization occurs inside cylindrical metal and quartz glass tubes and a numerical model was established to study the fluidized heat transport inside the quartz tube. Cong et al. [10] obtained the total heat transfer coefficient through logarithmic mean temperature difference method and conducted an experimental study on the heat transfer of a gas–solid two-phase mixture. Grewal and Saxena [10] analyzed the effects of particle size, shape, density, and specific heat; tube size; bed depth; heat flux density; and distributor design on the heat transfer coefficient by measuring serval particles. The experimental results showed that the heat transfer coefficient increases with the increasing of gas velocity and decreases with the further increase and the turning point is 0.5 m/s. Yin et al. [11] conducted an experimental study on the heat transfer characteristics of dusty gas through buried tubes in a granular bed. Their study utilized a solid corundum ball as the filtration medium and analyzed the influence of dust concentration and flue gas velocity on the bed temperature distribution through a comprehensive heat transfer coefficient of the bed. The experimental correlations between the bed heat transfer coefficient and dust concentration and flue gas velocity were proposed. Yin et al. [12] proposed an ammonia absorption cooling and heating dual-supply system based on off-peak electricity heat storage. This system can use the waste heat of flue gas for heating and cooling instead of off-peak electricity. Chen et al. [13] studied the collection mechanism and heat-transfer characteristics of a packed granular filter by using a three-dimensional randomly packed granular filter model.

In the present work, vertical heat exchange tubes were arranged in a granular bed with 3–5-mm hollow corundum balls as filler particles to reduce heat storage. The total heat transfer coefficient of the granular bed was used to characterize the heat transfer capability of the particle bed, and the heat transfer experimental equipment was built. The experiments were conducted at 1073.15 K, and the influence of inlet gas temperature and cooling water flow rate on the heat transfer process was studied. The temperature distribution in the bed was simulated through the computational fluid dynamics' (CFD) method, and the simulation results were compared with the experimental results.

2. Mathematical Model

The heat transfer process for a granular bed with buried tubes is complex [14]. This process includes the convection heat transfer between high-temperature flue gas and filled particles, heat transfer within filled particles, heat transfer between gas films on a particle surface, heat transfer between contact particles and air film, heat transfer between particles and tube walls, heat transfer within the tube wall, and the convection heat transfer of the cooling water inside a tube wall. To simplify the calculation process, an equivalent heat transfer coefficient method of a particle bed is used on the basis of logarithmic mean temperature difference formula. Macroscopically, the heat in high-temperature gas is exchanged with cooling water through the filling particles and heat exchange tubes. This formula is expressed as follows [15]:

$$Q = h_{ht} A_{ht} \Delta T = c\dot{m}(t_{w,out} - t_{w,in}).\tag{1}$$

The total heat transfer coefficient of the granular bed refers to the comprehensive heat transfer coefficient between high-temperature flue gas and cooling water. The influencing factors include the convective heat transfer coefficient of the inner and outer surfaces of the buried tube, the thermal conductivity of the buried tube, and the heat transfer characteristics.

The logarithmic mean temperature difference [16] is defined as follows:

$$t_{max} = t_{g,in} - t_{w,in}\tag{2}$$

$$t_{min} = t_{g,out} - t_{w,out}\tag{3}$$

$$\Delta T = \frac{(t_{max} - t_{min})}{\ln\left(\frac{t_{max}}{t_{min}}\right)}\tag{4}$$

The total heat exchange area is the sum of the total surface area of the heat exchange tubes and the surface area of the filled particles; this area can be expressed as follows:

$$A_{ht} = n\pi dl + \frac{m}{m_1} \times 4\pi r^2.\tag{5}$$

The formula for the total heat transfer coefficient of the granular bed is presented as follows:

$$h_{ht} = \frac{c\dot{m}(t_{w,out} - t_{w,in})}{\left[\pi ndl + \frac{m}{m_1} \times 4\pi r^2\right]\Delta T}\tag{6}$$

3. Experimental Design

The buried tube granular bed heat transfer experimental equipment included a cooling water pipeline system, a buried tube granular bed heat exchanger, a secondary heat exchanger, a granular bed system, and a programmable logic controller (PLC) control panel system. The equipment is illustrated in Figure 1. The combustion air and the gas in the combustion chamber were ignited by an electronic igniter to reach the experimental temperature. The high-temperature gas flowed through the buried tube granular bed and secondary heat exchangers, exchanged heat with the buried tube in the heat exchanger, and then discharged. Armored thermocouples and sensors were set between the cooling water inlet and outlet, the flue gas inlet and outlet, and the granular bed and secondary heat exchangers. The real-time cooling water flow rate, bed pressure drop, and temperatures of flue gas and cooling water were measured by the PLC control panel system.

The buried tube granular bed and secondary heat exchangers had a square structure, with section sizes of 1120 × 1000 and 1200 × 450 mm, respectively. The diameter of the heat exchange tube was 32 mm, and the thickness of the tube wall was 3 mm. To improve the heat transfer process in the buried tube granular bed heat exchanger, 60 tubes were arranged in a staggered way, for a total of 8 rows, as depicted in Figure 2.

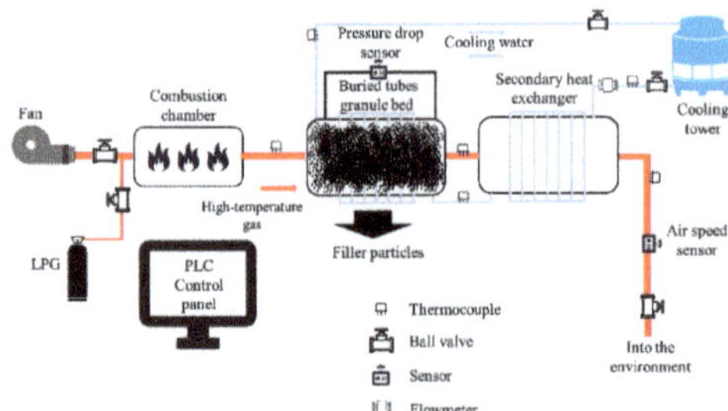

Figure 1. Schematic of the experimental flow and location of measuring points.

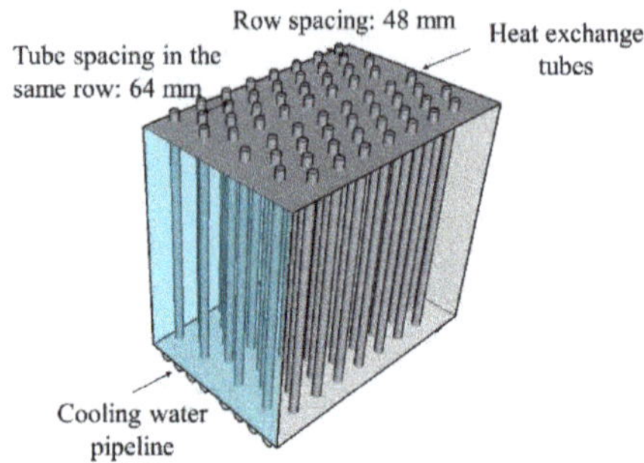

Figure 2. Schematic of the arrangement of heat exchange tubes.

4. Experiment and Result Analysis

Filled particles were added to the designed thickness, and an experiment was performed to study the heat transfer process. Moreover, the influences of gas temperature and cooling water flow rate on heat transfer were studied.

4.1. Analysis of Heat Transfer Experimental Results

The experiment was conducted at 1073.15 K. The environmental air temperature was 318.15 K, and the cooling water flow rate was 4.5 m^3/h. The flue gas temperature, cooling water temperature, bed pressure drop, and waste heat recovery rate are demonstrated in Figure 3. The temperatures of flue gas and cooling water remained stable, thereby indicating that the equipment operated stably. At the initial stage of the experiment, the heat storage of the corundum particles was incomplete, and the waste heat recovery rate of the equipment increased gradually. After 100 min, the heat storage of the particles was completed, and the recovery of waste heat stabilized gradually, at more than 72%. The bed pressure drop initially rose slowly and remained stable. At 520–550 min, 35 kg of particles were slowly and uniformly discharged, while the bed pressure drop decreased from 2000 Pa to 700 Pa.

4.2. Influence of Inlet Flue Gas Temperature on Waste Heat Recovery

After the stable operation of the equipment, the combustion air and gas flow were adjusted to change the inlet gas temperature. The cooling water flow rate was 4.5 m^3/h. The waste heat recovery rate of the equipment and the heat transfer coefficient of the granular bed are plotted in Figure 4.

The results show that the inlet flue gas temperature increased from 1096.65 to 1286.45 K, and the waste heat recovery rate of the equipment increased by 1.7% with the increase in the heat transfer coefficient of the granular bed by 26.6%. The heat of the gas brought into the equipment increased with the inlet gas temperature. The waste heat recovery rate of the equipment and the heat transfer coefficient of the granular bed also increased gradually and remained stable after the inlet gas temperature reached a critical value.

In Figure 4b, the variation curve of the heat transfer coefficient of the granular bed was fitted, and the experimental correlation formula was proposed as follows:

$$k_{ht} = \frac{755.22}{4.19e^{-0.01t} + 0.22} - 3363.45 \tag{7}$$

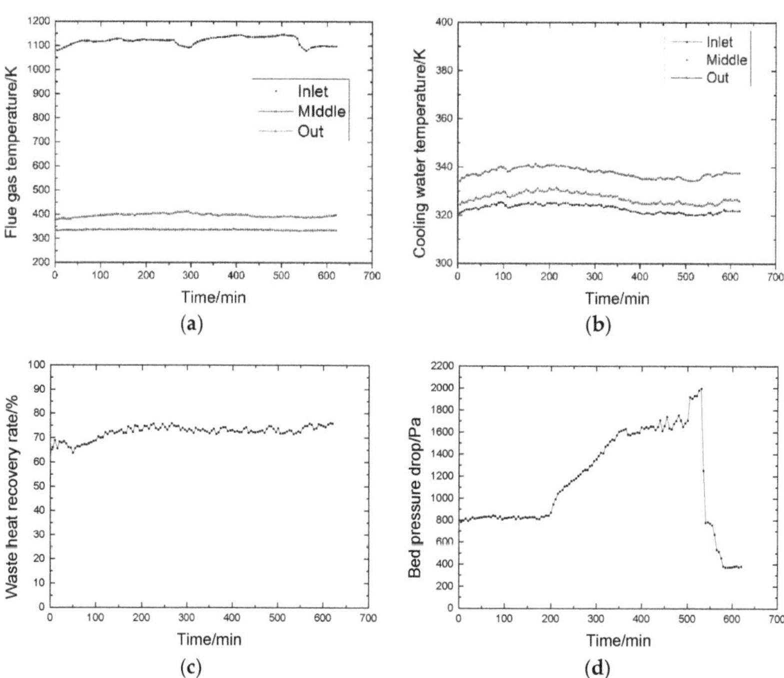

Figure 3. Variation curves of heat exchange in the buried tube granular bed. (**a**) Variation curve of flue gas temperature. (**b**) Variation curve of cooling water temperature. (**c**) Variation curve of waste heat recovery rate. (**d**) Variation curve of bed pressure drop.

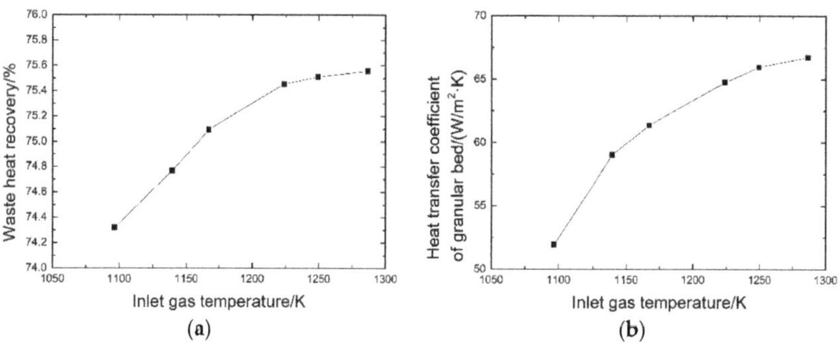

Figure 4. Influence curves of flue gas temperature on heat transfer. (**a**) Influence on waste heat recovery. (**b**) Influence on the heat transfer coefficient.

4.3. Influence of Cooling Water Flow on Waste Heat Recovery

After the stable operation of the experimental equipment, the cooling water pipeline valve was manually adjusted to change the cooling water flow rate. The flue gas inlet temperature was 1093.15 K and the flue gas flow was 350 Nm³/h. The initial cooling water flow rate was 2.6 m³/h. The influence curve of the waste heat recovery rate is exhibited in Figure 5.

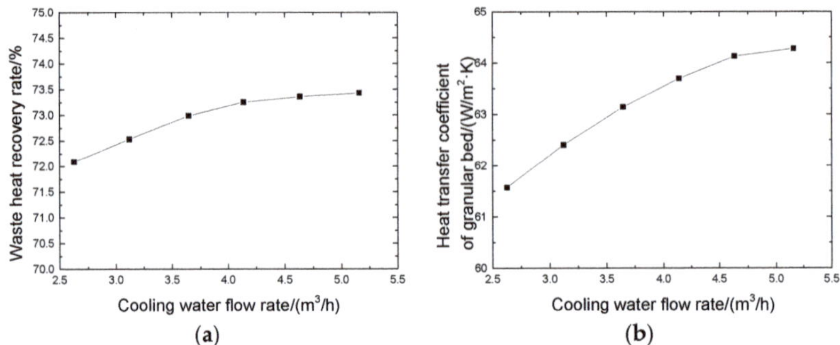

Figure 5. Influence curves of cooling water flow on waste heat recovery. (**a**) Influence on the recovery rate of waste heat. (**b**) Influence on the heat transfer coefficient.

The experimental results showed that the recovery of waste heat and the heat transfer coefficient of the granular bed increased with the cooling water flow rate. The cooling water flow increased from 2.6 m³/h to 5.1 m³/h, and the waste heat recovery rate of the equipment increased by 1.9% with the increase in the heat transfer coefficient of the granular bed by 4.4%. The increase in the cooling water flow rate promoted the convection between the cooling water inside the heat exchange tube and the tube wall. Thus, the total heat transfer coefficient of the granular bed and heat transfer process was promoted. In accordance with the experimental data displayed in Figure 5b, the changing curve of the heat transfer coefficient of the granular bed was fitted, and the experimental correlation formula was obtained as follows:

$$k_{ht} = \frac{36.69}{7974.34\mathrm{e}^{-2q} + 10.78} + 60.92. \tag{8}$$

5. Numerical Simulation and Analysis

The CFD method was adopted to build a grid model using the Integrated Computer Engineering and Manufacturing (ICEM), and the grid was imported into Fluent for related settings. The granular layer was handled as a porous medium. At present, the equivalent heat transfer coefficient of porous media is generally calculated by the macroscopic induction method to study the heat transfer process of porous media. The Nikitin equation considering the influence of solid particle contact thermal resistance, gas thermal conductivity, and radiation heat transfer was used in this paper. The Nikitin equation is expressed as follows:

$$k_e = k_g\left[1 + 3.91(1-\varphi)k_g^{0.1}\ln\frac{k_s}{k_g}\right]\left[1 + \frac{7\rho_g}{\rho_s+\rho_g}\left(\frac{L}{d}\right)^{0.55}\right]^{-1} + \frac{3.46\sigma T^3[3\varphi\xi_g + (1-\varphi)\xi_s]}{1+(1-\varphi)(1-\xi_s)} + k_c \tag{9}$$

The fluent porous medium model was used to conduct the simulation, with the relevant parameters listed in Table 1. The hydraulic diameter and turbulence intensity were determined by the size of tubes in the bed. The viscous resistance coefficient and the inertial resistance coefficient of porous media were determined by Ergun equation. The thermal conductivity of porous media was determined by the Nikitin equation, with the relevant parameters listed in Table 1.

Table 1. Fluent simulation parameter setting.

Parameter/Unit	Value
Hydraulic diameter/mm	16
Turbulence intensity/%	13
Material equivalent diameter/mm	5
Viscous resistance coefficient of porous media	4551.54
Inertial resistance coefficient of porous media	18,148.15
Thermal conductivity of porous media/W/(m^2·K)	39.59

To verify the grid independence, three grid sizes in the same simulation conditions were set up in this work. The inlet gas temperature was 800 K and the velocity was 1.5 m/s. The cooling water inlet temperature was 300 K and the total flow rate was 0.86 m^3/s. The grid independence verification is shown in Table 2. The errors of the three grid sizes were less than 3%, which indicated that the grid independence was verified.

Table 2. Grid independence verification.

Grid Sizes	Inlet Gas Temperature/K	Outlet Gas Temperature/K	Relative Error/%
80,342	800	665.32	
2,764,566	800	664.21	1.67
4,537,529	800	663.83	2.25

To analyze the temperature distribution in the granular bed intuitively, three planes were taken in the radial and vertical directions of the particle bed, as presented in Figure 6.

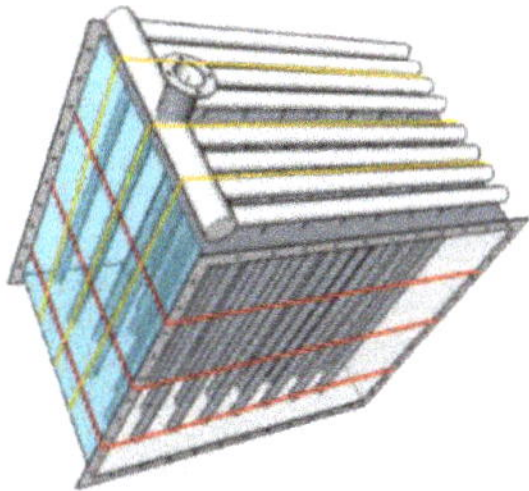

Figure 6. Schematic of the section position of the granular bed.

5.1. Heat Transfer under Different Inlet Flue Gas Temperatures

The inlet flue gas temperatures were adjusted and other experimental conditions were provided as follows: The inlet flue gas velocity was 1.5 m/s; the cooling water flow rate in a single heat exchange tube was 0.025 m/s, that is, the overall cooling water flow rate was 4.3 m^3/h; and the inlet water temperature was 300 K. The simulation results of the temperature distribution in the granular bed are illustrated in Figure 7. The heat transfer coefficient and waste heat recovery rate of the granular bed heat exchanger with changes in inlet flue gas temperature were calculated using Tecplot, as depicted in Figure 8. The flue gas brought more heat into the bed, and the total heat transfer coefficient and the waste heat recovery rate of the bed increased to different degrees with the increase in the inlet flue gas temperature. The increasing trend slowed down with the rise in the inlet flue gas temperature, and the heat transfer process in the bed remained stable after reaching the critical value.

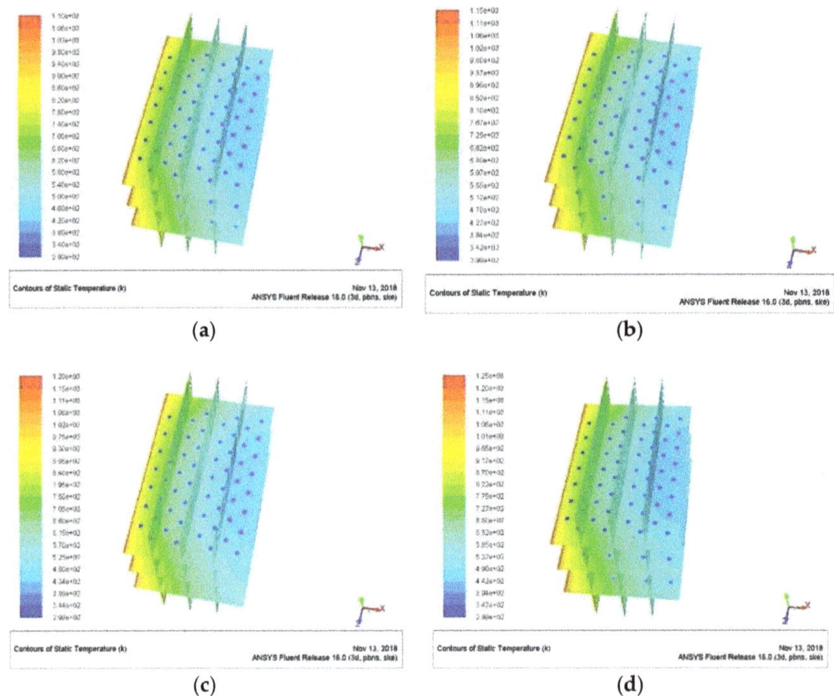

Figure 7. Temperature distribution under different inlet flue gas temperatures in granular bed. (**a**) Inlet flue gas temperature was 1100 K. (**b**) Inlet flue gas temperature was 1150 K. (**c**) Inlet flue gas temperature was 1200 K. (**d**) Inlet flue gas temperature was 1250 K

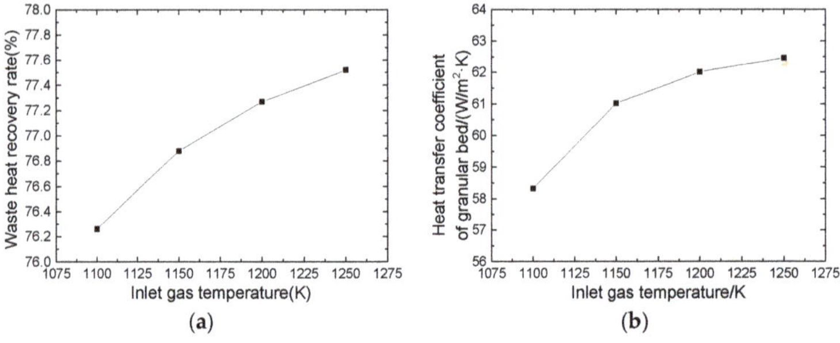

Figure 8. Influence of flue gas temperature on the heat transfer in granular bed. (**a**) Influence on the waste heat recovery rate. (**b**) Influence on the heat transfer coefficient.

To analyze the difference between the experimental and simulation results, the effect of flue gas temperature on the heat transfer coefficient of the granular bed was studied. The temperatures of the inlet flue gas were 1100, 1150, 1200, and 1250 K and the speed of inlet flue gas was 1.5 m/s. The flow rate of the cooling water was 4.3 m³/h and the temperature of the inlet cooling water was 300 K. The comparison curves between the experimental and simulated results of the granular bed heat transfer change with the inlet flue gas temperature are plotted in Figure 9.

The relative errors of the heat transfer coefficient and the waste heat recovery were less than 2%, thereby indicating that the simulation and experimental results were reasonable.

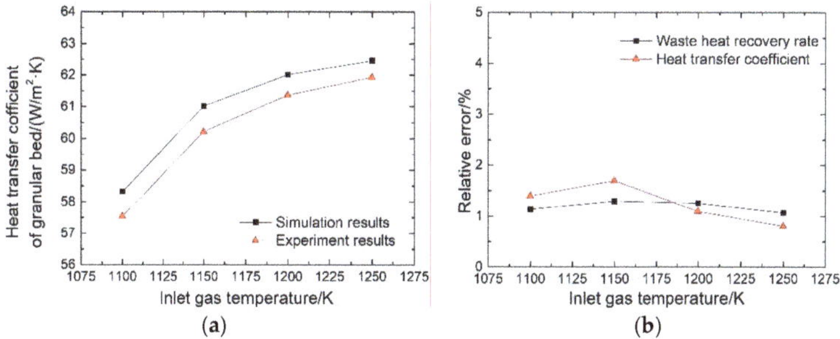

Figure 9. Comparison of experimental and simulation results. (**a**) The influence of gas temperature on the coefficient. (**b**) Variation curve of relative error.

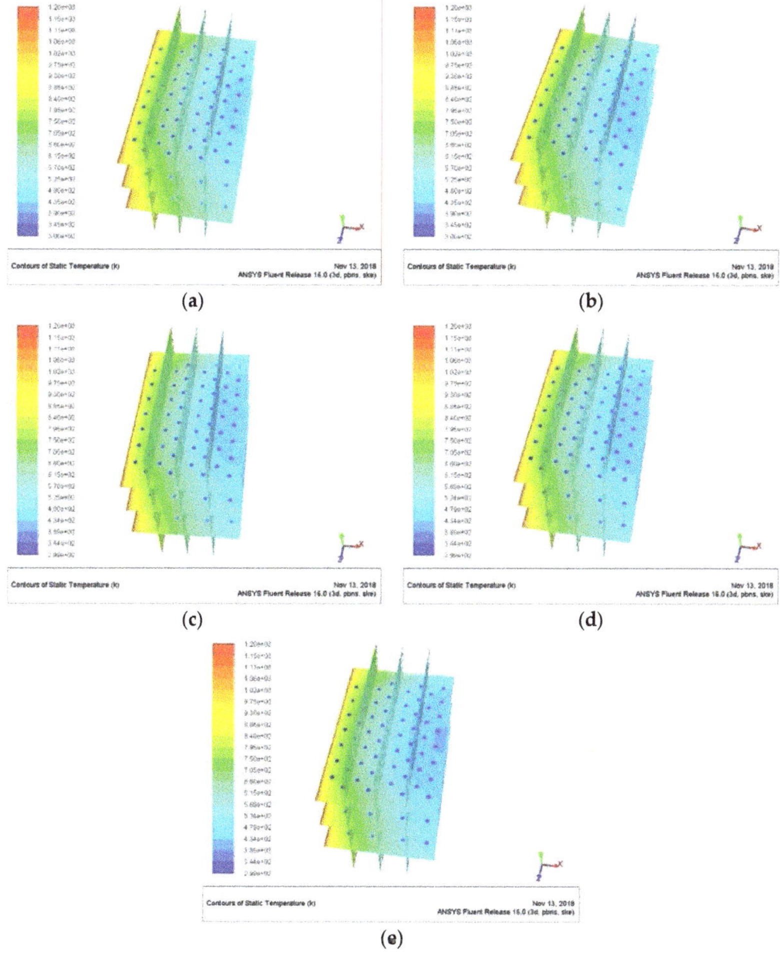

(a)

(b)

(c)

(d)

(e)

Figure 10. Temperature distribution under different cooling water flow rates. (**a**). Cooling water flow rate was 2.6 m³/h. (**b**) Cooling water flow rate was 3.5 m³/h. (**c**) Cooling water flow rate was 4.3 m³/h. (**d**) Cooling water flow rate was 5.2 m³/h. (**e**) Cooling water flow rate was 6.1 m³/h.

5.2. Heat Transfer under Different Cooling Water Flow Rates

The cooling water flow rate was adjusted and the other experimental conditions were presented as follows: The inlet flue gas temperature was 1200 K, the inlet flue gas velocity was 1.5 m/s, and the inlet cooling water temperature was 300 K. The simulation results are demonstrated in Figure 10.

The heat transfer coefficient and waste heat recovery rate of the granular bed heat exchanger with changes in inlet flue gas temperature were calculated using Tecplot, as exhibited in Figure 11. The increase in the cooling water flow rate slightly promoted the heat transfer in the bed, and the increasing trend slowed down with the rise in the cooling water flow rate. After reaching the critical value, the heat transfer process in the bed remained stable.

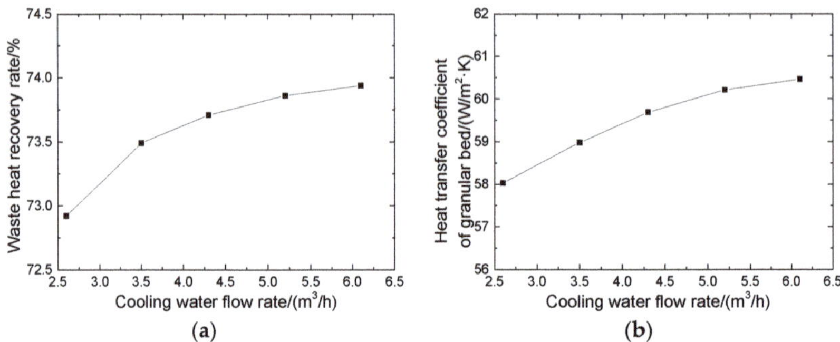

Figure 11. Influence of cooling water flow rate on the heat transfer in granular bed. (**a**) Influence on the waste heat recovery rate. (**b**) Influence on the heat transfer coefficient.

6. Conclusions

A buried tube granular bed with 3–5-mm hollow corundum balls as filler particles was developed to reduce heat storage. The experiment was performed at 1073.15 K, and the main conclusions are presented as follows:

(1) The experimental equipment operated stably at 1073.15 K. The waste heat recovery rate increased gradually when the heat and the recovery rate stored in the particles stabilized at higher than 72% after storage.

(2) The waste heat recovery rate of the equipment increased by 1.7%, and the heat transfer coefficient of the granular bed increased by 26.6% with the variation in the inlet gas temperature from 1096.65 K to 1286.45 K. The experimental correlation of the heat transfer coefficient of the granular bed with the inlet gas temperature was proposed.

(3) The waste heat recovery rate of the equipment increased by 1.9%, and the heat transfer coefficient of the granular bed increased by 4.4% with the variation in the cooling water flow rate from 2.6 m³/h to 5.1 m³/h. The experimental correlation of the heat transfer coefficient of the granular bed with the cooling water flow rate was proposed. With the increase of cooling water flow rate and flue gas inlet temperature, the increase rate of waste heat recovery rate and heat transfer coefficient of granular bed slowed down.

(4) The heat transfer in the granular bed was simulated. The influence of gas temperature on the heat transfer in the granular bed was studied, and the relative error between the experimental and simulation results was less than 2%.

Author Contributions: F.X. and S.Y. designed the manuscript; S.Y. and F.X. drafted the manuscript; F.X., X.W., and L.T. collected the data and revised the manuscript; L.W. and Y.D. checked the content and revised the manuscript. All authors made contributions to the study and the writing of the manuscript. All authors have read and agreed to the published version of the manuscript.

Energies **2020**, *13*, 3589

Funding: The authors gratefully acknowledge the financial support from the National key research and development plan of China (No. 2018YFB0605901).

Conflicts of Interest: The authors declare no conflict of interest.

Nomenclature

Name	Significance
h_{ht}	Total heat transfer coefficient of the granular bed, $W/(m^2{\cdot}K)$
A_{ht}	Total heat transfer area, m^2
ΔT	Logarithmic mean temperature difference, K
c	Specific heat capacity of cooling water, $J/(kg{\cdot}K)$
$\dot{m}$	Cooling water mass flow rate, kg/s
$t_{w,in}$	Temperature of inlet cooling water, K
$t_{w,out}$	Temperature of outlet cooling water, K
$t_{g,in}$	Inlet gas temperature, K
$t_{g,out}$	Outlet gas temperature, K
n	Number of heat exchange tubes in the granular bed
d	Diameter of heat exchange tubes, m
l	Length of a single heat exchange tube, m
m	Total mass of the filled particles in the granular bed, kg
m_1	Mass of a single filled particle, kg
r	Radius of a single filled particle, m
σ	Blackbody radiation constant, $W/(m^2{\cdot}K^4)$
k_c	Contact thermal resistance between particles, $m^2{\cdot}K/W$
ξ_g	Gas blackness in porous media
ρ_s	Density of particles, kg/m^3
r	Radius of a single filled particle, m

References

1. Nasr, K.; Ramadhyani, S.; Viskanta, R. An Experimental Investigation on Forced Convection Heat Transfer from a Cylinder Embedded in a Packed Bed. *J. Heat Transf.* **1994**, *116*, 73–80. [CrossRef]
2. Pivem, A.C.; Marcelo, J.S. Laminar Heat Transfer in a Moving Porous Bed Reactor Simulated with a Macroscopic Two-energy Equation Model. *Int. J. Heat Mass Transf.* **2012**, *55*, 1922–1930. [CrossRef]
3. Zumbrunnen, D.A.; Viskanta, R.; Incropera, F.P. Heat transfer through granular beds at high temperature. *Wärme-und Stoffübertrag.* **1984**, *18*, 221–226. [CrossRef]
4. Dayal, R.; Gambaryan-Roisman, T. A novel numerical method for radiation exchange in granular medium. *Heat Mass Transf.* **2016**, *52*, 2587–2591. [CrossRef]
5. Shen, C.; Yang, S.L.; Zhang, D.W.; Yu, P.; Yang, J.Z. Experimental Research on Heat Transfer Performance of Parallel Flow Heat Pipe Exchanges. *J. Eng. Thermophys.* **2018**, *39*, 1339–1343.
6. Zhang, Z.F.; Ma, T.Z.; Zhao, J.Q.; Huo, X.H. Studies on Heat Transfer from a Vertical Immersed Tube in a Large Particle Fluidized Bed. *J. Eng. Thermophys.* **1988**, *9*, 156–160.
7. Royston, D. Heat transfer in the flow of solids in gas suspensions through a packed bed. *Ind. Eng. Chem. Process. Des. Dev.* **1971**, *10*, 145–150. [CrossRef]
8. Doherty, J.A.; Verma, R.S.; Shrivastava, S. Heat transfer from immersed horizontal tubes of different diameter in a gas-fluidized bed. *Energy* **1986**, *11*, 773–783. [CrossRef]
9. Zhang, S.C.; Wang, Z.F. Experimental and Numerical Investigations on the Fluidized Heat Absorption inside Quartz Glass and Metal Tubes. *Energies* **2019**, *12*, 806. [CrossRef]
10. Cong, T.N.; He, Y.R.; Chen, H.S.; Ding, Y.L.; Wen, D.S. Heat transfer of gas-solid two-phase mixtures flowing through a packed bed under constant wall heat. *Chem. Eng. J.* **2007**, *130*, 1–10. [CrossRef]
11. Grewal, N.S.; Saxena, S.C. Maximum heat transfer coefficient between a horizontal tube and a gas-solid fluidized bed. *Ind. Eng. Chem. Process. Des. Dev.* **1981**, *20*, 108–116. [CrossRef]
12. Yin, S.W.; Li, J.; Shi, G.S.; Xue, F.Y.; Wang, L. Experiment Study on Heat Transfer Characteristics of Dusty Gas Flowing through a Granular Bed with Buried Tubes. *Appl. Therm. Eng.* **2019**, *146*, 296–404. [CrossRef]

13. Chen, J.L.; Li, X.F.; Huai, X.L.; Wang, Y.W.; Zhou, J.Z. Numerical Study of Collection Efficiency and Heat-Transfer Characteristics of Packed Granular Filter. *Particuology* **2019**, *46*, 75–82. [CrossRef]
14. Yin, S.W.; Shi, Y.L.; Tong, L.G.; Wang, L.; Ding, Y.L. Performance Simulation and Benefit Analysis of Ammonia Absorption Cooling and Heating Dual-Supply System Based on Off-Peak Electricity Heat Storage. *Energies* **2019**, *12*, 2298. [CrossRef]
15. Yin, S.W.; Xue, F.Y.; Wang, X.; Wang, L.; Tong, L.G. Experimental Study on Purification and Waste Heat Recovery Characteristics of Dusty Gas in Buried Tube Granular Bed. *J. Eng. Thermophys.* **2019**, *40*, 1928–1935.
16. Wang, N.H.; Meng, F.Z. Influence of Logarithmic Mean Temperature Difference on Heat Transfer Area of Heat Exchanges. *Petro-Chem. Equip.* **1999**, *28*, 13–15.

Article

Pathways for Low-Carbon Transition of the Steel Industry—A Swedish Case Study

Alla Toktarova [1,*], Ida Karlsson [1], Johan Rootzén [2], Lisa Göransson [1], Mikael Odenberger [1] and Filip Johnsson [1]

[1] Department of Space, Earth and Environment, Chalmers University of Technology, SE-412 96 Gothenburg, Sweden; ida.karlsson@chalmers.se (I.K.); lisa.goransson@chalmers.se (L.G.); mikael.odenberger@chalmers.se (M.O.); filip.johnsson@chalmers.se (F.J.)

[2] Department of Economics, University of Gothenburg, SE-405 30 Gothenburg, Sweden; johan.rootzen@economics.gu.se

* Correspondence: alla.toktarova@chalmers.se

Received: 30 June 2020; Accepted: 20 July 2020; Published: 27 July 2020

Abstract: The concept of techno-economic pathways is used to investigate the potential implementation of CO_2 abatement measures over time towards zero-emission steelmaking in Sweden. The following mitigation measures are investigated and combined in three pathways: top gas recycling blast furnace (TGRBF); carbon capture and storage (CCS); substitution of pulverized coal injection (PCI) with biomass; hydrogen direct reduction of iron ore (H-DR); and electric arc furnace (EAF), where fossil fuels are replaced with biomass. The results show that CCS in combination with biomass substitution in the blast furnace and a replacement primary steel production plant with EAF with biomass (Pathway 1) yield CO_2 emission reductions of 83% in 2045 compared to CO_2 emissions with current steel process configurations. Electrification of the primary steel production in terms of H-DR/EAF process (Pathway 2), could result in almost fossil-free steel production, and Sweden could achieve a 10% reduction in total CO_2 emissions. Finally, (Pathway 3) we show that increased production of hot briquetted iron pellets (HBI), could lead to decarbonization of the steel industry outside Sweden, assuming that the exported HBI will be converted via EAF and the receiving country has a decarbonized power sector.

Keywords: iron and steel industry; techno-economic pathways; decarbonization; CO_2 emissions; carbon abatement measures

---

1. Introduction

In Sweden, the industrial sector is responsible for over a third of the total energy demand. In 2017, the iron and steel industry was the largest industrial consumer of fossil fuels (natural gas, oil, coal and coke) and the resulting CO_2 emissions corresponded to 38% of the total industrial CO_2 emissions in Sweden [1]. In line with the global effort of keeping the temperature increase to well below 2 °C, Sweden introduced a nationwide climate policy framework which entered into force in 2018. Through this new framework, Sweden has formally committed to net zero greenhouse gas emissions by 2045 compared to the level in 1990, translating into at least 85% reduction in emissions with the remaining emission reduction to be taken by bio-carbon capture and storage (CCS), land-use change and measures in other countries. After 2045 Sweden is to achieve net negative emissions [2]. The Swedish steel-producing sector is facing the challenge of changing current energy carriers and implementing low carbon technologies to meet these targets.

To reach substantial cuts in emissions from the energy-intensive industries has proven to be challenging [3]. Bataille et al. [4] categorize the general decarbonization options for the energy-intensive industries by a decision-tree with three main branches, (i) dematerialization or recycle/reuse,

(ii) substantial changes of existing processes, (iii) maintaining existing processes with CCS or using an alternative heat source. Johansson et al. [5] investigate measures for the Swedish steel industry that enable becoming climate neutral and conclude that in order to reach deep emissions cuts, efficient energy use must be combined with alternative technologies such as fuel replacement and CCS. Wang et al. [6] investigate the deployment of biomass in the Swedish integrated steel plants applying an energy and mass balance model. The findings by Wang et al. show that using biomass to replace coal in one single blast furnace (the blast furnace located in Luleå), would decrease the CO_2 emissions of the entire Swedish steel industry by 17.3%. Yet, this would require 6.19 TWh of biomass, which correspond to about 4% of current (2017) annual biomass harvests from the Swedish forest industry, while there are several sectors that will compete over the biomass resource. Furthermore, Mandova et al. [7] use a techno-economic model to estimate the carbon dioxide emissions mitigation potential of bio-CCS in primary steelmaking across the European Union (EU). They demonstrate that up to 20% of the EU CO_2 emission reduction target can be met entirely by biomass deployment, and up to 50% by bio-CCS. Lechtenbohmer et al. [8] investigate electrification of the energy-intensive basic materials industry in the EU by means of an explorative method and conclude that electrification of the production of basic materials is technically feasible, yet, can have major implications on the interaction between the industries and the electric systems.

Fischedick et al. [9] have developed a techno-economic model to assess the potential of alternative processes for primary steel production, e.g., blast furnace with CCS (BF/CCS), hydrogen direct reduction, and direct electrolysis of iron ore. The study is made for Germany and the model is run for scenarios up to the year 2100. According to the study, the 80% emission reduction target defined by European Commission (EC) for the iron and steel industry can only be met with early implementation of alternative technologies such as hydrogen direct reduction and iron ore electrolysis, together with a strong climate policy and additional material efficiency measures.

The findings by Fischedick et al. [9] are confirmed by Arens et al. [10] who analyze four future pathways to a low-carbon steel production industry in Germany up to 2035 with emphasis put on estimating technical options, specific energy consumption and CO_2 emissions in the German steel industry. Even though Arens et al. [10] have a different time perspective than Fischedick et al. [9] they conclude that, in order to reduce carbon dioxide emissions from steel production to near zero, alternative steelmaking processes (hydrogen direct reduction, steel electrolysis) need to be developed while CO_2 reduction in short-term (heat recovery, scrap usage and the use of by-products to produce base chemicals) also need to be realized. Although the above works give important knowledge on the available options for abatement of carbon emissions from steel production, there is a lack of studies which shows how a transition from today's steel industry to a near zero-emitting steel industry could be allocated in time.

Therefore, using Sweden as an example, this study aims to further investigate the potential development of the iron and steel industry to become carbon neutral (by 2045, as it is the Swedish target year for carbon neutrality) with respect to the dynamics of the transition, i.e., which technology options to use and when it is reasonable to assume these can be implemented in the form of decarbonization pathways. We consider recent developments in the Swedish iron and steel industry as well as a general literature review on emission reductions options in the iron and steel industry. In addition, barriers and risks associated with developed pathways are put forward and discussed.

The outline of the paper is as follows: Section 2 gives an overview of the Swedish steel industry. Section 3 presents the method and assumptions. Section 4 presents the results. The paper ends with discussion and conclusions in Sections 5 and 6.

2. CO_2 Abatement in the Steel Industry

Sweden is one of the EU's leading producers of ores and metals; ore extraction is about 48 Mton annually of which 83% is refined into iron ore pellets. Furthermore, 77% of the iron ore extractions are exported, which correspond to about 17 Mton [11]. In Sweden, two different steel production technologies are currently applied: the ore-based steelmaking process using blast furnaces/basic oxygen furnaces (BF/BOF), and the scrap-based steel production applying electric arc furnaces (EAF) [12]. These processes have a different structure of the main inputs and energy intensity. The average annual production of crude steel in 2017 was around 4.9 Mt. Two-thirds of the steel production stems from the BF/BOF technology, which currently takes place in two locations (Luleå and Oxelösund) by one single company (SSAB, Stockholm, Sweden). SSAB is accountable for more than 90% of the CO_2 emissions from Swedish steel production, and about 80% of these emissions originate from iron ore reduction. Within the BF/BOF process, iron ore is reduced to pig iron using reducing agents in a blast furnace. Furthermore, in a basic oxygen furnace (BOF) pig iron together with ferrous scrap is processed and transformed into crude steel. As a first step toward carbon-neutral steel production, SSAB has decided to replace the blast furnace in Oxelösund with an electric arc furnace by 2025 [13], when the current blast furnace is scheduled to be retired due to age. EAF requires ferrous scrap and electricity as major inputs. Oxygen and natural gas are used to generate complementary chemical heat for the melting. Based on the configuration of the EAF plant, the availability of scrap and the desired quality of the end product, this process may require some quantities of pig iron from the BF or, optionally, direct reduced iron (DRI). Secondary steelmaking with EAF results in producing lower steel quality compare to virgin steel since scrap steel retains contaminants, such as copper. Steel produced in an EAF tends to be of lower quality than virgin steel because it retains whatever contaminants that were present in the scrap steel, such as copper. Although the EAF is less energy- and CO_2-intensive, high-quality virgin steel demand will remain.

The specific technological decarbonization options for the steel industry are found in Table 1, including information on CO_2 intensity, costs and technology readiness level (TRL) [14].

Table 1. Specifications of current commercially available and new transformative low CO_2 production processes for steel production in greenfield production facilities.

Process	TRL Status	CO_2 Emissions, Tonne CO_2/ Tonne Steel	Capital Expenses, €/Tonne	References
Primary steel production				
Blast furnace with basic oxygen furnace (BF/BOF)	Commercial (TRL 9)	1.6–2.2	386–442	[15,16]
Top gas recycling blast furnace (TGRBF/BOF)	TRL 7	1.44–1.98	632	[17–19]
CO_2 capture technology [1]	TRL 6–9	CO_2 capture efficiency (%): 90	25–85	[17,20–23]
Smelting reduction (SR/BOF)	Commercial (TRL 9)	1.2–2.25	393	[15,21]
Direct reduction using electric arc furnace (DR/EAF)	Commercial (TRL 9)	0.63–1.15	414	[15,18,24]
Hydrogen direct reduction using electric arc furnace (H-DR/EAF)	TRL 1–4	0.025	550–900	[25–27]
Electrowinning (EW)	TRL 4–5	0.2–0.29	639	[9,25,28]
Secondary steel production				
Electric arc furnace (EAF)	Commercial (TRL 9)	0.6	169–184	[15,29,30]
Electric arc furnace/biomass (EAF/biomass)	TRL 6–8	0.005	169–184	[26,31]

[1] Capture emission points: BF, TGRBF.

To assess the techno-economic potential of the CO_2 emissions reduction in the steel industry the following CO_2 emission reduction measures were selected and investigated: top gas recycling blast furnace (TGRBF); carbon capture and storage (CCS); substitution of pulverized coal injection (PCI) with biomass; steelmaking process with hydrogen direct reduction of iron ore (H-DR) and an electric arc furnace (EAF); and a secondary steel production route with EAF, where fossil fuels are replaced with biomass. Furthermore, the abatement measures are combined in three pathways to investigate the potential development implementation of these technologies over time towards zero-emission steelmaking.

The top gas recycling blast furnace concept relies on both removing the CO_2 from the top gas and reinjection of the remaining gas to the blast furnace. This technology enables a decrease in carbon dioxide emissions from the blast furnace since the demand for coke reduces and an opportunity of CO_2 storage. The TGRBF could be modified to an existing blast furnace [32].

As long as the blast furnace process uses coke and coal as fuels, CO_2 emissions are unavoidable, but they could be reduced by means of biomass-derived fuels and reductants applications. The following potential biomass applications can be specified: replacement of fossil fuels in sintering or pelletizing; substitute for coke as a reducing agent and fuel in the blast furnace; substitute for pulverized coal injected (PCI) as a fuel in the blast furnace; substitute for coal-based char utilized for recarburizing the steel; and reduction of pre-reduced feeds [33]. The biomass substitution rate varies between applications. Since the replacement of PCI with biomass is the most feasible application [34], this option is investigated in the present study.

However, in order to achieve deep CO_2 emission cuts down to zero or beyond zero, the steel industry must either capture the CO_2 emissions or shift to another means of iron reduction (hydrogen direct reduction, steel electrolysis). The deployment of CCS in a steel plant is in this work considers the integration of post-combustion capture, which can reduce carbon dioxide emissions from existing plants without major modifications. According to Eurofer [15] a full-scale deployment of the TGR and CCS technologies is assumed possible after 2020. The potential for CO_2 reduction is around 5–10% from TGR alone, 50–60% with TGR technology combined with carbon storage (TGRBF + CCS), and over 80% with TGR with biomass-based BF and carbon storage (TGRBF charcoal + CCS) [35,36].

Currently, the main focus for CO_2 mitigation of the steel industry in Sweden is to develop the hydrogen direct reduction of iron ore. In the present study, we assume hydrogen replaces coke as the main reductant in the reduction process and hydrogen is produced via electrolysis. Iron ore is converted into direct reduced iron (DRI) during the H-DR process and further compressed to hot briquetted iron (HBI), since HBI is less reactive than DRI and allows the problems associated with shipping and handling to be overcome. The principal market for HBI pellets is the electric arc furnace (EAF), but HBI also finds use as a feedstock in basic oxygen furnace (BOF). HBI pellets produced by the hydrogen direct reduction (H-DR) steelmaking process could decrease CO_2 emissions from ironmaking by 90% compared with iron production in a blast furnace, and by 80% compared with a direct reduction of iron using natural gas, as a reducing agent. The hydrogen direct reduction steelmaking process is expected to be feasible from 2040 [37]. The alternative secondary steel-making process is based on the conventional EAF, however, the chemical energy and carbon required to complement the electrical energy is taken from biomass.

3. Method

3.1. Techno-Economic Pathways Concept

In this study, the concept of techno-economic pathways is used (Figure 1). The pathways are characterized as series of techno-economic investments connecting current steel industry configurations to a desirable low-carbon future [38]. The pathways reveal sectoral-level changes through technological characteristics.

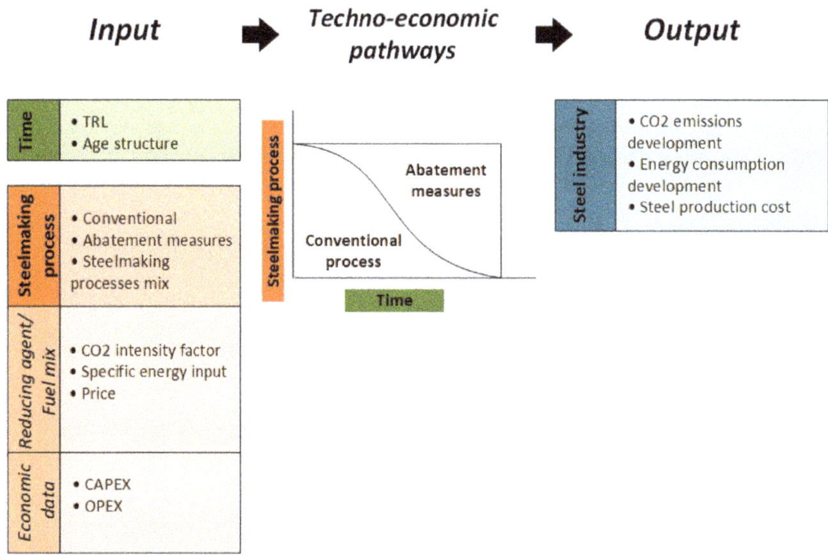

Figure 1. Schematic overview of the techno-economic pathways concept used in this work.

The pathway analysis follows the following steps:

- The assessment of CO_2 abatement measures in the steel industry serves as inputs for the techno-economic pathways as it establishes an upper limit for the emission reduction potential.

- The selection and combination of the CO_2 abatement measures are made in line with governmental climate goals and the visions of the steel companies, as well as a comprehensive literature review.

- The pace of retiring the conventional steelmaking technologies and replacing these with technologies which apply CO_2 abatement measures is in accordance with the age structure of the existing capital stock and assumptions concerning the average technical lifetime of steelmaking technologies (capital stock turnover). The assumed average technical lifetime of steelmaking technologies is set to 50 years. A timeline of the development CO_2 abatement measures throughout technology readiness levels phases, i.e., from concept design to technological maturation and deployment, has been generated based on [39] and industry reports prognosis.

- Based on the technology readiness level timeline, we estimate a timeline for investments in abatement measures to replace current processes, prompting a shift in innovative technology diffusion patterns. The technology readiness level is established by means of a literature review including industry and government agency reports (i.e., these are the ones listed in Table 1). The following outputs from the developed pathways are analysed: development of CO_2 emissions and energy consumption over time and steel production cost.

Three mitigation pathways (Pathways 1–3) as defined in Table 2 are investigated for the Swedish steel industry applying a selected combination of the CO_2 abatement measures listed in Table 1. As a reference, we also compare these pathways to current steel process configuration, for which 65% of the steel production is based on the conventional primary steelmaking process using blast furnace and basic oxygen converter (BF/BOF) and 35% of the steel is produced in conventional electric arc furnaces (EAF). For Pathways 1 and 2 the total annual production of the Swedish steel industry is assumed to remain at 4.9 Mtonne (average steel production of Year 2017) between 2020 and 2045. For Pathway 3, an ore based metallic production growth is assumed, i.e., hot briquetted iron pellets produced via H-DR process. The export of HBI pellets is arbitrarily assumed to reach 6 Mt in 2045, which, since the iron content in HBI pellets is higher than in iron ore pellets, corresponds to approximately 50% of LKAB's iron ore pellets export in 2017. Table 2 shows the combination of CO_2 abatement measures assumed for primary and secondary steelmaking and production rate level in the investigated pathways. The share of primary steelmaking is assumed to decrease compared to the current level for all pathways due to replacement of one of the blast furnaces by EAF in 2025 [13]. However, for Pathways 2 and 3, from 2040 the share between primary and secondary steelmaking is assumed to be on the current level [26]. The configurations do not include processes of steel casting, hot rolling, cold rolling and coating due to their relatively less energy consumption and carbon emissions.

Table 2. Overview of the process configurations as well as production rate assumption for pathways investigates.

Pathway	Primary Steelmaking	Commercially Available [1]	Secondary Steelmaking	Commercially Available [1]	Production Rate
Pathway 1	TGRBF/BOF + CCS + biomass	2030	EAF/biomass	2025	Constant
Pathway 2	H-DR/EAF	2040	EAF/biomass	2025	Constant
Pathway 3	H-DR/EAF	2040	EAF/biomass	2025	Increased

[1] Year when assumed commercially available.

3.2. Data

The assessment of the energy consumption and CO_2 emissions is based on the specific energy consumption and carbon dioxide emission intensity per ton steel, as outlined for the investigated process configurations in Table A1 in Appendix A. Emissions arise from the combustion of biomass are discarded from the emission estimates (i.e., assuming that the biomass is sustainable from a carbon accounting point of view). The Swedish climate goal to get 100% of renewable electricity by 2040 [2] and the current (2017) CO_2 emission grid factor is already low, equaling 0.069 kgCO$_2$/kWh [40]. The CO_2 emission associated with electricity is assumed to fall linearly from 0.069 kgCO$_2$/kWh in the current year to zero by the year 2040. (see Appendix A, Table A2). An economic analysis based on steel production cost is conducted for technologies used in the investigated pathways. The total steel production cost is determined as the sum of capital and variable operating costs, where variable operating cost includes the cost of reducing agent, fuel and other costs associated with running the steel process (see Appendix A, Table A3).

3.3. Modeling Pathways

The annual energy consumption for steel production ($Q_{t,i}$) in year t for Pathway i is calculated by applying the specific energy consumption of fuel and reducing agent combined with total annual crude steel production:

$$Q_{t,i} = \sum_j \sum_r s_{r,j} x_{t,i,r} P, \ \ t \in T, \ i \in I \tag{1}$$

where $s_{r,j}$ is specific energy consumption of fuel and reducing agent j in the technology r (kWh/t). P denotes total annual crude steel production (tonne) and $x_{t,i,r}$ represents the share of the production from the technology r in year t for pathway i (%).

For each pathway, the annual CO_2 emissions from the steel production $E_{t,i}$(tonne CO_2) in year t are given by:

$$E_{t,i} = \sum_r e_r x_{t,i,r} P, \ \ t \in T, \ i \in I \tag{2}$$

where P denotes total annual crude steel production (tonne), e_r is CO_2 emissions intensity of the steelmaking technology r (tonne CO_2), and $x_{t,i,r}$ represents the share of the production from the technology r in year t for pathway i (%).

The CO_2 emission intensity of the steel production e_r is expressed as:

$$e_r = \sum_j f_j s_{r,j}, \quad r \in R \tag{3}$$

where f_j denotes CO_2 emission factor of fuel and reducing agent j (kg CO_2/kWh), $s_{r,j}$ is specific energy consumption of fuel and reducing agent j in the technology r (kWh/t).

The steel production costs, C_r^T, for each of the steel production technologies r are calculated as:

$$C_r^T = V_r + F_r + A_r, \quad r \in R \tag{4}$$

where V_r is the variable operating cost, F_r is the fixed operating costs and A_r is the annualized capital costs which are calculated as:

$$A_r = \frac{CAPEX_r (1+i)^n i}{(1+i)^n - 1}, \quad t \in T, \quad r \in R \tag{5}$$

where i is the interest rate assumed to be 5%, n is the economic lifetime assumed to be 20 years and $CAPEX_r$ is the capital cost of the steel production technologies r.

3.4. Sensitivity Analysis

Several European steel production companies have announced a transition towards electrification of steel production [26,41–43]. The level of the steel decarbonization from electrification option relies heavily on the level to which the electricity grid is decarbonized. Sweden already has a low CO_2 emission grid factor compared to other countries. Therefore, a sensitivity analysis is performed for the carbon dioxide intensity of steel production via the H-DR/EAF process used in Pathway 2, 3 by varying CO_2 emission grid factors. The results are compared to the CO_2 intensity of the TGRBF/CCS/Biomass process used in Pathway 1 to estimate mitigation potential and feasible implementation time of these processes depending on CO_2 emission grid factors. The sensitivity of CO_2 emissions in steelmaking processes to the CO_2 emission grid factor is calculated based on Equation (2) by applying future European CO_2 emission grid factors estimated by the International Energy Agency (IEA) [44] (and given in Table A1). The IEA projection is done based on Sustainable Development Scenario [45], which is aligned with the Paris Agreement of limiting global temperatures to well below 2 °C. The estimates of carbon dioxide emission per tonne of steel produced are done for the period 2020–2040 to identify when in Europe deep emission reduction via hydrogen direct reduction steelmaking should take place in order to meet targets.

4. Results

This section first presents the three pathways in terms of the development of energy consumption over time and comparison of the total steel production cost. This is followed by the of CO_2 emissions along the pathways. Finally, the results of the sensitivity analysis are discussed.

4.1. Future Productivity—Outline of Pathways

Figure 2 gives the three production pathways for the Swedish steel industry showing the timing of replacement of current technology.

Pathway 1 (Figure 2a) represents a shift to the top gas recycling blast furnace with carbon capture and biomass for the conventional primary steel production and to the EAF with biomass for secondary steel production. From 2025, the total production level of BF/BOF equals 2.1 Mt/year (42% of the total steel production in Sweden) due to the blast furnace shutdown in Oxelösund and replacing it by the EAF (SSAB, 2018). By the year 2030 the current primary steel production technology is replaced by a combination of TGRBF and CCS technologies and the replacement of the coal for PCI with biomass. As regards CO_2 capture technology, post-combustion technology is assumed.

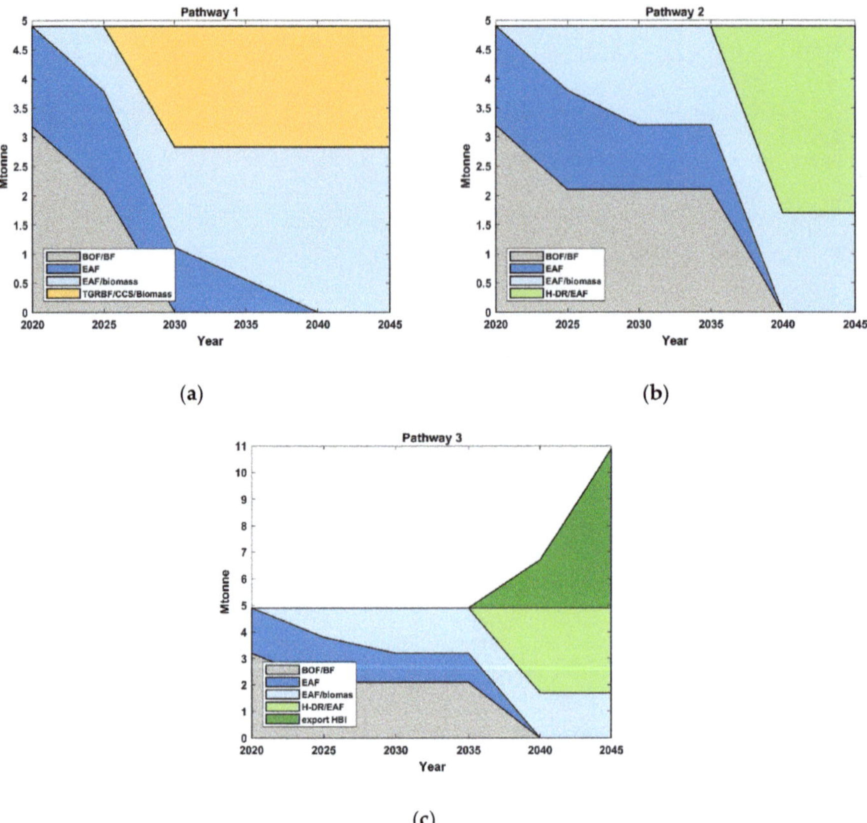

Figure 2. Production processes mix for the Swedish steel industry in Pathway 1 (**a**), Pathway 2 (**b**) and Pathway 3 (**c**) from 2020 to 2045. Note the different scale of the y-axis of Figure 2c.

In Pathways 2 and 3 (Figure 2b,c), Sweden's two blast furnaces are replaced by the hydrogen direct reduction (H-DR/EAF) steelmaking process, which is assumed to be implemented by 2040 (HYBRIT, 2016). Between 2025 and 2040, steel production is assumed to be done by the EAF with biomass at a level corresponding to about 58% of the current total production, which is due to the retirement of one blast furnace in 2025.

For Pathway 3 (Figure 2c), the export of iron ore pellets is assumed to be replaced by the export of hot briquetted iron (HBI) pellets from 2040. The increased production of HBI in Sweden can replace current iron making in other regions and consequently lead to a reduction of CO_2 emissions from ironmaking.

4.2. Energy and Fuel Demand

Figure 3 shows energy consumption for both primary and secondary steelmaking technologies. In Pathways 1, 2 and 3 (Figure 3a–c), the replacement of the iron ore-based steel plant with an EAF results in a coal consumption reduction in 2025. In Pathway 1 (Figure 3a), further coal demand decline is observed in 2030 since the injected pulverized coal into the blast furnace is replaced by biomass.

Due to the reinjection of the top gas components CO and H_2 to the blast furnace as a reducing agent of iron ore, total coke consumption for primary steel production in Pathway 1 is lowered by 27% compared to the conventional BF. In 2030, an increase in natural gas consumption by 44% is observed compared to current steel industry configuration, despite the reduction in natural gas consumption using biomass in EAFs. In the TGRBF/CCS, natural gas is utilized for the preheating of the steam, as well as for the supplemental thermal energy demand of the CCS technology [29].

For Pathway 2 (Figure 3b), the demand for fossil fuel-based energy carriers, such as coke, coal, oil and natural gas, decreases by almost 100% in 2040 compared to the demand with current steel process configuration, due to the transition to the hydrogen direct reduction technology. However, from 2025 to 2040 the demand for fossil fuel-based energy carries is higher compared to Pathway 1. The electricity use increases significantly, implying an electricity need of around 12 TWh per year in 2045. For Pathway 3 (Figure 3c), the energy consumption level is similar to Pathway 2 until 2040 when the electricity consumption increases dramatically to reach a level of 33 TWh per year in the year 2045.

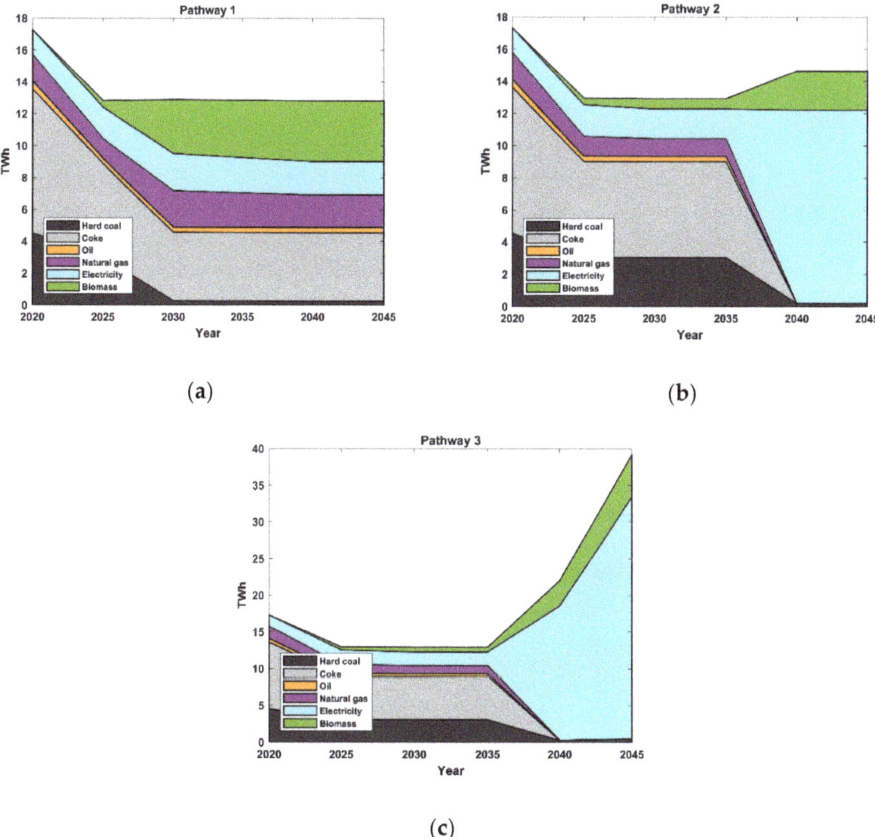

Figure 3. Energy use for the Swedish steel industry in Pathway 1 (**a**), Pathway 2 (**b**) and Pathway 3 (**c**) from 2020 to 2045. Note the different scale of the y-axis of Figure 3c.

4.3. Steel Production Costs

Figure 4 shows the production costs (Equation (4)) of 1 tonne of steel from primary and secondary steelmaking technologies applied in the investigated pathways, where capital expenditure (CAPEX) for the steel production technologies calculated as annuity payments (cf. Equation (5)). Nearly 60%

of current steel production costs consist of raw materials (i.e., iron ore, ferroalloys, scrap and fluxes), fuels and reductant, while CAPEX only contributes to around 20% of the total cost. Thus, since steel production costs are strongly influenced by different market drivers, mainly raw material cost and energy prices, which vary from location to location, the production cost figures obtained are indicative. Figure 4a shows steel production cost for primary steelmaking via the conventional process (BF/BOF), TGRBF/CCS with biomass (Pathway 1) and H-DR/EAF (Pathways 2,3), since the same primary steelmaking technology is used in Pathways 2 and 3, the production costs for these pathways are the same. As for secondary steelmaking, conventional EAF is compared to EAF with biomass implemented in Pathways 1–3 (Figure 4b).

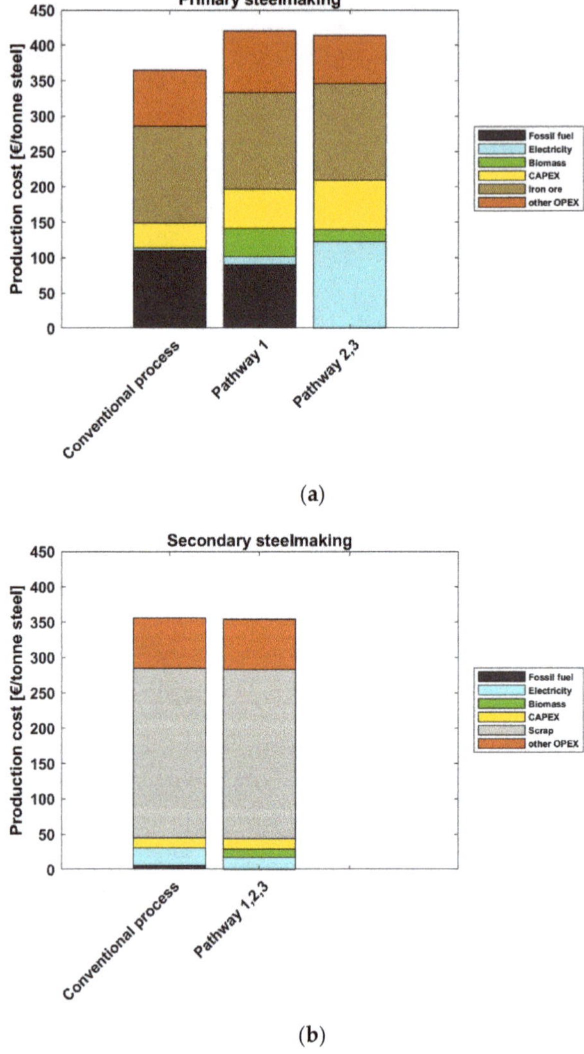

Figure 4. Steel production cost for primary steelmaking technologies (**a**) and secondary steelmaking technologies (**b**) applied in the pathways investigated in this work.

Primary steelmaking with CO_2 emissions reduction, such as applied in Pathways 1–3, implies steel production cost increase by 12–13% compared to conventional primary steelmaking. Capital expenditures for Pathway 1 and Pathways 2, 3 increase by 55% and 97%, respectively, compared to capital expenditures for the conventional process. The cost of electricity is the dominant cost for Pathways 2 and 3 and makes up 30% of total production cost. In this work, an average electricity price for Sweden between 2012 and 2019 of 35 EUR per MWh is used and it is assumed this electricity price level remains constant up to 2045. This, since little is known about the future costs of electricity, but cost can be reduced due to increased share of renewables. Yet, in order to achieve this electricity price level, there is flexible operation of the electrolyser so that periods of high electricity prices are avoided, is likely required. However, such operation strongly depends on electricity system composition and might lead to additional capital expenses of hydrogen storage and electrolyser capacities. Based on our assumptions, secondary steelmaking using EAF, where coke and natural gas are replaced with biomass, offers production cost similar to conventional EAF (Figure 4b).

Figure 5 shows the development of the average steel production cost over time for investigated pathways. Pathways 2 and 3 have identical production cost development, since the same steelmaking technologies are invested in along these pathways. All three pathways show a slight increase in production cost due to investments and increased fuel prices. The production cost in Pathway 1 increases by 5% in 2030 and by 8% in 2040 compared to the current production cost due to the investments in new production technology. The average steel production cost in Pathway 2 is relatively stable up to 2040. From 2040, the steel production cost of Pathway 2 is 16% higher compared to the 2020 cost.

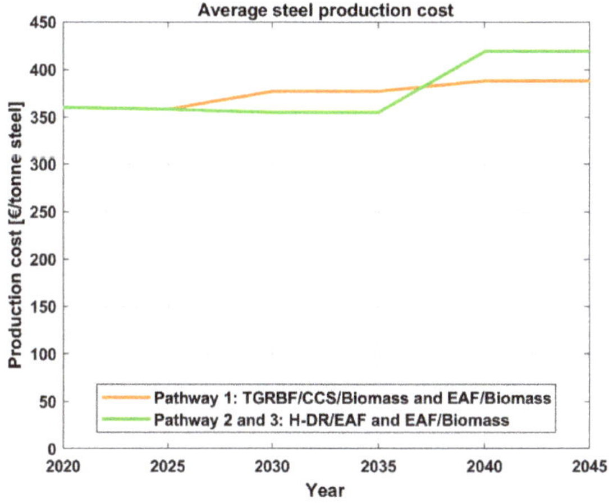

Figure 5. Development of the average steel production cost for the three pathways investigated.

4.4. The Pathways in Relation to the CO_2 Emission Targets

Figure 6 shows the development of the CO_2 emission intensity of steel production for the three pathways. Steel production via processes with substantial electricity demand, such as H-DR/EAF (Pathways 2, 3) and EAF (Pathways 1–3), results in low CO_2 intensity of steel production due to the low CO_2 emission grid factor of the Swedish electricity system. For primary steelmaking in Pathway 1, the decrease in the CO_2 emission grid factor between 2030 and 2045 results in the reduction of steel CO_2 intensity only by 2%.

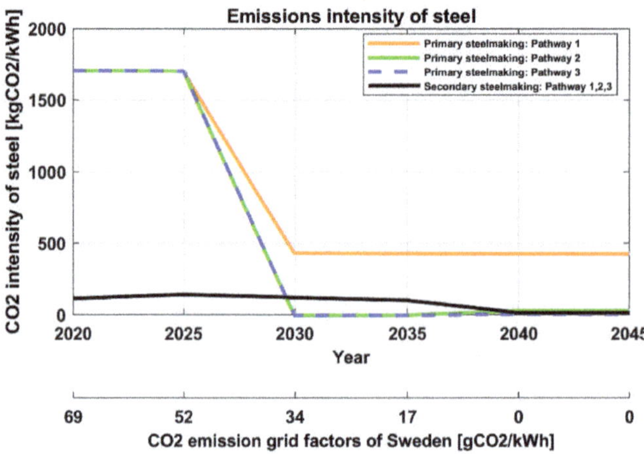

Figure 6. Development of CO_2 emission intensity of the steel production (primary and secondary steelmaking) for the three pathways.

Figure 7 shows the development of CO_2 emissions over time for the Swedish steel industry for the three pathways. As shown in Figure 7, Pathway 1 yields up of 83% emissions reduction in 2045, i.e., applying CCS in combination with biomass substitution in the blast furnace as well as a replacement iron ore-based steel plant with an EAF. Furthermore, already in 2030, an 80% reduction in CO_2 emissions is obtained. Pathways 2 and 3, including electrification, enable further emission reductions compared to implementing CCS and utilization of biomass. As can be seen in Figure 7, none of the pathways can achieve zero CO_2 emissions due to emissions emerging in lime production and the addition of carbon to make steel, which is an essential component in steelmaking.

From 2040, there is a slight increase in CO_2 emissions for Pathway 3 resulting from the large growth in HBI pellet production for export, which could support international emissions reduction efforts not accounted for here.

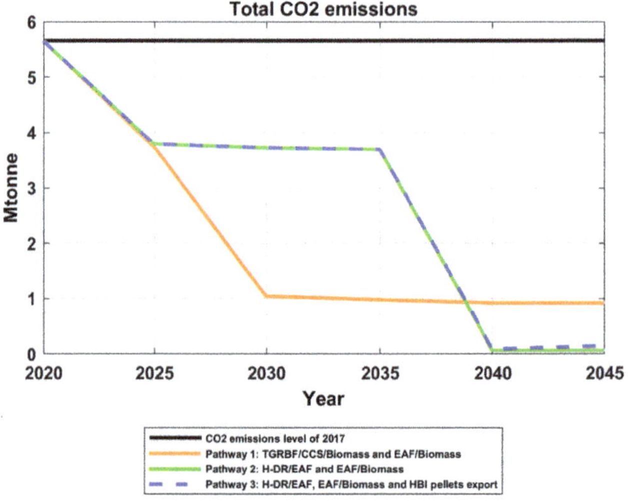

Figure 7. Development of CO_2 emissions for the Swedish steel industry pathways from 2020 to 2045.

4.5. Sensitivity Analysis

Figure 8 illustrates the results of the CO_2 emission intensity of the primary steelmaking processes applied in Pathways 1 and 2, depending on the CO_2 emission grid factors, including the timeline for the European electricity mix as estimated by IEA. The right y-axis shows the timing of future European CO_2 emission grid factors estimated by IEA [44]. For future European CO_2 emission grid factor using CCS in combination with biomass substitution in the blast furnace provides higher CO_2 emissions reduction potential compared to hydrogen direct reduction steelmaking up until 2025. With given future European emission grid factors, the hydrogen direct reduction (H-DR/EAF) steelmaking process allows the reduction of emissions from conventional steel production by 50% in 2020, however, already in 2030, this option provides the greatest CO_2 emissions reduction potential compared to investigated abatement measures for primary steelmaking. The complexity of the plant infrastructure is one of the central issues in capturing CO_2 from steel production. Carbon dioxide emissions are distributed over a large area from different point sources (the lime kilns, sinter plants, coke ovens, hot stoves, BF, and BOF) with potentially different emission rates and flue gas compositions. Since in this study we assumed TGRBF as a capture point, it is not possible to reach near-zero emissions from primary steelmaking in Pathway 1. Applying CCS to all stacks in an integrated steel plant is possible in theory and would lead to near-zero CO_2 emissions. It should be mentioned that the current Swedish CO_2 emission grid factor is 69 g CO_2/kWh which would make hydrogen direct reduction the best solution to cut CO_2 emission from steelmaking deeply already at the present electricity production mix. The sensitivity analysis shows that the decarbonization of electricity supply is decisive for achieving near zero CO_2 emission cuts in the steel industry.

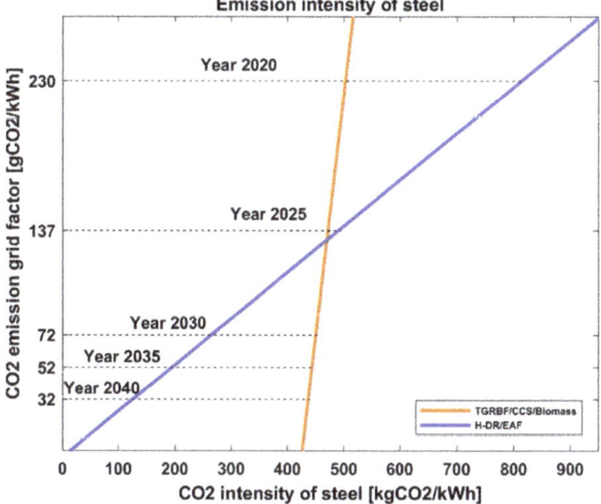

Figure 8. CO_2 emissions intensity of primary steelmaking in Pathway 1 (**orange**) and Pathways 2, 3 (**blue**) as function of future European CO_2 emission grid factor. The horizontal dotted lines indicate the development of the European CO_2 emission grid factors as estimated by IEA [44].

5. Discussion

The aim of this study was to explore how different choices, with respect to technological development in the Swedish steel industry, impact energy use, CO_2 emissions and cost over time. However, it should be noted that the steel production pathways assessed in this study are exploratory and not intended as projections.

It should also be pointed out that this study does not consider the variation of scrap availability and demand in the investigated pathways. For all investigated alternative pathways scrap consumption should increase from 2025 due to replacement BF/BOF by EAF. A global increase in scrap availability due to stocks building up in emerging economies is expected [46] while the availability in the EU will stabilize, as steel stock saturates [47]. In addition, it is important to prioritize innovation and technological development related to delivering the highest quality of steel from recycling (EAF) (see e.g., [48]).

Furthermore, the uncertainty of the steel production cost results obtained in this study may be larger than quantified by our analysis. The primary steelmaking process in Pathways 2, 3, hydrogen direct reduction, allows for flexible operation of the steel plant. The flexibility in the steelmaking process benefits form periods of low electricity price, and this becomes particularly important for electricity systems with a high share of variable renewables. However, it also brings investments in storage technologies and additional investments in production capacities (electrolyzer, direct reduction shaft, EAF). This study did not assess these consequences of flexibility. Furthermore, the introduction of the carbon price, by means of carbon credits and/or carbon tax can be estimated to increase the competitiveness of steel production via alternative processes (Pathways 1–3). Feliciano-Bruzual C. [49], shows that the price of carbon emission in the range of 40–190 €/t CO_2 could make charcoal substitution economically competitive.

Finally, in two of the pathways the study assumed Swedish steel production will remain constant at the 2017 level until 2045. Steel is a globally traded good and steel demand internationally is affected by several factors, e.g., state of the global economy, and therefore development in a region, such as the Swedish steel industry, is difficult to predict. However, change of future demand and production levels obviously will have major impacts on the results for energy use and CO_2 emissions.

Only a relatively small share of the steel produced in Sweden has a domestic end-use, i.e., most (>85%) of the steel produced in Sweden is exported. Still, even though mitigating CO_2 emission by using less steel has a limited potential on national basis such efforts will: (i) limit the use of steel; (ii) maximize upgrading, recycling and reuse of steel already in use; (iii) switch to lower- CO_2 materials; and (iv) use less steel for same function. These aspects will be important to decrease carbon dioxide emissions related to steel production and to reach the long-term emission reduction goals.

Steelmaking firms seeking to invest in high-cost high-risk (but low-CO_2) technology face a dilemma. On the one hand, it is difficult to motivate and find a business case for investments away from traditional and established technologies, especially in the currently uncertain policy regime, on the other hand, a failure to invest in a shift to less carbon-intensive technology is incompatible with the Paris Agreement. Thus, it is worth pointing out, which is also done in other work [4], that a current policy mix targeting the basic material industry will need to be accompanied by complementary policy interventions and/or private initiatives to secure financing and lower the financial risk in investments for decarbonization up to 2045.

6. Conclusions

This paper explores pathways for deep CO_2 emission cuts in the Swedish steel industry up to 2045, with respect to technological development, energy use, carbon dioxide emissions and cost over time. The alternative pathways, e.g., TGRBF with CCS and biomass, H-DR/EAF and EAF/biomass, are compared to the current (2017) Swedish steelmaking technologies.

The technological assessment has shown that in 2030, it should be reasonable to assume that CO_2 emission reductions of 80% compared to current process configurations can be achieved applying TGRBF/CCS with biomass along with electric arc furnace with biomass as CO_2 mitigation options. Using biomass instead of PCI for the primary steelmaking process, would result in a biomass demand from the steel industry in 2045 equal to 6% of the current total current biomass consumption in Sweden. At present, biomass is hardly used at all in the steel industry. Even though there is potential for increased utilization of biomass instead of PCI in the Swedish steel industry in the mid to long term [50], the available biomass is subject to competition, since other sectors are also aiming to increase their use of biomass to achieve their emission reduction goals.

Pathway 2 shows that electrification of primary steel production, in terms of using hydrogen as a reducing agent in H-DR/EAF technology, can result in a 10% reduction in total Swedish carbon dioxide emissions. The main challenge of the electrification in Pathway 2 is the resulting electricity demand of almost 14 TWh in 2045.

The results from this work suggest that the increased production of HBI pellets, as assumed in Pathway 3, can lead to reduction in CO_2 emissions from the steel industry outside Sweden, assuming that the exported HBI will be converted via EAF and the receiving country has a decarbonized power sector. Such a pathway leads to new investments in Swedish steel production capacities and an additional electricity demand of 25.6 TWh (current electricity demand of steel industry is 7.4 TWh).

Author Contributions: Conceptualization, A.T., L.G. and F.J.; methodology: A.T., L.G. and F.J.; software, A.T.; validation: A.T.; formal analysis, A.T., I.K. and J.R.; investigation, A.T. and I.K.; resources, A.T., data curation: A.T.; writing – original draft preparation, A.T.; writing – review and editing, A.T., L.G., I.K, J.R., M.O. and F.J; visualization, A.T.; project administration, F.J. and M.O.; funding acquisition, F.J. All authors have read and agreed to the published version of the manuscript.

Funding: This work was financed by the Mistra Carbon Exit Research programme.

Acknowledgments: Financial support from Mistra is gratefully acknowledged.

Conflicts of Interest: The authors declare no conflict of interest.

Appendix A

Table A1. Specific energy consumption per tonne of steel, CO_2 emission factors and price of reducing agent and fuel mix applied in the investigated pathways.

Specific Energy Input, kWh/Tonne	Primary Steelmaking			Secondary Steelmaking		CO_2 Intensity Factor, kg CO_2/kWh [4]	Reducing Agent/Fuel Mix Price, €/MWh [5]
	Conventional	Pathway 1	Pathway 2,3	Conventional	Pathway 1,2,3		
	BF/BOF [1]	TGRBF +CCS/Biomass [2]	H-DR/EAF [3]	EAF [1]	EAF + Biomass [3]		
Biomass	0	1319	560	0	380	0	30
Coke	2835	2067	0	0	0	0.385	28
Electricity	108	333	3488	700	494	var	35
Hard coal	1381	62	42	64	42	0.342	10
Natural gas	408	997	0	219	0	0.202	25
Oil	159	159	0	0	0	0.277	42

[1] All values are from [51]. [2] All values are from [18], except for biomass and oil consumption values. The biomass consumption value is assumed based on [34]. The oil consumption value is from [51]. [3] All values are from [37]. [4] All values from [52], The grid CO_2 emission factor depends on the year (Table A1). [5] All values are from [53], except for biomass price, electricity price and coke price. The biomass price is from [54], average electricity price for years 2012–2019 is from [55], and coke price is from [56].

Table A2. CO_2 emission grid factor of Sweden and Europe for the years 2020–2040.

CO_2 Emission Grid Factor, g CO_2/kWh	Sweden [1]	European Union [2]
2020	69^2	230
2025	52	137
2030	34	72
2035	17	52
2040	0	32

[1] Own calculations. The CO_2 emission grid factor for Year 2020 is assumed to be equal CO_2 emission grid of 2017.
[2] All values are from [44].

Table A3. Capital and operating expenses of steelmaking technologies for investigated pathways.

	Primary Steelmaking			Secondary Steelmaking	
	Conventional	Pathway 1	Pathway 2,3	Conventional	Pathway 1,2,3
	BF/BOF	TGRBF + CCS/Biomass	H-DR/EAF	EAF	EAF + Biomass
CAPEX, €/tonne	442 [1]	692 [2]	874 [3]	184 [1]	184 [4]
Total OPEX, €/tonne	216	224	205	311	310
Iron ore, €/tonne [5]	136	136	136		
Scrap, €/tonne [6]				239	239
Other OPEX, €/tonne [1]	80	88	69	72	71

[1] Values are from [16]. [2] Value is based on [19,21]. [3] Value is from [9]. Hydrogen storage capacity is assumed for 14 days with CAPEX 0.09€/kWh. [4] Value is assumed to be equal to the CAPEX of the conventional EAF. [5] Iron ore demand is from [51] and iron ore price is from [53]. [6] Scrap demand is from [51] and scrap price is from [56].

References

1. SCB Greenhouse Gas Emissions and Removals. Available online: https://www.scb.se/en/finding-statistics/statistics-by-subject-area/environment/emissions/greenhouse-gas-emissions-and-removals/ (accessed on 17 June 2020).
2. Ministry of the Environment and Energy. *The Swedish Climate Policy Framework*; Ministry of the Environment and Energy: Stockholm, Sweden, 2018.
3. Ballester, C.; Furió, D. Effects of renewables on the stylized facts of electricity prices. *Renew. Sustain. Energy Rev.* **2015**, *52*, 1596–1609. [CrossRef]
4. Bataille, C.; Åhman, M.; Neuhoff, K.; Nilsson, L.J.; Fischedick, M.; Lechtenböhmer, S.; Solano-Rodriquez, B.; Denis-Ryan, A.; Stiebert, S.; Waisman, H.; et al. A review of technology and policy deep decarbonization pathway options for making energy-intensive industry production consistent with the Paris Agreement. *J. Clean. Prod.* **2018**, *187*, 960–973. [CrossRef]
5. Johansson, M.T.; Söderström, M. Options for the Swedish steel industry–Energy efficiency measures and fuel conversion. *Energy* **2011**, *36*, 191–198. [CrossRef]
6. Wang, C.; Mellin, P.; Lövgren, J.; Nilsson, L.; Yang, W.; Salman, H.; Hultgren, A.; Larsson, M. Biomass as blast furnace injectant–Considering availability, pretreatment and deployment in the Swedish steel industry. *Energy Convers. Manag.* **2015**, *102*, 217–226. [CrossRef]
7. Mandova, H.; Patrizio, P.; Leduc, S.; Kjärstad, J.; Wang, C.; Wetterlund, E.; Kraxner, F.; Gale, W. Achieving carbon-neutral iron and steelmaking in Europe through the deployment of bioenergy with carbon capture and storage. *J. Clean. Prod.* **2019**, *218*, 118–129. [CrossRef]
8. Lechtenböhmer, S.; Nilsson, L.J.; Åhman, M.; Schneider, C. Decarbonising the energy intensive basic materials industry through electrification–Implications for future EU electricity demand. *Energy* **2016**, *115*, 1623–1631. [CrossRef]
9. Fischedick, M.; Marzinkowski, J.; Winzer, P.; Weigel, M. Techno-economic evaluation of innovative steel production technologies. *J. Clean. Prod.* **2014**, *84*, 563–580. [CrossRef]
10. Arens, M.; Worrell, E.; Eichhammer, W.; Hasanbeigi, A.; Zhang, Q. Pathways to a low-carbon iron and steel industry in the medium-term—the case of Germany. *J. Clean. Prod.* **2017**, *163*, 84–98. [CrossRef]
11. LKAB. *Annual and Sustainability Report*; LKAB: Luleå, Sweden, 2017.
12. Jernkontoret Companies and Plants. Available online: https://www.jernkontoret.se/en/the-steel-industry/companies-and-plants/ (accessed on 17 June 2020).
13. SSAB First in Fossil-Free Steel. Available online: https://www.ssab.com/company/sustainability/sustainable-operations/hybrit (accessed on 17 June 2020).
14. Klar, D.; Frishammar, J.; Roman, V.; Hallberg, D. A Technology Readiness Level scale for iron and steel industries. *Ironmak. Steelmak.* **2016**, *43*, 494–499. [CrossRef]
15. Association, E.E.S. *A Steel Roadmap for a Low Carbon Europe 2050*; The Boston Consulting Group (BCG): Boston, MA, USA, 2013.

16. Wörtler, M.; Schuler, F.; Voigt, N.; Schmidt, T.; Dahlmann, P.; Lüngen, H.B.; Ghenda, J.-T. Steel's contribution to a low-carbon Europe 2050: Technical and economic analysis of the sector's CO_2 abatement potential. *Lond. BCG Retrieved April* **2013**, *20*, 2015.

17. Carpenter, A. CO_2 abatement in the iron and steel industry. *IEA Clean Coal Cent.* **2012**, 67–70.

18. GHG, I. *IEAGHG Iron and Steel CCS Study (Techno-Economics Integrated Steel Mill)*; IEA: Paris, France, 2013.

19. Lee, S.; Pollitt, H.; Fujikawa, K. *Energy, Environmental and Economic Sustainability in East Asia: Policies and Institutional Reforms*; Routledge: Abingdon, UK, 2019.

20. Haines, M.; Kemper, J.; Davison, J.; Gale, J.; Singh, P.; Santos, S. *Assessment of Emerging CO_2 Capture Technologies and Their Potential to Reduce Costs*; International Energy Agency (IEA): Paris, France, 2014.

21. Kuramochi, T.; Ramírez, A.; Turkenburg, W.; Faaij, A. Comparative assessment of CO_2 capture technologies for carbon-intensive industrial processes. *Prog. Energy Combust. Sci.* **2012**, *38*, 87–112. [CrossRef]

22. Torp, T.A. Drastic Reduction of CO_2 Emissions from Steel Production with CO_2 Capture and Storage (CCS)—ULCOS Project. In Proceedings of the Technical Committee Meeting, Rome, Italy, 27 June 2005.

23. McKinsey, C. *Pathways to a Low Carbon Economy—Version 2 of the Global Greenhouse Gas Abatement Cost Curve*; McKinsey Company: New York, NY, USA, 2009.

24. Kirschen, M.; Badr, K.; Pfeifer, H. Influence of direct reduced iron on the energy balance of the electric arc furnace in steel industry. *Energy* **2011**, *36*, 6146–6155. [CrossRef]

25. Axelson, M.; Robson, I.; Wyns, T.; Khandekar, G. *Breaking Through-Industrial Low-CO_2 Technologies on the Horizon*; Institute for European Studies, Vrije Universiteit Brussel: Bruxelles, Belgium, 2018.

26. LKAB; SSAB. Vattenfall Fossil-Free Steel—Hybrit. Available online: http://www.hybritdevelopment.com/ (accessed on 5 April 2020).

27. Vogl, V.; Åhman, M.; Nilsson, L.J. Assessment of hydrogen direct reduction for fossil-free steelmaking. *J. Clean. Prod.* **2018**, *203*, 736–745. [CrossRef]

28. Siderwin Development of new methodologies for Industrial CO_2-free steel production by electrowinning. Available online: https://www.siderwin-spire.eu/ (accessed on 17 June 2020).

29. Otto, A.; Robinius, M.; Grube, T.; Schiebahn, S.; Praktiknjo, A.; Stolten, D. Power-to-steel: Reducing CO_2 through the integration of renewable energy and hydrogen into the German steel industry. *Energies* **2017**, *10*, 451. [CrossRef]

30. Xylia, M.; Silveira, S.; Duerinck, J.; Meinke-Hubeny, F. Weighing regional scrap availability in global pathways for steel production processes. *Energy Effic.* **2018**, *11*, 1135–1159. [CrossRef]

31. Bianco, L.; Baracchini, G.; Cirilli, F.; Di Sante, L.; Moriconi, A.; Moriconi, E.; Agorio, M.M.; Pfeifer, H.; Echterhof, T.; Demus, T. Sustainable electric arc furnace steel production: GREENEAF. *BHM Berg-und Hüttenmännische Monatshefte* **2013**, *158*, 17–23. [CrossRef]

32. Birat, J.P.; Maizière-lès-Metz, D. *Steel Sectoral Report Contribution to the UNIDO Roadmap on CCS1-Fifth Draft JP*; Arcelor Mittal Global R and D: Maizières-lès-Metz, France, 2010.

33. Adams, S.; Schnittger, S.; Kockar, I.; Kelly, N.; Xu, H.; Monari, F.; Edrah, M.; Zhang, J.; Bell, G. *Impact of Electrolysers on the Network*; Scottish & Southern Electricity Networks: Perth, Scotland, 2016.

34. Mathieson, J.G.; Rogers, H.; Somerville, M.A.; Jahanshahi, S.; Ridgeway, P. Potential for the Use of Biomass in the Iron and Steel Industry. In Proceedings of the Chemeca 2011: Engineering a Better World: Sydney Hilton Hotel, NSW, Australia, 18–21 September 2011.

35. Borlée, J. Low CO_2 Steels-ULCOS Project (Ultra Low CO_2 Steelmaking). In Proceedings of the IEA Deployment Workshop, Paris, France, 8–9 October 2007; Volume 9.

36. Chunbao Charles, X.U.; Cang, D. A brief overview of low CO_2 emission technologies for iron and steel making. *J. Iron Steel Res. Int.* **2010**, *17*, 1–7.

37. HYBRIT. *HYBRIT—A Swedish Prefeasibility Study Project for Hydrogen Based CO_2—Free Ironmaking*; SSAB: Stockholm, Sweden, 2016.

38. Rosenbloom, D. Pathways: An emerging concept for the theory and governance of low-carbon transitions. *Glob. Environ. Chang.* **2017**, *43*, 37–50. [CrossRef]

39. Forum, O.E. Ocean energy strategic roadmap. *Build. Ocean Energy Eur.* **2016**. Available online: https://webgate.ec.europa.eu/maritimeforum/en/node/3962 (accessed on 22 July 2020).

40. IEA Key energy statistics. 2018. Available online: https://www.iea.org/countries/sweden (accessed on 17 June 2020).

41. Arcelor Mittal. Arcelor Mittal Commissions Midrex to Design Demonstration Plant for Hydrogen Steel Production in Hamburg. Available online: https://corporate.arcelormittal.com/media/news-articles/2019-sep-16-arcelormittal-commissions-midrex-to-design-demonstration-plant (accessed on 5 April 2020).
42. Salzgitter AG. Salzgitter AG und Tenova Unterzeichnen Absichtserklärung für das SALCOS-Projekt: CO_2-Arme Stahlproduktion auf Wasserstoffbasis. Available online: https://www.salzgitter-ag.com/de/newsroom/pressemeldungen/pressemeldung-der-salzgitter-ag/2019-04-03-1/salzgitter-ag-und-tenova-unterzeichnen-absichtserklrung-fr-das-salcosprojekt-co2arme-stahlproduktion-auf-wasserstoffbasis.html (accessed on 22 July 2020).
43. Voestalpine the Three Pillars of Decarbonization. Available online: https://www.voestalpine.com/blog/en/innovation-en/the-three-pillars-of-decarbonization/ (accessed on 5 April 2020).
44. IEA Tracking Power. 2019. Available online: https://www.iea.org/reports/tracking-power-2019 (accessed on 24 June 2020).
45. IE World Energy Model. Scenario Analysis of Future Energy Trends. Available online: https://www.iea.org/reports/world-energy-model/sustainable-development-scenario#abstract (accessed on 24 June 2020).
46. Pauliuk, S.; Milford, R.L.; Müller, D.B.; Allwood, J.M. The steel scrap age. *Environ. Sci. Technol.* **2013**, *47*, 3448–3454. [CrossRef] [PubMed]
47. Consortium, E. *Treating Waste as a Resource for the EU Industry: Analysis of Various Waste Streams and the Competitiveness of Their Client Industries*; ECSIP Consortium: Rotterdam, The Netherlands, 2013.
48. Allwood, J.; Azevedo, J.; Clare, A.; Cleaver, C.; Cullen, J.; Dunant, C.F.; Fellin, T.; Hawkins, W.; Horrocks, I.; Horton, P. *Absolute Zero: Delivering the UK's Climate Change Commitment with Incremental Changes to Today's Technologies*; University of Cambridge: Cambridge, UK, 2019.
49. Feliciano-Bruzual, C. Charcoal injection in blast furnaces (Bio-PCI): CO_2 reduction potential and economic prospects. *J. Mater. Res. Technol.* **2014**. [CrossRef]
50. Mellin, P.; Wei, W.; Yang, W.; Salman, H.; Hultgren, A.; Wang, C. Biomass availability in Sweden for use in blast furnaces. *Energy Procedia* **2014**, *61*, 1352–1355. [CrossRef]
51. Remus, R.; Monsonet, M.A.A.; Roudier, S.; Sancho, L.D. *Best available techniques (BAT) reference document for iron and steel production*; Publications Office of the European Union: Brussels, Belgium, 2013.
52. World Steel Association. Worldsteel association CO_2 emissions data collection. *User Guide* **2008**, *6*, 21.
53. WorldBank Commodity Markets. Available online: https://www.worldbank.org/en/research/commodity-markets (accessed on 17 June 2020).
54. Mandova, H.; Leduc, S.; Wang, C.; Wetterlund, E.; Patrizio, P.; Gale, W.; Kraxner, F. Possibilities for CO_2 emission reduction using biomass in European integrated steel plants. *Biomass Bioenergy* **2018**, *115*, 231–243. [CrossRef]
55. NordPool. Day-Ahead Prices. Available online: https://www.nordpoolgroup.com/Market-data1/Dayahead/Area-Prices/ALL1/Yearly/?view=table (accessed on 17 June 2020).
56. Moya, J.A.; Boulamanti, A. *Production costs from energy-intensive industries in the EU and third countries*; Publications Office of the European Union: Luxembourg, 2016.

Article

Roadmap for Decarbonization of the Building and Construction Industry—A Supply Chain Analysis Including Primary Production of Steel and Cement

Ida Karlsson [1,*], Johan Rootzén [2], Alla Toktarova [1], Mikael Odenberger [1], Filip Johnsson [1] and Lisa Göransson [1]

[1] Department of Space, Earth and Environment, Chalmers University of Technology, SE-412 96 Gothenburg, Sweden; alla.toktarova@chalmers.se (A.T.); mikael.odenberger@chalmers.se (M.O.); filip.johnsson@chalmers.se (F.J.); lisa.goransson@chalmers.se (L.G.)
[2] Department of Economics, University of Gothenburg, SE-405 30 Gothenburg, Sweden; johan.rootzen@economics.gu.se
* Correspondence: ida.karlsson@chalmers.se

Received: 2 July 2020; Accepted: 3 August 2020; Published: 10 August 2020

Abstract: Sweden has committed to reducing greenhouse gas (GHG) emissions to net-zero by 2045. Around 20% of Sweden's annual CO_2 emissions arise from manufacturing, transporting, and processing of construction materials for construction and refurbishment of buildings and infrastructure. In this study, material and energy flows for building and transport infrastructure construction is outlined, together with a roadmap detailing how the flows change depending on different technical and strategical choices. By matching short-term and long-term goals with specific technology solutions, these pathways make it possible to identify key decision points and potential synergies, competing goals, and lock-in effects. The results show that it is possible to reduce CO_2 emissions associated with construction of buildings and transport infrastructure by 50% to 2030 applying already available measures, and reach close to zero emissions by 2045, while indicating that strategic choices with respect to process technologies and energy carriers may have different implications on energy use and CO_2 emissions over time. The results also illustrate the importance of intensifying efforts to identify and manage both soft and hard barriers and the importance of simultaneously acting now by implementing available measures (e.g., material efficiency and material/fuel substitution measures), while actively planning for long-term measures (low-CO_2 steel or cement).

Keywords: construction; building; supply chain; decarbonization; roadmap; heavy industry; CO_2 emissions; carbon abatement; emissions reduction; climate transition

---

1. Introduction

Sweden has committed to reducing greenhouse gas (GHG) emissions to net-zero by 2045 and to pursue negative emissions thereafter, in line with its obligations to the Paris agreement [1,2]. It is clear that the future development over several decades of the economic, social, and technical dynamics that govern demand for energy and materials, and the associated greenhouse gas emissions, are likely to be speculative. Nevertheless, as there is an urgent need to start a transformation towards deep decarbonization, decisions must be made now as to how to best manage the transition, while taking the future into account [3]. This includes starting with the current situation to map mitigation measures to see which measures that can be applied already at present and those which will require longer lead times to be implemented.

Seeing that the energy and climate performance of the user phase of the built environment in Sweden keeps improving, the climate impact of the construction process has increasingly come in to focus [4]. Emissions arising from manufacturing, transporting, and processing of construction materials to buildings and infrastructure account for approximately one-fifth of Sweden's annual CO_2 emissions [5–7]. However, current estimates of the climate impact from building and construction processes in Sweden is associated with a significant degree of uncertainty. Environmentally extended input-output data has provided estimates for the year 2015. These determine territorial emissions associated with building construction to be 6.6 Mt CO_2e, increasing to 11.6 Mt CO_2e when including imports [7,8]. Territorial emissions linked transport infrastructure construction is similarly estimated at 1.5 Mt CO_2e increasing to 1.9 Mt CO_2e including imports [7,9]. The imported emissions are associated with greater uncertainty as they are estimated by calculating differences in emissions from trading partners compared to emissions in Sweden, giving the limitation of not capturing differences between different industries in the importing countries [10]. On the other hand, a process-based bottom-up life cycle analysis (LCA) approach, combining statistics detailing new net area from newbuilds and refurbishments with LCA data per building type, provides a lower estimate of 5.4 Mt CO_2e emissions associated with building construction in 2015 [8,11].

Indeed, as demonstrated in literature, there is evidence that life-cycle assessments based on process data and environmental extended input–output (EEIO) tend to lead to very different results, where EEIO LCAs often lead to higher emissions and process LCAs to lower emissions [12]. There are several reasons for these discrepancies, with EEIO LCAs suffering from inherent homogeneity and linearity assumptions, along with aggregation errors due to several different industries being comprised into one input-output sector [12,13]. The combination multiple economic subsectors with quite different emissions profiles into one sector, along with the assumption that the market price linearly correlates with higher emissions results in systematic overestimations [14]. On the other hand, process LCA suffers from an inherent 'truncation error' due to indirect impacts (e.g., capital goods) or excluding upstream processes along the supply chain due to the need for a system boundary leading to systemic underestimation [14–16]. Comparative building case studies demonstrate 20–73% higher embodied carbon emissions for EEIO LCA versus process LCAs [12,17–19].

In view of the differences in the LCA approaches, several studies regard EEIO methods most useful in assessments of entire economies or industries [13,20,21]. We conclude that, to enable analysis into the ongoing development in the construction sector and the opportunities for the sector to contribute to the national climate targets, better estimates are needed, including the main components making up those emissions, from different materials to transport of those materials and construction processes.

The focus of this study was on the path towards net-zero emissions in 2045, which necessitates not only looking at current emissions and the components therein but also require comprehensive assessments into current, as well as prospective future, abatement options and potentials. In literature, one can find an array of sector-specific or industry level studies focused on future carbon abatement options (see, e.g., Reference [22–26]) for steel, Reference [27–29] for cement/concrete, and Reference [30–33] for heavy transport and construction equipment). A comprehensive review of 40 energy-intensive industry roadmaps was recently performed by Gerres et al. [34]. This review remarked that roadmaps with a focus on subsector specific technology assessments often disregard the cross-sectorial dimensions of the abatement options considered, while top-down approaches tend to provide limited details on technological and economic feasibility. Gerres et al. found little consensus on how deep decarbonizations of industry are to be achieved but could identify a few key areas of importance and agreement, including alternative feedstock and carbon capture in the cement industry, carbon neutral steelmaking, and decarbonization of low temperature heat in the petrochemical industry. The authors finally noted that carbon capture, transport and storage (CCS), the electricity system, and the hydrogen economy, i.e., external system transformations, must be considered when evaluating decarbonization pathways.

In addition to sector-specific abatement studies, we have found recent evidence, particularly in grey literature of synthesis reports, reports which integrate the perspectives from different industries [35–41]. The target of these reports is predominantly either a European or a global level, emphasizing the cross-sectorial potential of reducing demand for products and services via circular economy, logistic optimization, and material efficiency measures while highlighting the potential and alternatives contributed by biomass, carbon capture, and electrification, including links to hydrogen.

Thus, we see that roadmaps detailing industry decarbonization on a sector by sector or multinational level are prevalent. However, focusing in on the building and construction sector, there are limited examples in literature of national assessments of future abatement options and potentials and the pathway towards close to zero emissions [42,43], with most studies pertinent to the UK [44–47].

In Sweden, within the government-initiated Fossil Free Sweden (http://fossilfritt-sverige.se/in-english/) initiative, business industries have drawn up roadmaps towards 2045, describing in varying details technological solutions, investment needs, and obstacles required to be removed. These provide some key information on abatement options within individual industry sectors with the construction sector roadmap capturing a cross-sectorial perspective [48,49]. Some initial assessments have also been made on emissions reductions and energy needs on a cumulative level for the year 2045 [50,51]. However, to explore critical factors on the pathway towards 2045, including impacts from upscaling and the risk of lock-in effects, there is a need for studies that take a broader perspective while combining a short and long-term perspective of abatement potential across the supply chain.

In this study, we used material and energy flow analysis combined with an extensive literature review to assess (i) the current status of emissions from the Swedish construction sector and (ii) the extent to which abatement technologies across the construction supply chain could reduce the GHG emissions if combined to its full potential based on implementation timelines linked to their technical maturity and expected readiness for implementation. The ambition was to analyze the current and future GHG emissions reduction potential by considering the development, over time, of emission abatement measures in different parts of the construction supply chain.

With support of scenarios, we created a roadmap exploring different future trajectories of technological developments in the supply chains for buildings and transportation infrastructure. By matching short-term and long-term goals with specific technology solutions, the roadmap made it possible to identify key decision points and potential synergies, competing goals, and lock-in effects. While the study is performed in a Swedish setting, and the updated estimate of current emissions are predominantly based on northern European LCAs, the analysis of abatement options, timelines, and pathways are relevant and applicable on a European, if not a global level.

2. Materials and Methods

This work combines quantitative analytical methods, i.e., scenarios and stylized models, with a participatory process involving relevant stakeholders in the assessment process. The participatory process served to identify the main abatement options but also to adjust decisions and assumptions regarding abatement portfolios and timelines to make these as realistic and feasible as possible. Stakeholders have thus provided input and feedback via workshops undertaken during the study development period. Stakeholders include industry representatives and experts along the supply chain: material suppliers, contractors, consultants, clients, and governmental agencies.

Estimates are provided of the magnitude of current and future GHG emissions reduction potential across the building and transport infrastructure construction supply chain by (i) estimating the current emissions, material, and energy flows associated with the sector; (ii) identifying possible GHG abatement options relevant to the construction works and their estimated abatement potentials; (iii) using (i) and (ii) to assess the impact of combining abatement measures along the construction supply chain; and (iv) crafting scenarios to highlight challenges and possibilities up to 2045 given different assumptions regarding future practices and technological development.

Current emissions from the Swedish building and construction industry is analyzed by comparing existing estimates with a mapping of the material and energy flow through the supply chain of building and transport infrastructure construction produced via a literature review of life cycle analyses and equivalent studies (where literature searches were conducted in Scopus and Web of Science with search string algorithms targeting a combination of LCA OR "life cycle analysis" OR "life cycle assessment" OR "carbon footprint" AND building* OR construction OR infrastructure with subsequent screening to identify studies of relevance for the scope of this study, e.g., transport infrastructure and buildings of equivalent design and construction techniques, as in Sweden.). In the technology roadmap of this work, we analyze the climate impact linked to construction of buildings and transport infrastructure, i.e., we do not include construction of for example utilities, such as waterworks, wastewater treatment plants, power plants, and power lines. Construction of buildings and transport infrastructure is equivalent to around 80% of construction investments in Sweden [52]. Focus of the analysis is on emissions from materials production and the construction phases (i.e., corresponding to life cycle stage A1–A5 [43]). The latter includes emissions from mass and material transport and the construction process. A schematic of the mapping is shown in Figure 1.

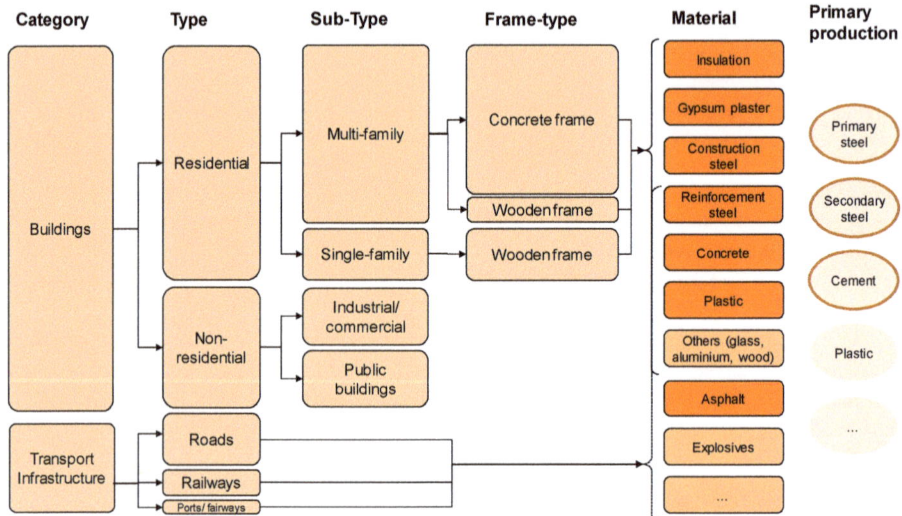

Figure 1. Schematic figure of material flow mapping for buildings and transport infrastructure construction in Sweden. The height of the category to frame-type boxes represent the approximate relative sizes of the associated emissions. Regarding materials, the dark orange boxes depict materials studied in detail, while the dark orange contours in the primary production column depict material production processes evaluated in detail. The analysis also includes emissions from mass and material transport and construction processes.

The Swedish Transport Administration (STA) provides a breakdown of the emission share from various materials and activities regarding new construction of state-owned transport infrastructure. However, this is not a complete picture of transport infrastructure as around half of the transport infrastructure investments in Sweden are made by regional and local government [53]. More detailed analysis has been performed by [9], including both state, municipal, and privately-owned transport infrastructure. The analysis by Liljenström et al. describes the emissions share of material production and on-site activities (transports and construction processes) for both new construction and reinvestments (defined as larger projects intended to restore the infrastructure to its original state by replacing a construction component (for example, the bounded base layer and tunnel lining) with the same, or a similar, type of construction component.) for road and rail infrastructure, ports, and fairways and airports. In this study, we slightly refined the emissions shares given by Liljenström et al. based on additional data [52,54], while excluding airports due to the minor emissions associated with airport construction (0.03 kt CO_2e [9]). We further used the total emissions for construction of transport infrastructure provided by the detailed bottom-up analysis performed by Liljenström et al. [9] and national environmentally extended input-output modeling [7], as these provide a coherent result of 1.9 Mt CO_2e emissions for the year 2015.

As this coherence does not apply for building construction, an estimate for the national emissions associated with building construction was developed using data on the emissions share from different lifecycle stages and materials sourced from the literature review combined with validated emissions levels of different components. Where available, the literature review was concentrated to LCA studies in a Northern European setting as to account for equivalent design and construction techniques, along with requirements stemming from climatic conditions. While LCA studies of buildings are prevalent, studies that describe and separate material inputs, material transports, and construction processes are more limited, particularly regarding non-residential buildings and refurbishments (see, e.g., reviews in Reference [55–57]). As LCA studies are limited for refurbishments, no detailed breakdown for refurbishments has been developed here. We instead use an adjustment factor to reflect emissions from transports and specific materials considered dominant in refurbishments in the few studies available.

The share of emissions for specific materials for construction of different building types was calculated based on the estimates in literature for these building types and the estimated share of emissions per building type. The total share of emissions for different material/activities for building construction were subsequently calculated using estimates for different life cycle stages for the various building types.

The compilation of material, energy, and emissions flow serves as the baseline when applying identified abatement potentials from the abatement options review. The inventory of GHG abatement options (described in detail in Section 2.2) is established by means of a comprehensive literature review, including industry and governmental agency reports (grey literature), together with input from supply chain stakeholders. (Literature searches were conducted via a combination of academic bibliometric databases (Scopus and Web of Science) and web browser searches was used to enable the sourcing of the relevant grey literature, which is not as evident in academic bibliometric databases. Search string algorithms targeted a combination of the material/activity in question together with "carbon emissions" OR CO_2 OR GHG OR "greenhouse gas emissions" AND abatement OR "emission* reduction" OR mitigation OR decarbonization.), The main types of abatement options considered in the assessment are material efficiency and optimization measures together with shifts in: material production processes, transport vehicles and construction equipment technologies, and fuel substitutions in both equipment and production plants. The options include certain reuse and recycling measures resulting in emissions reductions, but not for the specific purpose of resource conservation. The inventory comprises both current best available technology and technologies assumed to be available over time to 2045.

A timeline is applied to test the potential implications to the climate impact when constructing the same assets while applying a combination of GHG abatement measures along the supply chain appraised to have reached commercial maturity at different points in time (over 5-year time periods until 2045). From this inventory, portfolios of abatement measures for the respective supply chain activities are constructed with selections of measures applied on a timeline up to the year 2045. The abatement measures are combined in pathways according to strategic choices [58], namely access to biofuels and renewable electricity, as well as enactment of material efficiency measures (as described in detail in Section 2.3).

The analysis assumes emission factors for electricity and district heating declining in accordance with scenario analysis from the Swedish Energy Agency, implying that GHG emissions related to electricity generation are close to zero in 2045 [59].

2.1. Pathway Generation and Quantification Approach

Total emissions from buildings and infrastructure construction in each time period t is calculated as:

$$E_{tot,t} = E_{b,t} + E_{ti,t}, \tag{1}$$

where $E_{b,t}$ is the emissions resulting from building construction, and $E_{ti,t}$ is the emissions resulting from transport infrastructure construction. The analysis includes emissions from materials production and the construction phase (i.e., corresponding to life cycle stage A1–A5 according to EN 15978 [60]), with the latter comprising emissions from mass and material transport, and the construction process (A4 and A5, respectively).

2.1.1. Emissions from Transport Infrastructure Construction

The transport infrastructure construction emissions, $E_{ti,t}$, are calculated as the sum of emissions from the material production stage and the construction activities as:

$$E_{ti,t} = \sum_m (E_{ti,m,t}) + \sum_{tc} (E_{ti,tc,t}), \tag{2}$$

where $E_{ti,m,t}$ is the emissions associated with material production of material m in timestep t; and $E_{ti,tc,t}$ is the emissions for construction activities tc in timestep t. Construction activities tc comprise mass and material transport and the construction process. Five material categories (m) are included in the analysis: concrete, reinforcement steel, construction steel, asphalt, and others.

The share of emissions from transport infrastructure construction coming from materials production and the construction process activities, respectively, in the base year, year 2015, is based on data from the Swedish Road Administration [53,61] and Liljenström et al. [9]. The emissions $E_{tc,2015}$ from the construction activities, tc, in the base year, year 2015, is calculated as:

$$E_{ti,tc,2015} = E_{ti,2015} * \sum_{i,c,tc} (e_{i,c} * e_{i,c,tc}), \tag{3}$$

where $E_{ti,2015}$ is the total emissions from transport infrastructure construction in 2015; $e_{i,c}$ is the share of emissions from transport infrastructure type i (i.e., road, railway, ports, and fairways) and construction type c (i.e., new construction and reinvestment); $e_{i,c,tc}$ is the share of emissions from transport infrastructure type i, construction type c and construction activities tc.

Correspondingly, emissions from material production are calculated as:

$$E_{m,2015} = E_{ti,2015} * \sum_{i,c,m} (e_{i,c} * e_{i,c,m}), \tag{4}$$

where $E_{m,2015}$ is the emissions from material production in 2015 for the specific material m; $E_{ti,2015}$ is the total emissions from transport infrastructure construction in 2015; $e_{i,c}$ is the share of emissions from transport infrastructure type i and construction type c; $e_{i,c,m}$ is the share of emissions from transport infrastructure type i, construction type c and material m.

2.1.2. Emissions from Building Construction

The building construction emissions are also calculated as the sum of emissions from the material production stage and the construction activities:

$$E_{b,t} = \sum_m (E_{b,m,t}) + \sum_{tc} (E_{b,tc,t}), \tag{5}$$

where $E_{b,m,t}$ is the emissions associated with material production of material m in timestep t; and $E_{b,tc,t}$ is the emissions for construction activities tc in timestep t. The analysis covers seven material categories, m, including: concrete, reinforcement steel, construction steel, insulation, gypsum and plaster, plastics and paint, and others (glass, aluminium, and wood).

For the base year of 2015, validated emissions for construction equipment (as per data from the national EEIO data reported in [7]) was used to extrapolate total building construction emissions:

$$E_{b,2015} = \frac{E_{cp,2015}}{e_{cp}}, \tag{6}$$

where $E_{b,2015}$ is the total annual emissions associated with building construction and refurbishment in 2015; $E_{cp,2015}$ is the emissions estimate for construction equipment in 2015 according to the national EEIO data; and e_{cp} is the emissions share estimated for construction processes.

The construction equipment data from the national EEIO data is considered reliable as construction equipment contribute to domestic emissions only and is used in construction and refurbishments (and not in operation of buildings). Once the total emissions estimate is produced, it is validated by means of comparing the resulting emissions for specific materials with available data to confirm its feasibility.

The share of emissions for the construction processes, material transports and material production were calculated using estimates for different life cycle stages for various building types

$$e_{lc} = \sum_{i=0}^{n} \left(e_i * e_{lc,i} \right), \tag{7}$$

where e_{lc} is the emissions share associated with the different life cycle stages lc (equivalent to A1–A3 for material production, A4 for material transport, and A5 for the construction process according to the European standard for "Sustainability of construction works - Assessment of environmental performance of buildings" (EN 15978)); e_i is the emission share for building type i; and $e_{lc,i}$ is the emissions share for life cycle stage lc and building type i. The analysis covers three building types, i, including: multi-family dwellings, single-family dwellings, and non-residential buildings.

The share of emissions for different materials for construction of different building types were calculated based on the estimates in literature for these building types and the estimated share of emissions per building type. Where available and most applicable (i.e., for multi-family dwellings), the building type was also divided into building typology and frame type, namely concrete frame and wood frame. The emissions share e_m associated with material production of the material m is thus calculated as:

$$e_m = \sum_{i=0}^{n} \left(e_i * e_{m,i} \right), \tag{8}$$

where e_i is the emission share for building type i; and $e_{m,i}$ is the emissions share for material m and building type i.

The initial estimated shares for both life cycle stages and materials were subsequently amended based on validated data for certain components in combination with adjustments for materials commonly used in refurbishments.

2.1.3. Material and Energy Demand

Emissions and energy intensity factors for materials, activities, and fuels were combined with the emissions figures to estimate material and energy demand. The emission intensity factors for materials, activities, and fuels, along with data for associated quantity and source of energy used for material production, were sourced in a literature review. Table 1 lists the details for the reference energy carriers, materials or material combinations used in the calculation of material and energy demand for the construction of buildings and transport infrastructure in the year 2015. Details on specific materials, material production processes, and energy sources can be found in Tables A1 and A2 in Appendix A.

Table 1. Emissions and energy intensity factors along with energy mix in the production of reference materials and energy carriers used in the construction of buildings and transport infrastructure in the base year of 2015.

Reference Materials (m)/ Activities (tc)	Unit	Emissions Intensity Ef_m (tCO$_2$/unit)	Energy Intensity Qf_m (MWh/unit)	Energy mix $q_{sh,m,s}$ (%)							Comment	References
				Fossil Fuels	Coal/Coke	Oil/Diesel	Gas	Fossil Waste	Biomass	Electricity		
Concrete	m^3	353	656.0	6%	22%			37%	17%	15%	18% cement share (corresponding to 420 kg cement per m^3 concrete) as the average of building and infrastructure concrete	[7,62,63]
Cement	t	0.82	1.35		25%			43%	20%	10%	Cement with 86% cement clinker and 14% alternative binders. Thermal energy in clinker production; electrical energy in cement production	[51,64,65]
Reinforce-ment steel	t	0.78	2.50		29%	1%	16%			54%	85% scrap-based, 15% primary steel	[24,62,66–75]
Construction steel	t	2.12	6.40		64%	2%	17%			17%	Galvanized steel, 100% primary steel	[69,71,72,76–79]
Asphalt	t	0.35	0.90	100%							Hot mix asphalt with 6.2% bitumen. Does not include transports and paving	[80–82]
Insulation	t	3.30	17.40	89%						11%	Varying depending on insulation material; Assuming 60% polystyrene and 40% mineral wool	[66,74,79,83,84]
Gypsum and plaster	t	0.30	1.50	87%						13%	Average of values for gypsum plasterboards	[62,79,85–87]
Plastic	t	2.50	20.00	89%						11%	Average of polyvinylchloride (PVC) and polyethylene (PE)	[37,62,67,74,75,79]
Aluminium	t	11.0	19.70	12%						88%	Primary aluminium	[67,74,75,88,89]
Glass	t	1.00	3.50	70%				30%				[67,74,85,90]
Timber	t	0.28	2.60	15%					70%	15%	Average of cross-laminated timber, glulam beams and sawn timber	[62,69,75,85,91]
Construction process	kWh	0.24	-		2%	64%			17%	17%	Based on 75% diesel use, 17% electricity use and 8% heat from district heating	[11,75,92–94]
Material transports	kWh	0.30	-			85%			15%		Low biofuel blended diesel (Diesel MK1) with the Swedish national biofuel share from 2015	[95]

The specific emission intensity figures were combined with emission shares to calculate the resulting material and energy demand. Accordingly, the material demand M_m for each material m for the base year of 2015 is calculated as:

$$M_m = \frac{\left(E_{b,2015} + E_{ti,2015}\right) * e_m}{Ef_m}, \tag{9}$$

where $E_{b,2015}$ is the total annual emissions associated with building construction and refurbishment; $E_{ti,2015}$ is the total annual emissions associated with construction of transport infrastructure; e_m is the emissions share associated with material m; and Ef_m is the emission intensity factor associated with material production of the material m.

The energy demand for material transports and construction processes for the base year, year 2015, is calculated as:

$$Q_{tc} = \sum_s \frac{\left(E_{b,2015} + E_{ti,2015}\right) * e_{tc,s}}{Ef_s}, \tag{10}$$

where Q_{tc} is the energy demand for construction activities tc; $e_{tc,s}$ is the emissions share associated with energy source s for construction activity tc; and Ef_s is the emission intensity factor for energy source s.

The total energy demand per energy source is calculated by using energy intensity factors combined with the energy mix data for production material processes and fuels used in transport and construction processes:

$$Q_{tot,s} = \sum_m \left(M_m * Qf_m * q_{m,s}\right) + \sum_{tc} Q_{tc} * q_{tc,s} \tag{11}$$

where $Q_{tot,s}$ is the total energy demand associated with energy source s; M_m is the material demand of each specific material m; Qf_m is the energy intensity associated with the production of material m; $q_{m,s}$ is the share of energy in the material production of material m of the energy source s; $Q_{tc,s}$ is the energy demand for construction activity tc and energy source s; and $q_{tc,s}$ is the energy share in the reference fuel used for construction activity tc of the energy source s. Three energy sources are detailed in the analysis: fossil fuels (coal, gas, oil, and fossil waste), biofuels, and electricity.

2.1.4. Pathway Generation

Pathways are subsequently created were portfolios of abatement measures for the respective supply chain activities are constructed with selections of measures applied on a timeline up to year 2045. In the pathway analysis, the production levels of each material in each time step was estimated based on the remaining material demand after implementation of abatement options affecting demand of each material:

$$M_{m,t} = \left(1 - A_{re,m,t}\right) * \left(1 - A_{ms,m,t}\right) * \left(1 - A_{me,m,t}\right) * M_m, \tag{12}$$

where $M_{m,t}$ is the material demand of material m in time step t; A is the total material demand reduction of material m in time step t associated with each of the following abatement measures: *re-* recycling, *ms*–material substitution, *me*–material efficiency measures; and M_m is the original material demand of each specific material m in the base year of 2015. An illustration of how this generic calculation is performed for concrete demand (and resulting demand for cement and Supplementary Cementitious Material, SCM) is illustrated in Figure 2.

In the pathway analysis, the emissions and energy demand associated with material production, material transports and construction processes were adjusted based on the abatement options selected and applied in the assessment for each supply chain activity, as described in Section 2.3. The energy intensity factors, and energy mixes, were adjusted based on abatement measures, including energy efficiency and hybridization, biofuel substitution, and electrification. The energy demand for $Q_{cp,t}$ construction process cp in timestep t is consequently calculated as:

$$Q_{cp,t} = \left(1 - A_{op,cp,t}\right) * \left(1 - A_{ee,cp,t}\right) * Q_{cp}, \tag{13}$$

where A is the total energy demand reduction for construction processes in time step t associated with each of the following abatement measures: *op*–optimization and *ee* energy efficiency (including from hybridization and electrification), Q_{cp} is the energy demand for construction processes in the base year of 2015.

The energy demand $Q_{mt,t}$ for material transport mt in timestep t is, consequently, calculated as:

$$Q_{mt,t} = \left(1 - A_{\overline{me},mt,t}\right) * \left(1 - A_{op,mt,t}\right) * \left(1 - A_{ee,mt,t}\right) * Q_{cp,} \tag{14}$$

where A is the total energy demand reduction for construction processes in time step t associated with each of the following abatement measures: $\overline{me}$–average of material efficiency measures for main materials (concrete, steel, asphalt), op–optimization and ee–energy efficiency (including from hybridization and electrification); Q_{mt} is the energy demand for material transports in the base year of 2015.

The energy demand per energy source in each time steps is consequently calculated as:

$$Q_{tot,t,s} = \sum_{m} (M_{m,t} * Qf_{m,t} * q_{m,t,s}) + \sum_{tc} Q_{tc,t} * q_{tc,t,s,} \tag{15}$$

where $Q_{tot,s,t}$ is the total energy use of energy source s in timestep t; $M_{m,t}$ is the material demand of material m in timestep t; $Qf_{m,t}$ is the energy intensity factor for production of material m in timestep t; $q_{m,s,t}$ is the share of energy source s for the production of material m in timesteps t; Q_{tc} is the energy demand for construction stage tc; $q_{tc,t,s}$ is the energy share for construction processes and material transports in timestep t for energy source s.

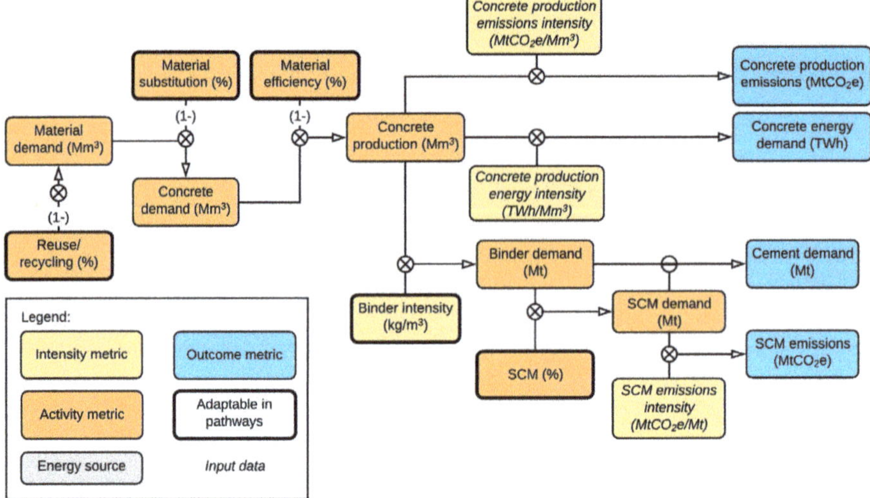

Figure 2. Schematic illustration of the calculation of concrete, cement, and Supplementary Cementitious Material (SCM) demand, along with associated energy demand and emissions for concrete manufacture and SCM. Boxes linked with an encircled x are multiplied, a box linked with an encircled x combined with 1-in brackets are reduced by the percentage figure in the box closest to the brackets, while a box linked via an encircled minus sign is subtracted. Boxes with thick outlines are metrics that are adaptable over time in the pathways depending on the abatement measures applied, while boxes with cursive texts are input data provided in Tables A1 and A2 in Appendix A. The initial material demand figure is only adaptable in the sensitivity analysis. Blue boxes are result figures. The cement demand figure is used as input for the cement production calculation, as displayed in Figure 3.

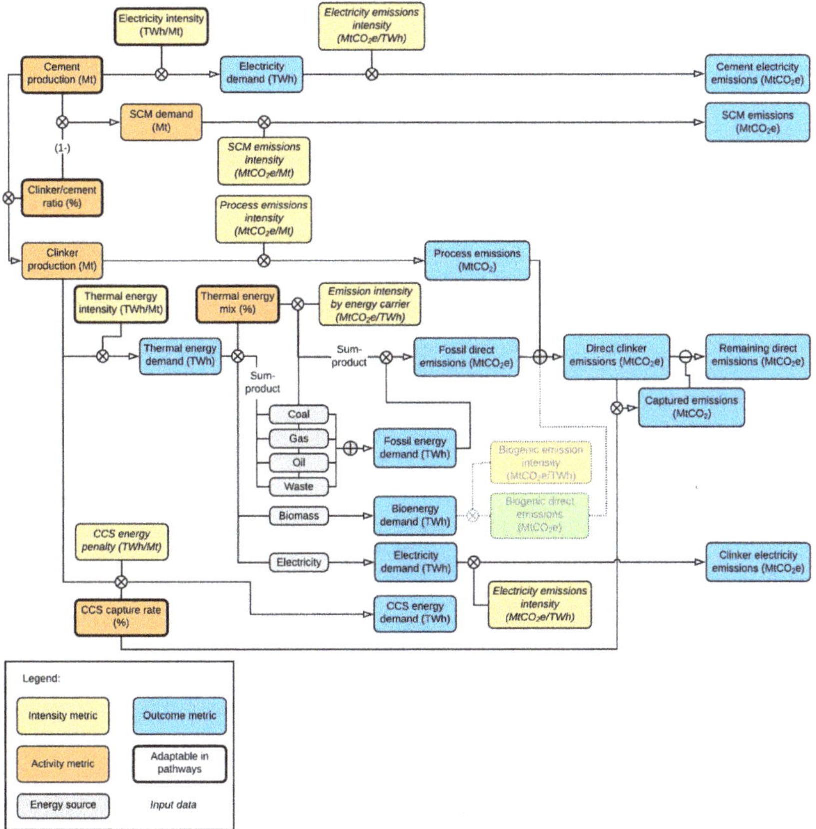

Figure 3. Schematic illustration of the calculation of emissions and energy demand per energy source for cement production. The cement production figure stems from the concrete calculation depicted in Figure 2. Boxes linked with an encircled x are multiplied, a box linked with an encircled x combined with 1-in brackets are reduced by the percentage figure in the box closest to the brackets, while a box linked via an encircled minus sign is subtracted, and boxes linked with an encircled plus sign are added up. Boxes with thick outlines are metrics that are adaptable over time in the pathways depending on the abatement measures applied, while boxes with cursive texts are input data provided in Table A1, Table A2, and Table A3 in Appendix A. Blue boxes are result figures.

For material production, emissions from direct energy use, together with process emissions, were also adjusted based on the level of carbon capture applied. The resulting emissions for each material m are calculated as:

$$E_{m,t} = M_{m,t} * \left((Ef_{pr,m} + Qf_{m,t} * \sum_s (q_{m,s,t} * Ef_s)) * (1 - CC_{m,t}) + Qf_{m,t} * q_{m,el,t} * Ef_{el,t} \right), \quad (16)$$

where $E_{m,t}$ is the emissions resulting from the production of material m in timestep t; $M_{m,t}$ is the material demand of material m in timestep t; $Ef_{pr,m}$ is the process emissions intensity factor to produce material m; $Qf_{m,t}$ is the energy intensity factor for production of material m in timestep t; $q_{m,s,t}$ is the share of direct energy sources s for the production of material m in timesteps t; Ef_s is the emissions intensity factor of direct energy source s; $CC_{m,t}$ is the share of direct and process emissions captured via carbon capture technologies in the production of material m in timesteps t; $q_{m,el,t}$ is the share of energy use from electricity in the production of material m in timestep t; $Ef_{el,t}$ is the emissions intensity factor of electricity in timestep t. Illustrations of how the generic calculation of materials emissions is performed for cement and primary steel production is displayed in Figures 3 and 4. The abatement options considered and applied are described in Sections 2.2.1 and 2.2.2, respectively. Below, an example calculation is made for construction steel in Pathway 1 for the year 2040, where 30% of the coal use is substituted for biofuel and 30% of the thermal emissions are captured:

$$E_{construction\ steel,2040}(\text{ktCO2e}) =$$

$$414\ (\text{kt}) * \left(\left(0 + 6.4\left(\tfrac{\text{GWh}}{\text{kt}}\right) * (0.34(\%) * 0.37\left(\tfrac{ktCO2e}{GWh}\right) + 0.02(\%) * 0.228\left(\tfrac{ktCO2e}{GWh}\right)\right.\right.$$
$$\left.\left. + 0.17(\%) * 0.248\left(\tfrac{ktCO2e}{GWh}\right)\right) * (1 - 0.30) + 6.4\left(\tfrac{\text{GWh}}{\text{kt}}\right) * 0.17(\%) * 0.115\left(\tfrac{ktCO2e}{GWh}\right)\right) = 372\ ktCO2e,$$

where the energy intensity factor and energy source shares are taken from Table 1 with the coal share reduced by 30%, and the emissions intensity factors are taken from Table A2 in Appendix A for the thermal energy sources and Table A3 in Appendix A for electricity.

For material transport and construction process, construction activities tc, the emissions in each timestep t is calculated as:

$$E_{tc,t} = Q_{tc,t} * \sum_{s} (q_{tc,s,t} * Ef_{s,t}), \tag{17}$$

where $E_{tc,t}$ is the emissions for construction stage tc in time step t, $Q_{tc,t}$ is the energy for lifecycle stage tc in timestep t; $q_{tc,s,t}$ is the energy share for construction stage tc of energy source s in timestep t; $Ef_{s,t}$ is the emissions factor for energy source s in timestep t.

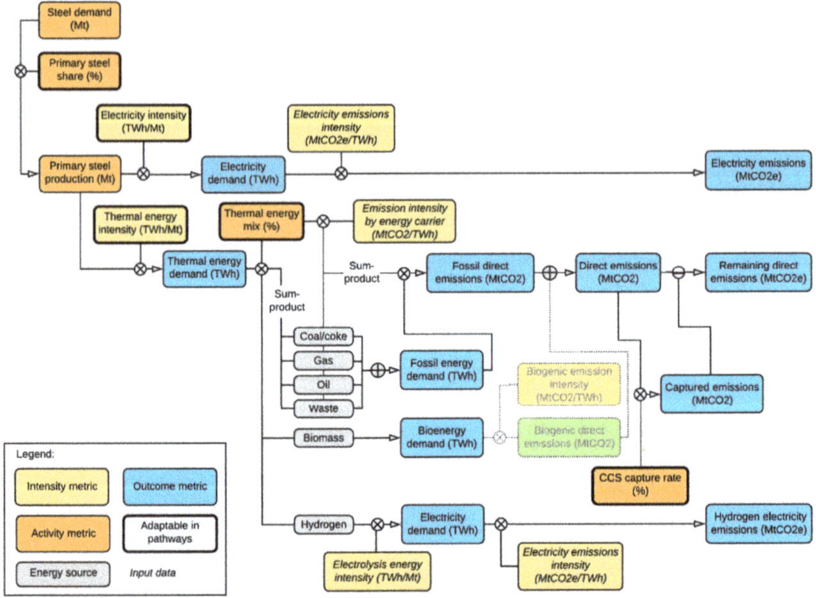

Figure 4. Schematic illustration of the calculation of emissions and energy demand per energy source for steel production. Boxes linked with an encircled x are multiplied, a box linked via an encircled minus sign is subtracted, and boxes linked with an encircled plus sign are added up. Boxes with thick outlines are metrics that are adaptable over time in the pathways depending on the abatement measures applied, while boxes with cursive texts are input data provided in Table A1, Table A2, and Table A3 in Appendix A. The initial material demand figure is only adaptable in the sensitivity analysis. Blue boxes are result figures.

2.2. Abatement Options

2.2.1. Cement/Concrete

The cement clinker production is responsible for the majority of GHG emissions related to concrete use with around 65% of the CO_2 emissions stemming from the calcination process and 35% emanating from the fuels used in the cement ovens, the so-called kilns. The main current emission abatement options comprise of replacing fuels in the cement kilns with waste- or bio-based fuels, reducing the amount of cement clinker by using Supplementary Cementitious Material (SCMs or so-called alternative binders), and optimizing the concrete recipes to use less cement [27,29,96,97]. Sweden is a frontrunner when it comes to alternative fuels [51] but is behind the rest of Europe in using alternative binders with a clinker share of 86% [98] compared to the European average of 73% [29]. In addition, the average cement/binder content used in concrete is higher in Sweden than in other countries, with around 420 kg binder per m^3 concrete compared to an average 400 kg binder per m^3 concrete in Europe overall [7,99,100]. It is worth noting that high levels of SCMs require process adjustments due to additional hardening times prolonging project timelines, while optimized concrete recipes impact site practices as multiple specific concrete mixes require further logistics and on-site coordination.

Other prominent abatement options include design optimization to slim constructions, increased prefabrication to reduce waste and minimized construction process emissions, and material substitutions towards wood-based solutions [37]. For building construction, the development of engineered wood products has increased the opportunities for building multi-floor building with a structural core of timber.

Indeed, engineered wood products have recently experienced annual growth rates between 2.5% and 15% [101], with a range of studies showing that buildings with wooden structures have a lower carbon footprint than buildings with other types of structures (see reviews in, e.g., Reference [57,102–105]).

However, even if current abatement options are combined to its full potential, transformative technologies are still required to reach the goal of close to or net zero emissions in the cement industry by 2045 [54]. Carbon capture technologies (CCS) with or without electrification of the cement kilns are key deep decarbonization alternatives. The Swedish cement industry roadmap is targeting climate neutrality by 2030, with the main focus being on biofuels together with CCS [98]. However, Cementa is also pursuing electrification together with Vattenfall through its CemZero project, with a pre-feasibility study released in 2018 [106].

2.2.2. Steel

Construction steel, often galvanized, is predominantly produced by primary steel, i.e., from iron ore in integrated steel plants, while reinforcement steel is mainly produced by scrap steel in secondary steelmaking plants, called electric arc furnaces (EAF), although depending on the availability of scrap steel, this varies globally [107]. Predominant current abatement options to reduce embodied emissions associated with steel are enhanced material efficiency and circularity measures [9,15,58,108]. The main opportunities lie in reducing waste during the construction process; reduce the amount of material in each building by avoiding over-specification and using higher-strength materials; and reusing buildings and building components [38,44,109]. With better sorting and separation, there is also a potential for increased scrap share for construction steel production [38,110].

Regarding the different production methods, EAFs mainly use electricity but also require fueling by natural gas (25–30%) and a smaller share of coal (<5%) [70–73]. With electricity as the main energy carrier, the emission intensity of the electricity used is an important factor [107,108]. Refurbishments and upgrades of current electric arc furnaces provide potential for decreased electricity consumption [70,111,112], and there is also potential for biomass to substitute fossil process energy in EAFs, both as a reducing agent and as fuel in reheating furnaces [70,73,113]. Fuel substitution from natural gas to bio-based syngas or biooil is similarly proposed in metallurgical processes [114].

For primary steel production, about 80% of the CO_2 emissions stem from the reduction of iron ore [22,23,115]. The main options for deep emission reduction in primary steel production are electrification with renewable electricity (either via hydrogen direct reduction or through electrowinning) [22,26,71,90,115–117], use of biomass to replace coke as fuel and reducing agent [26,76,113,118–122], and/or use of carbon capture and storage (CCS) [22,26,40,117,123,124]. Partial CO_2 capture is a mature and low-cost technology that can be implemented in the coming 10–15 years without major changes to the existing process and which can be combined with biomass substitution [123,125,126].

2.2.3. Other Materials

At present, polystyrene and mineral wool are the most frequently used for insulating buildings [127], with mineral wools in general having a lower carbon footprint, which is why material substitution together with recycling is a current abatement measure for insulation [74,104]. Other abatement measures include fuel change together with energy efficiency measures for production of both mineral wool and polystyrene insulation [37,128,129]. Steam cracking is responsible for a large share of the carbon footprint (~40%) of plastics production [37] (which is also a raw material in polystyrene insulation), which is why deep abatement options for plastics production include electrification or carbon capture in cracking and polymerization [37,90]. Other abatement measures for plastics include material efficiency measures and recycling either by mechanical or chemical means [38,41,47].

In the production of gypsum for plasterboards, the most prominent abatement measure is the use of recycled gypsum which can be combined with electrification or biofuel substitution in the heating furnaces used in the gypsum production [130].

Main abatement options for asphalt include biofuel substitution, lowered temperatures, and increased recycling rates [80,131,132].

2.2.4. Material Efficiency

Material efficiency is a key abatement measure for all construction materials, and a measure that generally deserves more attention in policy and climate mitigation discussions. Evidence (see, e.g., Reference [37,38,44,47,133]) suggests that, on average, one-third of all material use could be saved if designs were optimized for material use rather than for cost reduction, since downstream production (and design) are generally dominated by labor costs and not material costs. For example, it is easier to use constant cross-sections across a structure than to design each beam and column individually since this leads to more rapid construction.

In addition, motivations to use excess material are driven by an asymmetry costs of product failure compared with the costs of over-specification, by over-specified components copied across projects to minimize costly design time, by cheaper manufacture of standard parts, and by the fact that many products experience higher loads prior to use (in installation or transport) than in use [46].

2.2.5. Construction Equipment and Heavy Transports

High potential abatement measures for heavy vehicles and machinery in the short to medium term include biofuel substitution, energy efficiency measures, hybridization, and optimization of logistics and fleet management. Over the longer term, deeper emissions reductions would result from electrification of construction equipment, crushing plants and heavy trucks. For the latter, options include plug-in hybrid or fuel-celled heavy-duty trucks/haulers potentially in combination with electric road systems. Model shifts for heavy transport to rail and ship is also an abatement measure with large potential. While such shifts are out of scope for this analysis, this is an important level towards a more transport-efficient society [134].

2.2.6. Summary of Abatement Options

A summary of all abatement options and their identified emission reduction potential are described in Figure 5. The graph illustrates the range of GHG emissions reduction potential recognized in literature for each of the abatement options explored, where the range may depend on the level of the abatement measure that is adopted. Full details of measures for all activities, including timelines, potentials, and references, are available in the Supplementary Material.

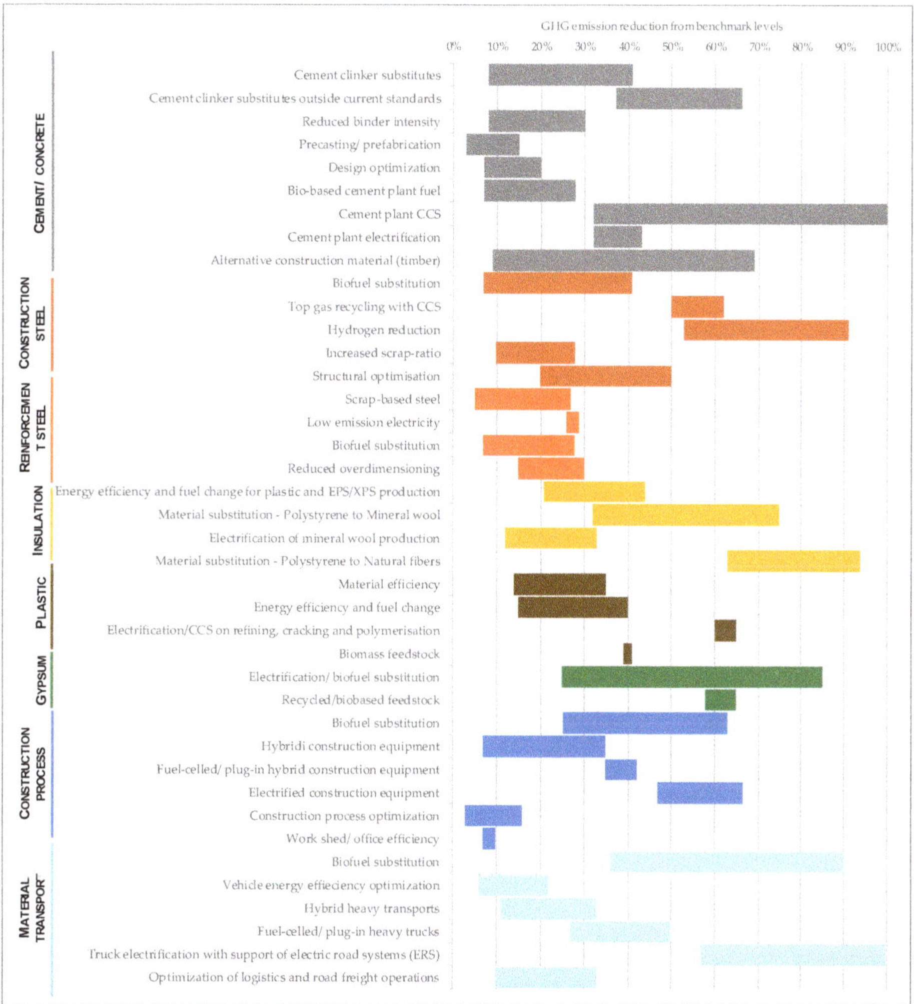

Figure 5. Range of greenhouse gas (GHG) emissions reduction potential for the abatement options identified in the literature review for the main emissions sources (color coded). The study analysis is based around reaching the medium-high range of the emission reduction potentials for each selected abatement measure when fully implemented. The Supplementary Material provides full details of measures for all activities, including timelines, potentials, and references.

2.3. Alternative Pathways

Four pathways have been devised for buildings and transport infrastructure, describing different future trajectories of technological developments in the supply chains of buildings and transportation infrastructure in Sweden, two with a focus on bio-based measures together with CCS and two with a focus on electrification:

- Pathway 1: Biofuels and CCS;
- Pathway 2: Electrification;
- Pathway 3: Biofuels, CCS and material efficiency; and
- Pathway 4: Electrification and material efficiency.

The second of the two within each focus explores the role material efficiency measures may play in the low-carbon transition. Details of the emissions reduction measures applied over the timeline for the different pathway scenarios are displayed in Table 2.

For cement, the bio/CSS pathway adopts post-combustion carbon capture with amine scrubbing, which is the technology tested by HeidelbergCement in Breivik in Norway [135]. In all pathways, a progressive realization of cement clinker substitution and cement demand reduction from optimization of concrete recipes is assumed.

For primary steel production, the bio/CCS pathways adopt process modification enabling top gas recycling combined with carbon capture and storage, while the electrification pathways pursue a hydrogen direct reduction (H-DR/EAF) steelmaking process. Current electric arc furnaces for scrap-based secondary steel production are being refurbished and upgraded at a continuous rate in all pathways, alongside partial bioenergy substitution in the bio/CCS pathways.

Separate pathways have also been devised for construction equipment and heavy transports, while other materials follow a common decarbonization pathway (based on, e.g., Reference [37,41,74,83,108,130,136]).

The pathway portfolios are predominantly based around reaching the medium-high range of the emission reduction potentials for each selected abatement measure when fully implemented (as per Figure 5) with measures and timelines largely compatible with roadmaps and pathways developed within the EU Commission long term climate strategy (combination of electrification and hydrogen scenarios), along with relevant industry roadmaps developed within the Fossil Free Sweden project [48,137].

Table 2. Details of abatement measures applied across pathways with percentage figures depicting the diffusion of the specific mitigation option.

Material/Process	Pathway	2025	2030	2035	2040	2045
Cement/concrete	All pathways	20% alternative binders (SCM) 5% reduced binder intensity 2% wood substitution	25% alternative binders (SCM) 12% reduced binder intensity 3% wood substitution	28% alternative binders (SCM) 15% reduced binder intensity 5% wood substitution	32% alternative binders (SCM) 22% reduced binder intensity 7% wood substitution	35% alternative binders (SCM) 28% reduced binder intensity 10% wood substitution
	Biofuel + CCS	40% biofuels	45% biofuels 45% CCS	50% biofuels 45% CCS	52% biofuels 80% CCS	55% biofuels 90% CCS
	Electrification	40% biofuels	45% electrification	45% electrification	90% electrification	100% electrification
	Material efficiency	8%	15%	20%	25%	30%
Reinforcement steel	Biofuel + CCS	100% secondary steel	10% energy efficiency 7% biofuel	14% biofuel	25% biofuel	35% biofuel
	Electrification	100% secondary steel	10% energy efficiency 7% electrification (plasma heating)	14% electrification	14% electrification 10% biofuel	14% electrification 21% biofuel
	Material efficiency	5%	10%	15%	20%	25%

Table 2. *Cont.*

Material/Process	Pathway	2025	2030	2035	2040	2045
Construction steel	Biofuel + CCS		20% biofuel	30% biofuel	30% CCS 30% biofuel	60% CCS 30% biofuel
	Electrification		20% biofuel	30% biofuel	50% electrification (hydrogen-reduction)	100% hydrogen-reduction
	Material efficiency	10%	15%	20%	25%	30%
Construction equipment	All pathways	5% optimization	10% optimization	10% optimization	10% optimization	10% optimization
	Biofuel + CCS	42% biofuel 9% hybridization 5% electrification	63% biofuel 14% hybridization 9% electrification	78% biofuel 23% hybridization 13% electrification	85% biofuel 31% hybridization 15% electrification	81% biofuel 31% hybridization 19% electrification
	Electrification	42% biofue 19% hybridization 5% electrification	75% biofuel 14% hybridization 9% electrification	76% biofuel 23% hybridization 24% electrification	59% biofuel 23% hybridization 41% electrification	50% biofuel 23% hybridization 50% electrification
Heavy transports	All pathways	5% efficiency/ optimization	10% efficiency/ optimization	15% efficiency/ optimization	20% efficiency/ optimization	25% efficiency/ optimization
	Biofuel + CCS	42% biofuel 5% electrification	63% biofuel 10% electrification	78% biofuel 15% electrification	80% biofuel 20% electrification	75% biofuel 25% electrification
	Electrification	42% biofuel 5% electrification	63% biofuel 20% electrification	70% biofuel 30% electrification	55% biofuel 45% electrification	40% biofuel 60% electrification
Insulation	All pathways	2% energy efficiency 20% material substitution	4% energy efficiency 50% material substitution 10% electrification	6% energy efficiency 70% material substitution 20% electrification	70% material substitution 30% electrification/CCS	70% material substitution 30% electrification/CCS
Gypsum/ plaster	All pathways	25% biofuel/material substitution	25% biofuel/material substitution 25% recycling	25% biofuel/electrification 50% recycling	50% biofuel/electrification 50% recycling	100% biofuel/electrification 75 recycling
Plastic	All pathways	20% energy efficiency/biofuel	40% energy efficiency/biofuel	40% energy efficiency/biofuel	50% electrification/CCS	100% electrification/CCS
	Material efficiency	5%	10%	15%	20%	25%
Asphalt	All pathways	66% biofuel 37% recycling 4% energy efficiency	66% biofuel 45% recycling 8% energy efficiency	85% biofuel 50% recycling 12% energy efficiency	85% biofuel 55% recycling 15% energy efficiency	85% biofuel 60% recycling 15% energy efficiency

Sensitivity Analysis

The main assumption in the model is a constant construction demand up until 2045. However, this assumption is uncertain and different sources provide diverse interpretations of how the demand for building and transport infrastructure construction will develop. For example, the Swedish Energy Agency, in its long-term prognosis, predicts the energy use of the building construction sector to increase until 2020 due to extensive construction of new housing and to then fall back to previous lower levels after 2025 [138]. This would imply reductions of around 20% to 2030 and 30% to 2045 based on 2020 levels.

On the other hand, Boverket estimates that, by 2025, Sweden needs 600,000 new dwellings, implying a level of construction not anticipated in the prognosis of the Swedish Energy Agency [7,139]. Further, a great need for renewed transport infrastructure has been identified to enable the climate transition of the transport network to be realized while meeting increased transport demands, including the anticipated but heavily discussed construction of a highspeed railway network [140,141].

Consequently, a scenario analysis has been performed to test the implications of reductions/increases in construction demand of ±20% to 2030 and ± 30% to 2045.

3. Results

3.1. Current Emissions from Building and Infrastructure Construction

As described in the Introduction the range of current estimates of GHG emissions linked to the construction of buildings in Sweden is notable (8.5 MtCO$_2$e based on a process-based bottom-up LCA approach, 8.1 MtCO$_2$e for territorial emissions, and 13.5 MtCO$_2$e including imports based on EEIO data) with potential variances including different system boundaries (e.g., agricultural properties not included in the bottom-up model) and possible overstating of the importance and emissions intensity of imports in the input-output analysis [7,8,11]. Further, a great majority of construction steel is imported [142], and, while the cement market is mostly domestic (85% of Swedish cement use) [143], the concrete market is turning more international, at least pertaining to precast elements [143–145].

To validate the estimates of the current GHG emissions, and to specify emissions components, further analysis into the existing estimates were combined with a literature review focused on relevant LCA studies detailing embodied emission sources for different construction types.

3.1.1. Estimate and Validation of Current Emissions from Building Construction

Around 2/3 of the construction emissions correspond to new buildings and 1/3 to refurbishments and maintenance. In addition, around 40–50% of the annual climate impact from building construction stem from construction of non-residential buildings, such as offices, schools, and other premises. A growing share of around 40–50% arise from multi-family dwellings and the remaining 10–15% from single family houses [7,52]. Multi-family buildings are predominantly constructed with concrete frames (85% in 2018), with smaller shares of timber frames (13%) and steel frames (2%) [146].

A detailed overview of the share of emissions components, and share of individual materials, related to new building construction for new builds of various building and frame types can be found in Tables A4 and A5 in Appendix A. The total share and amount of emissions for different material/activities for building construction were calculated using estimates for different life cycle stages for various building/frame types from the literature review, with the initial estimates shown in Table 3.

Table 3. Initial and updated annual emissions estimates per lifecycle stage for building construction in the base year of 2015.

Emissions Estimate	Building Materials (A1–A3)	Transport (A4)	Construction Process (A5)
Initial estimate share of embodied emissions (%)	85%	5%	12%
Initial estimate amount of embodied emissions (Mt CO_2)	6.5	0.4	0.9
Updated estimate share of embodied emissions (%)	80%	9%	11%
Updated estimate amount of embodied emissions (Mt CO_2)	6.3	0.7	0.9

Worth noting about these estimates is that ground preparation is often not included in LCA studies, which would increase the share of construction processes. On the other hand, the estimates do not include refurbishments, which would increase the share of material transports and certain materials.

The estimates can be compared to the approximate sector division for the building and real estate sector (territorial emissions including emissions associated with real estate management during building use) from the Swedish EEIO analysis. The sector division include 3.9 Mt CO_2 from building materials (only domestically produced materials), 0.9 Mt CO_2 from construction equipment, and 1.5 Mt CO_2 from transports, while the share of the transport emissions estimate belonging to construction versus real estate management is not entirely clear [7]. The level of transport emissions nonetheless is significantly higher than the level and share of emissions allocated to transports resulting from the LCA studies in the literature review (~24% versus 5% or 0.4 Mt CO_2), noting that the latter figure does not encompass refurbishments. Further, the process-based bottom-up approach estimates transport emissions from building construction of 0.9 Mt CO_2 (17% of building construction emissions), while also including people transport in this estimate [8]. The initial emissions share and estimate for material transports is consequently adjusted upwards.

The initial and updated emissions estimates from materials are displayed in Table 4.

Table 4. Initial and updated annual emissions estimates per material for building construction in the base year of 2015. The initial estimates are based on a combination of emissions share data per lifecycle stage for construction of new buildings together with data on construction of different building/frame types, while the updated estimates are the data used in the model after adjustments for refurbishments and validation.

Building Materials (A1–A3)	Concrete	Reinforcement Steel	Construction Steel	Insulation	Gypsum and Plaster	Plastic and Paints	Others (Glass, Aluminium, Timber)
Initial estimate share of building material emissions (%)	44%	10%	11%	8%	6%	4%	17%
Initial estimate amount of material emissions (Mt CO_2)	2.8	0.6	0.7	0.5	0.4	0.3	1.1
Updated estimate share of building material emissions (%)	40%	10%	11%	10%	6%	4%	19%
Updated estimate amount of material emissions (Mt CO_2)	2.5	0.6	0.7	0.6	0.4	0.3	1.2

The sector division in the Swedish EEIO analysis further details an approximate 2.4 Mt CO_2 from the mineral industry (predominantly cement) [7]. Regarding cement, emissions from Cementa were 2.5 Mt CO_2 in 2015 [147], which corresponds to 85% of Swedish cement use [143]. In total, emissions from Swedish cement use were thus about 2.9 Mt CO_2 in 2015. However, while the cement market is mostly domestic, the concrete market is turning more international, particularly pertaining to precast concrete. There is a lack of data and reporting to determine the extent of concrete imports, but an estimate can be made based on the import-export balance of concrete, cement and gypsum products of SEK 1.8 billion [143].

If these imports are considered to correspond to concrete elements and the concrete costs 60–70 EUR/ton (about SEK 600–700/ton) [148,149], this would correspond to concrete imports of 2.5–3 Mt per year, corresponding to emissions of about 0.4–0.5 Mt CO_2, giving a total emissions estimate of 3.3 Mt CO_2e from Swedish concrete use. With around 75% of concrete being used in building construction [150], the emissions from concrete use in building construction would correspond to around 2.5 Mt CO_2. The emissions estimates of concrete and material production overall are adjusted accordingly in the model.

Additional upwards adjustments are based on literature detailing refurbishments which report the main embodied emissions resulting from insulation, windows and metals for new ventilation, and heating systems [55,151,152].

3.1.2. Estimate of Current Emissions from Building and Transport Infrastructure Construction

The total climate impact of building and transport infrastructure construction in Sweden is estimated to around 9.8 Mt CO_2 per year, with building construction responsible for 80% and transport infrastructure for 20%. This can be compared with the national territorial GHG emissions of 51.8 MtCO_2e in 2018 [6]. As can be seen in Figure 6, this carbon impact derives predominantly from concrete and steel together with diesel use in construction processes and material transports.

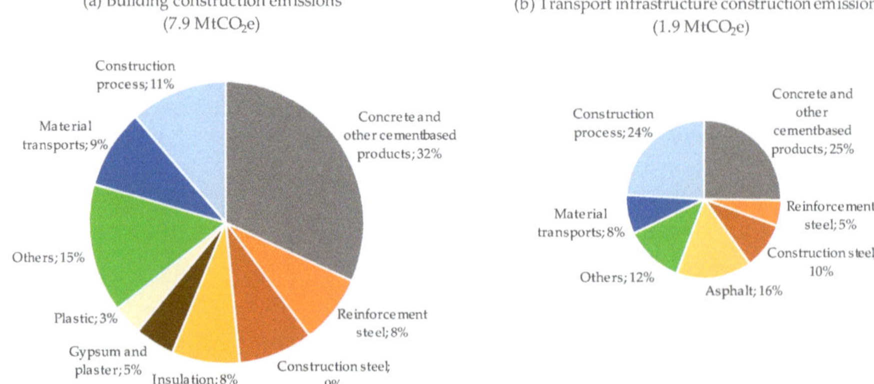

Figure 6. Carbon impact from (**a**) construction of buildings and (**b**) construction of transport infrastructure with the size of the pie charts reflecting the relative magnitude of emissions.

3.1.3. Validation of Building and Transport Infrastructure Construction Emissions Estimate

The total estimated emissions from buildings and transport infrastructure construction of 9.8 Mt CO_2e is in the middle of the range of estimates of 8.1–13.5 Mt CO_2e, as reported by Naturvårdsverket and Boverket [7,8].

Focusing in on concrete, the resulting concrete emissions estimate for building and transport infrastructure combined is 3.0 Mt CO_2e, which corresponds well to the estimate of concrete use in Sweden discussed in the building construction Section 3.1.1 (considering the exclusion of utilities in this analysis).

Another validation can be made regarding steel use. A great majority of steel used in construction is imported [143]. Swedish steel imports were 3.2 Mt in 2015 [142,153] with research demonstrating that around 25–50% of steel consumption goes to the construction industry [154,155]. This would correspond to the use of 0.8–1.6 Mt steel in constructions, matching the model estimate of 1.4 Mt steel (based on the equivalent emissions intensities for reinforcement and construction steel).

3.2. Pathway Results

The main results from the pathway analysis, i.e., energy use per energy carriers and carbon emission reductions, for the construction of buildings and transport infrastructure up until 2045, are depicted in Figures 7 and 8. The results show that it is possible to reduce CO_2 emissions associated with construction of buildings and transport infrastructure by at least 50% to 2030 (51–62%), and reach close to zero emissions by 2045 (90–94%) with the electrification and material efficiency pathways demonstrating the highest reductions. The energy use is also reduced in all pathways albeit with more variance between the pathways (6–19% to 2030 and 16–37% to 2045). In addition, regarding energy use, it is the electrification and material efficiency pathway which demonstrates the highest reductions.

The analysis demonstrates that currently, construction of buildings and transport infrastructure use approximately 32 TWh energy, accounting for around 8% of total Swedish energy use [156]. All the pathways demonstrate a reduction in total energy use over time, with the reduction varying from 6–19% to 2030 and 16–37% to 2045.

When comparing the total energy use, the electrification pathways demonstrate a total energy use of around 6–8% lower than the biofuel pathways in by 2045. This is mainly a result of the lowered energy requirements from electric propulsion compared to combustion engines for construction equipment and heavy-duty trucks combined with the energy penalty for post-combustion carbon capture for cement production.

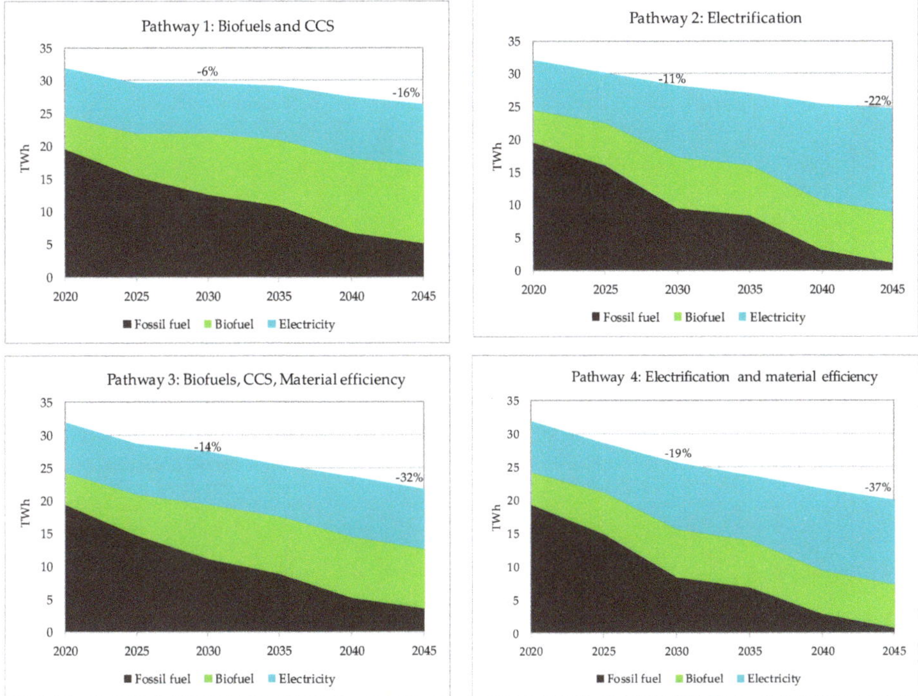

Figure 7. Energy use for each energy carrier over time for the buildings and transport infrastructure pathways.

A focus on material efficiency has the potential to reduce total energy use by 8–10% to 2030 and 18–20% by 2045 for both the biofuel and electrification pathways (noting that the reduction potential would be even higher compared to a reference scenario).

Regarding biofuels, they are at current mainly used in the transport sector, and in asphalt, timber and cement production. Over time, the use is set to expand with the overall share of biofuels increasing from 15% of total energy use at current to around 30% in the electrification pathways and to 40% in the biofuel pathways by 2045. This would mean an increase from 5 TWh to 9 TWh, which can be compared with the current total bioenergy use of 89 TWh in 2017 [156].

Electricity use remain almost constant in the biofuel pathways, while increasing from 7 TWh up to 13–16 TWh in 2045, reaching a share of around 40% in the biofuel pathways and 65% in the electrification pathways.

As can be seen in Figure 8, all pathways reach close to zero emissions in 2045, with total emissions reduction of 90–94%, with the highest emission reduction potential in the electrification pathways. Up until 2030, we see potential emissions reductions of 51–56% for Pathways 1 and 2, indicating that the emissions reduction goal of 50% set by the Construction and Civil Engineering sector in in its own roadmap [49] could be met if the measures suggested in this roadmap would be implemented. Before 2030, most emissions reductions stem from increased use of alternative binders combined with reduced binder intensity in concrete (25%), as well as optimization and energy efficiency measures on the construction sites combined with biofuel substitution in construction equipment and material transports (36–40%). The biofuel substitution partly ensues as a result of the Swedish reduction duty regulation, which specifies increasing emissions reductions in line with a growing share of renewable content in diesel fuel [157]. The emission reduction up until 2030 is also supported by the use of reinforcement steel produced only from recycled steel combined with measures, such as improved electricity emissions factors together with material and fuel substitutions regarding insulation materials.

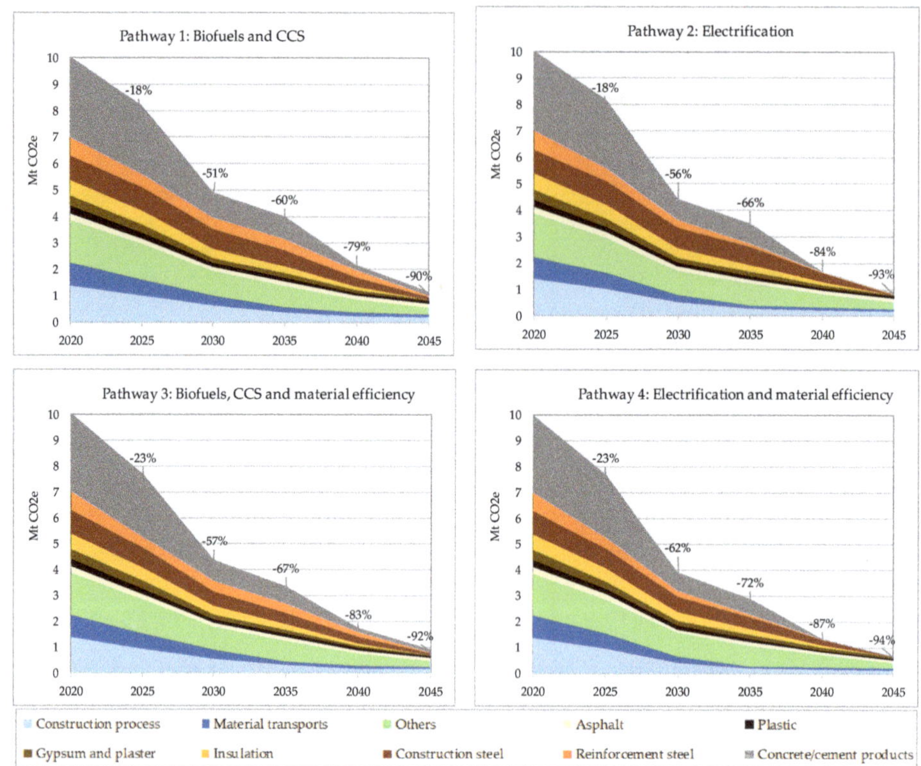

Figure 8. Results on CO_2 emissions for the buildings and transport infrastructure pathways from 2020 to 2045.

A focus on material efficiency provides for additional reductions, particularly in the medium term. An additional 12% brings the total emissions reductions down to around 56–62% by 2030, implying a difference of 0.5–0.6 Mt CO_2 emissions per year.

After 2030, deeper emissions reductions come about as a result of continued biofuel substitution combined with hybridization and electrification for construction equipment and trucks (contributing to a large share of the emissions reductions in 2030–2035). Fuel substitution also plays a role in primary and secondary steelmaking in 2030–2035.

In the biofuel+CCS pathways, this fuel substitution is combined with CCS in primary steelmaking, as well as in cement kilns (contributing to around 40% of the emissions reductions in 2040–2045, respectively).

In the electrification pathways, plasma heating is instead used to create the necessary temperatures in secondary steelmaking, cement kilns, in cracking, and polymerization for plastic production, as well as mineral wool production (contributing to around 45% of the emissions reductions in 2040–2045 combined). Electrification in the primary steelmaking in the form of hydrogen reduction also contributes considerably in the electrification pathway (40% in 2045).

In view of the remaining carbon budget, up to 2045 the material efficiency pathways could reduce the total cumulative amount of CO_2 emitted from construction of buildings and transport infrastructure over the years 2020 to 2045 by 10% compared to its corresponding biofuel/electrification pathways, equivalent to 10–11 MtCO$_2$.

Sensitivity Analysis

The results from the sensitivity analysis into impacts on emissions and energy use from reduced/increased construction demand (denoted as low/high) are shown in Figure 9. The percentage values indicate the reduction from the reference level (cf. the pathway reduction results in Figures 7 and 8).

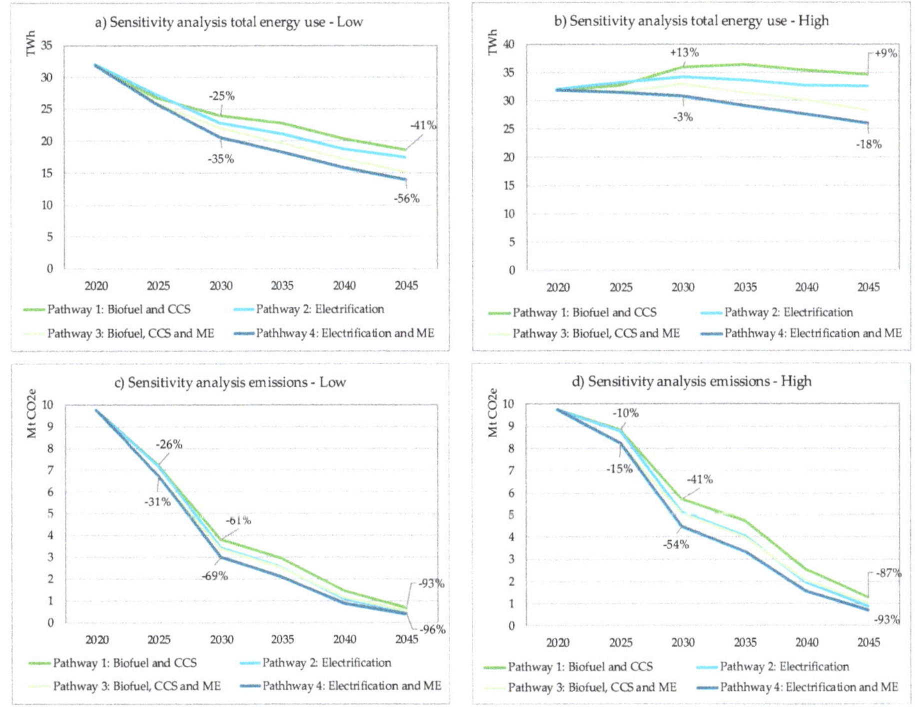

Figure 9. Sensitivity analysis based on reduced or increased construction demand for the different pathways with (**a,b**) depicting the impact on total energy demand, and (**c,d**) depicting the impact on emissions, for low and high construction demand, respectively.

The sensitivity analysis demonstrates that great variances are found in between scenarios of reduced or increased construction demand regarding total energy use. The main implications are seen with regards to an increased construction demand potentially leading to the energy demand remaining relatively stagnant (Pathway 2 and 3) or even growing by 13% to 2030 as in the biofuel+CCS pathway.

In this case, the biofuel demand increases to 11 TWh in 2030 and to 15 TWh in 2045 (compared to 4 TWh in 2015). For the electrification pathway, the electricity demand increases to 13 TWh in 2030 and 21 TWh in 2045 (compared to 8 TWh in 2015). In contrast, reduced construction demand combined with material efficiency measures could reduce to total energy demand by 41–56% implying a total energy demand of between 14–18 TWh in 2045.

Regarding emissions, the sensitivity analysis demonstrates that even with an increased construction demand, it will be possible to reduce CO$_2$ emissions by at least 85% to 2045 compared to 2020 with the combination of measures proposed in this roadmap. However, it also demonstrates that the emissions reduction in the short and medium terms could be slowed down further and possibly impact the potential to reach the sector goal of 50% emissions reduction to 2030 [49]. On the other hand, a slowdown in construction rates could make emissions reductions of over 60% to 2030 possible.

4. Discussion

Cement and steel, together with diesel use in construction processes and material transports, account for the majority of the CO_2 emissions associated with building and infrastructure construction (cf. Figure 6). While the analysis in the study has served to improve the current estimate of the climate impact of building and transport infrastructure construction, it is still associated with a degree of uncertainty. To provide well-grounded decision support for the climate transition ahead, it is important that sufficient resources and competence are allocated so that development of emissions can be properly evaluated and so that the effects of planned measures and policies can be assessed before implementation.

Based on the updated estimate, this roadmap served to illustrate how the basic materials industry and supply chains for buildings and transport infrastructure construction are affected, in terms of energy and material use and associated greenhouse gas emissions, by different technical choices. The study also aimed to illustrate the timing of measures needed to reach intermediary and long-term emission reduction targets. The results show that it is possible to reduce CO_2 emissions associated with construction of buildings and transport infrastructure by at least 50% to 2030, applying already available measures, and reach around 90% emissions reductions by 2045, while the energy use may be reduced by varying degrees (6–19% to 2030 and 16–37% to 2045), indicating that strategic choices with respect to process technologies and energy carriers may have different implications particularly on the energy use over time. It is worth noting that no pathway reaches zero carbon emissions, which is why it is important to further investigate the potential for and accounting of negative emissions (e.g., carbon capture of biogenic emission) and carbon sinks (e.g., use of long-lived wood products in construction) as to enable an approach towards net-zero emissions by 2045. The measures proposed in this roadmap could (and perhaps should) also be backed by strategies to avoid building by exploring alternatives and by repurposing assets, as well as reduce the floor area per capita by smarter floor plans and increased shared spaces [38,43,158].

This study, in alignment with previous analysis, as reported in, e.g., Gerres et al. [34], demonstrates the importance of ensuring sufficient availability of sustainable biomass/bioenergy, electricity and hydrogen. The urgency in upscaling these energy sources becomes particularly evident as experience shows that planning, permitting, and construction of both support infrastructure (renewable energy supply, electricity grid expansion, hydrogen storage, CCS infrastructure) and piloting and upscaling to commercial scale of the actual production involve long lead times. Strategic planning for key support infrastructure therefore needs to be initiated as early as possible, even if not all uncertainties will be fully resolved.

As there are already today known measures and technologies which can reduce emissions to zero, from circularity and material efficiency measures, biofuel or biomaterial substitution, electrification (direct or indirect) with renewable electricity, and/or carbon capture and storage, the challenge to meet climate targets is not only a technological challenge but relative to economics and financial risk, particularly since the current climate policy is too weak [159]. Indeed, large scale demonstration of key processes is required to obtain confidence in technologies, gain experience, and reduce financial risk, but technologies are available at high maturity levels. This would also serve to reduce the uncertainties inherent in the span of emission reduction potential from different abatement measures found in literature (ref. Figure 5).

A key message from this work (as illustrated in Figure 10) is the importance of simultaneously focusing on short- and long-term abatement measure. With this statement, we that that the pursuit of 'low-hanging fruits' (e.g., material substitution and efficiency measures) cannot be an excuse for not acting to lay the foundation for the high-cost long lead-time measures (zero-CO_2 basic materials) required to reach deep decarbonization. Vice versa, we cannot let the promise, e.g., of low-CO_2 steel or cement, be an excuse not to act to unlock the potential for measures that already exists today. Successful decarbonization of the supply chains for buildings and infrastructure, including the production of basic materials, will involve the pursuit—in parallel—of emission abatement measures with very different characteristics. Consequently, to facilitate the transition, the support tools box will need to encompass a variety of policies and strategies.

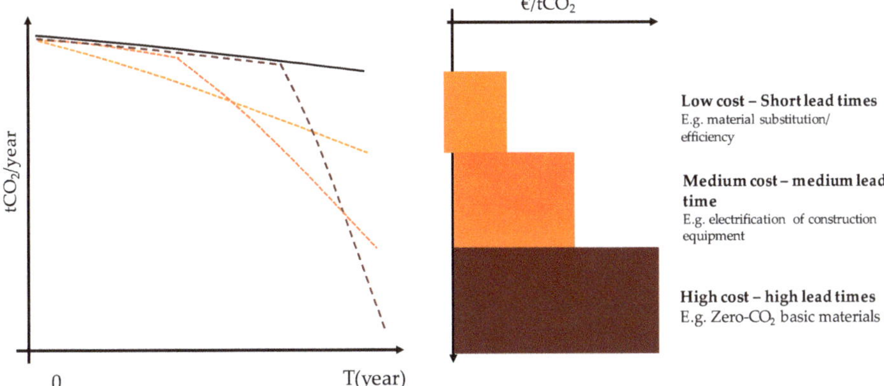

Figure 10. Successful decarbonization of the supply chains for buildings and infrastructure in less than three decades will require the parallel pursuit of emission abatement measures with very different characteristics. Figure adapted from Vogt-Schilb and Hallegatte (2014) [160].

The results thus illustrate the importance of intensifying efforts to identify and manage both soft (organization, knowledge sharing, competence) and hard (technology and costs) barriers and the importance of both acting now by implementing available measures (e.g., material efficiency and material/fuel substitution measures) and actively planning for long-term measures (low-CO$_2$ steel or cement). Unlocking the full abatement potential of the range of emission abatement measures that are described in this study will require not only technological innovation but also innovations in the policy arena and efforts to develop new ways of co-operating, coordinating, and sharing information between actors in the supply chain. Key priorities include, e.g.,

- Continuous efforts around process optimization, material efficiency, and material substitution [43] to reduce the climate impact from basic materials and construction, particularly in the short to medium terms. This includes efforts in all planning process, and among all actors, to:

  - avoid building (where possible),
  - re-using old assets,
  - recycle building materials and components,
  - optimize material use, and
  - shift to low-CO$_2$ materials and services.

- Development of integrated industrial climate strategy including adaptation of legislation, and innovative schemes to share the risk and costs associated with developing and implementing new process technology and infrastructures (see, e.g., Reference [161,162]).
- Strategic planning for support infrastructure. Lead times related to planning, permitting, and construction of both support infrastructure (renewable electricity supply, electricity grid expansion, hydrogen storage, CCS infrastructure) and piloting and upscaling to commercial scale of the actual production units will influence the speed of change [163,164]. Historical transition processes provide valuable lessons around the importance of going beyond the physical planning, ensuring transparency, broad participation, and fairness (e.g., acceptable distributional effects)

early on, as well as planning for agility and endurance in the face of the unforeseen (e.g., delays, changing market conditions). Similar planning processes, including identification of designated strategic areas/zones, have previously been carried out for wind and hydro power [165,166].

- Ensuring focus on logistical optimization (via, e.g., digitalization), sufficient availability of sustainably produced second-generation biofuels and support for hybridization and electrification of heavy transport and construction equipment (as called for in, e.g., Reference [30,31,54,167–169]).
- Using public procurement as a tool to spur innovation, creating markets for low-CO_2 products and opening up for economies of scale [35]. Public procurers in governmental agencies, municipalities, and county councils, with their significant purchasing power, can play an important role as drivers and by setting examples. In addition, private actors can help to legitimize public strategies and increase the volume of demand for low-CO_2 products [163]. It is imperative that embodied carbon emissions start weighing as heavily as project costs, timescales, functionality, and aesthetics do regarding client priorities [170]. At the same time, the applicability of procurement requirements for carbon reduction depends on how well these requirements are aligned with industry culture, policies, and capabilities in the local context (see, e.g., Reference [171]).
- Capacity building and information spreading to change the culture and established practices of the conservative, cost-driven, and risk averse construction industry [46] via, for example:

  - Establishment of an (public or private) umbrella organization with the responsibility to oversee and support the low-CO_2 transition.
  - Securing new competence by including low-CO_2 building and construction as a central part of the in upper secondary school and higher education.
  - Training of active practitioners (engineers, architects …).

It is of course also important to continue to find ways to sharpen existing climate policies, such as the EU-ETS and renewable policies, most important being to make them as long term as possible [35]. There is no guarantee that investments in the development and implementation of hydrogen direction reduction in the steel industry, CCS in the cement industry, nor other low-carbon technologies for industrial applications will pay off [172,173]. However, choosing not to, or failing to act within the next few years, to create the economic, organizational, and infrastructural conditions that could facilitate a shift towards low-CO_2 production and practices will severely compromise the chances of a successful decarbonization of the steel and cement-industries, as well as the supply chains for buildings and transport infrastructure, up to the year 2045.

Although the findings reported in this paper draw primarily on Swedish experiences, with some of the conclusions valid only under certain conditions and circumstances, it is clear that many of the challenges that have been raised here, which must be overcome to achieve a transition to zero-CO_2 production and practices in the supply chains for buildings and infrastructure, are universal [43,174,175]. Whereas rapid improvements of the climate performance of the use phase (i.e., related to heating and cooling) of the existing and new building stocks is a key priority in many parts of the world, it is equally important to take measures to reduce the climate impact of the construction process and the production and supply of building materials.

From a global perspective, this is important, not the least, since there are still many regions of the world where much the of the buildings and the infrastructure to provide shelter from the elements, mobility for people and goods, and infrastructures for the supply of water, electricity, and heat remains to be built. Estimates suggest that more than half of the urban infrastructure that will exist in 2050 has yet to be built [175,176] and that total global floor area of buildings will double within the next three or four decades [43,174].

5. Conclusions

In this paper, material and energy flow analysis is combined with a literature review to improve and validate the estimate of the current climate impact from building and construction processes in Sweden. The result is an estimate of around 9.8 Mt CO_2 per year, close to 20% total Swedish GHG emissions, deriving predominantly from concrete and steel, together with diesel use in construction processes and material transports.

From the current estimate, the work provides a roadmap with an analysis of different pathways of technological developments in the supply chains of the buildings and construction industry, including primary production of steel and cement. The analysis combines quantitative analysis methods, including scenarios and stylized models, with participatory processes involving relevant stakeholders in the assessment process. By applying a combination of circularity and material efficiency measures, biofuel or biomaterial substitution, electrification (direct or indirect) with renewable electricity, and carbon capture and storage, this roadmap demonstrates that the CO_2 emissions associated with construction of buildings and transport infrastructure could be reduced by over 50% to 2030 and by over 90% to 2045. At the same time, strategic choices with respect to process technologies, energy carriers, and the availability of biofuels, CCS, and zero CO_2 electricity may have different implications on energy use and CO_2 emissions over time.

Supplementary Materials: The following are available online at http://www.mdpi.com/1996-1073/13/16/4136/s1, Inventory of abatement options and potential.

Author Contributions: Conceptualization: I.K., J.R.; methodology: I.K.; software: I.K.; validation: I.K.; formal analysis: I.K., J.R., A.T.; investigation: I.K., A.T.; resources: I.K., data curation: I.K.; writing—original draft preparation: I.K.; writing–review and editing: I.K., J.R., A.T., M.O., F.J., L.G.; visualization: I.K.; project administration: F.J., M.O., L.G.; funding acquisition: F.J. All authors have read and agreed to the published version of the manuscript.

Funding: This research was funded by Mistra.

Acknowledgments: Close collaboration with both public and private stakeholders involved in the Mistra Carbon Exit research program have been important in the development of this study.

Conflicts of Interest: The authors declare no conflict of interest.

Abbreviations

The following abbreviations are used in this manuscript:

Parameter	Definition	Unit
E	Emissions	tCO_2
e	Emissions share for specific emissions sources (e.g., lifecycle stages, materials, etc.)	%
M	Material demand/production	metric tonne (t)
Q	Energy demand	kWh
q	Energy share for specific energy sources in material production or fuels	%
Ef	Emissions intensity factor per unit (t for material production and kWh for transport/ construction processes)	tCO_2/unit
Qf	Energy intensity factor per unit (t for material production and l for transport/ construction processes)	kWh/unit
A	Abatement measures reducing material/energy demand, i.e., re - recycling, ms-material substitution, me - material efficiency measures, op–optimization of logistics and construction process, ee–energy efficiency measures (including from hybridization and electrification)	%
CC	Emissions from material production abated via carbon capture.	%

Energies **2020**, 13, 4136

The following indexes are used in this manuscript:

Index	Definition	Index Components
lc	Lifecycle stages	Material production, material transports, construction process
tc	Lifecycle stages after material production	Material transports, construction process
m	Materials	Concrete, reinforcement steel, construction steel, asphalt, insulation, gypsum and plaster, plastics and paint, others
b	Building type	Multi-family dwellings, single-family dwellings, non-residential buildings
i	Infrastructure type	Road, railway, ports and fairways
p	Construction phase	New construction, reinvestment
s	Energy sources	Fossil fuels (coal, gas, oil, fossil waste), biofuels, electricity
t	Timesteps	2025, 2030, 2035, 2040, 2045

Appendix A

Table A1. Emissions and energy intensity factors along with energy mix for specific materials or material production processes.

Specific Materials/ Production Processes	Emissions Intensity Ef_m (tCO$_2$/t)	Energy Intensity Qf_m (kWh/t)	Energy Mix $q_{sh,m,s}$ (%)							Comment	References
			Fossil Fuels	Coal/Coke	Oil	Gas	Fossil Waste	Biomass	Electricity		
Concrete production	2*10^{-6}	36	50%						50%	Only concrete production (i.e., excluding emissions from cement or SCM in concrete)	[62,68,69]
Clinker production process emissions	0.51	-									[27,29,177]
Clinker production carbon capture	-	1220								Energy penalty based on post combustion via amine scrubbing (90% capture rate)	[40,106,149,178–181]
SCM (alternative binders)	0.10	2							100%	Supplementary Cementitious Materials. Average of fly ash, blast furnace slag (GGBS), limestone and calcined clay	[40,149,182–184]
Primary steel production	1.90	5400		77%	2%	9%			12%		[23,37,71,72,111]
Electrolysis energy intensity	-	2600							100%		[112,185]
Secondary steel production	0.40	1200		3%		28%			69%	Assuming European average electricity emissions factor	[23,71,72]
Metallurgy	0.14	400				100%				Assuming reheating furnaces run on gas	[37,72,186]
Galvanizing	0.08	600				100%				Assuming furnaces run on gas	[76–78]
Polystyrene-based insulation	4.70	22200	98%						2%	Average of expanded polystyrene (EPS) and extruded polystyrene (XPS)	[66,74,83,84,128,187]
Mineral wool insulation	1.30	7700	75%						25%	Average of rock and glass wool	[66,74,83,128,187]

Table A2. Emissions intensity factors for energy sources.

Energy Sources	Emissions Intensity Ef_s (kgCO$_2$/kWh)	Comment	References
Coal/coke	0.370	Average of coking coal and bituminous coal (including upstream emissions)	[188,189]
Oil	0.228	Fuel oil (including upstream emissions)	[188,189]
Fossil diesel	0.338	Component in Swedish standard Diesel MK1 (85% fossil diesel and 15% biofuel in 2015)	[95]
Gas	0.248	Natural gas (including upstream emissions)	[188,189]
Fossil waste	0.288	Average of tyres and plastic waste	[190,191]
Biofuel	*0.299*	Average of forest and agricultural residues; biogenic emissions (not included in emissions calculation due to assumption of carbon neutrality from sustainable forest management); may however contribute to negative emissions when carbon capture is applied	[190]
Electricity (Sweden)	0.047	Swedish average electricity emissions factor in 2015 (including upstream emissions). Used for construction processes, cement and concrete and non-mineral materials	[95,192]
Electricity (Europe)	0.314	European average electricity emissions factor in 2015. Used for steel and other metals production	[193]
District heating	0.069	Swedish national average from 2017; 23% fossil fuels, 68% biofuels, and 9% electricity (from heat pumps)	[189]

Table A3. Predicted emissions intensity factors for electricity and district heating.

Energy Sources	Year	Emissions Intensity Ef_s (kgCO$_2$/kWh)	Comment	References
Electricity (Sweden)	2025	0.034	According to a linear reduction to the figure in 2045 from the emission factor in 2015.	
	2030	0.025		
	2035	0.017		
	2040	0.008		
	2045	0.003	According to the average figure in 2045 from the scenario analysis Four energy futures from the Swedish Energy Agency	[194]
Electricity (Europe)	2025	0.261	Calculated according to estimated EEA projections	[195]
	2030	0.230		
	2035	0.172	According to a linear reduction from the estimated figure in 2030 down to zero emissions in 2050	
	2040	0.115		
	2045	0.057		
District heating	2025	0.064	According to a linear reduction to the figure in 2045 from the emission factor in 2015	
	2030	0.059		
	2035	0.055		
	2040	0.050		
	2045	0.045	According to the average figure in 2045 from the scenario analysis Four energy futures from the Swedish Energy Agency	[194]

Table A4. Overview of the share of emission components from life cycle analysis (LCA) literature for new builds of various building and frame types.

Building Type	Building Sub-Type	Building/Frame Type	Façade Type	Building Materials (A1-A3)	Transport (A4)	Construction Process (A5)	Comments	References
Non-residential	Offices	Reinforced in-situ cast concrete	Concrete	89%	2%	9%		[196]
		Reinforced in-situ cast concrete	Plaster/ wood panel	86%	2%	12%		[197]
		Hybrid precast/in-situ cast concrete and timber	N/A	95%	3%	2%		[198]
		Reinforced in-situ cast concrete	Plaster	93%	-	7%	Transports included in material emissions	[199]
	Industrial	Concrete/steel	Brick	97%	-	3%		[200]
		Prefab concrete/ steel	Steel	97%	1%	2%		[201,202]
Residential	Multi-family dwellings	In-situ cast concrete	Plaster	81–84%	3–4%	13–16%		[92,93]
		Prefab concrete	Plaster	79%	9%	13%		[92]
		Prefab concrete	Plaster	86%	8%	16%		[75]
		Hybrid prefab concrete/wood	Plaster	74%	5%	13%		[203]
		Wooden volume element	Plaster	79%	8%	17%		[92]
		Cross-laminated timber	Plaster	75%	9%	17%		[92]
		Cross-laminated timber	Wood panel	78%	6%	16%	Including ground preparation of 8%	[204]
	Single-family dwellings	Wooden	Wood panel	82%	2%	16%	A5 including 14% from waste	[205]
		Wooden	Brick	85%	2%	14%	A5 including 12% from waste	[205]
		Masonry	Brick	86%	3%	12%	A5 including 10% from waste	[205]
		Wooden	Wood panel	96%	4%	0%		[206]

Table A5. Overview of the share of emission components from different materials in LCA literature for new builds of various building and frame types.

Building Type	Building Sub-Type	Building/Frame Type	Concrete	Reinforcement Steel	Construction Steel	Insulation	Gypsum and Plaster	Plastic and Chemicals	Others (Aluminium, Glass, Timber)	Comments	References
Non-residential	Offices	Steel-reinforced in-situ cast concrete	35%	4–12%	33–39%	6–11%	3–4%	4%	11%		[196]
		Steel-reinforced in-situ cast concrete	70%*	*	11%	4%	3%	-	12%	*Reinforcement steel included in concrete emissions	[197]
		Hybrid precast/in-situ cast concrete and timber	20%	-	32%	9%	<1%	-	-		[198]
		Reinforced in-situ cast concrete	48%	*	22%	14%	14%	-	2%	*Reinforcement steel included in construction steel	[199]
	Municipal	Wood frame and panel	34%	2%	15%	14%	9%	3%	21%	Pre-school	[207]
	Industrial	Prefab concrete/steel	35%	20%	28%	1%	1%	4%	17%		[208]
Residential	Multi-family dwellings	In-situ cast concrete	58–65%	6%	0–2%	6–9%	2–5%	3–6%	14%		[92,93]
		Prefab concrete	43%	23%	6%	10%	3%	7%	8%		[92]
		Prefab concrete	62%	11%	2%	10%	4%	1%	10%		[75]
		Wooden volume element	13%	-	-	14%	36%	11%	26%		[92]
		Cross-laminated timber	12%	2%	9%	16%	15%	9%	37%		[92]
		Cross-laminated timber	34%	10%	8%	11%	5%	-	54%		[204]
	Single-family dwellings	Wooden	15%	0%	3%	4%	3%	2%	73%		[205]
		Wooden	40%	0%	9%	5%	7%	1%	38%		[206]

References

1. Regeringskansliet Riksdagen Antar Historiskt Klimatpolitiskt Ramverk. Available online: http://www.regeringen.se/pressmeddelanden/2017/06/riksdagen-antar-historiskt-klimatpolitiskt-ramverk/ (accessed on 21 July 2018).
2. UNFCCC Paris Agreement. Conference of the Parties on Its Twenty-First Session. 2015, Volume 32. Available online: https://unfccc.int/resource/docs/2015/cop21/eng/l09r01.pdf (accessed on 22 July 2018).
3. Bataille, C.; Waisman, H.; Colombier, M.; Segafredo, L.; Williams, J.; Jotzo, F. The need for national deep decarbonization pathways for effective climate policy. *Clim. Policy* **2016**, *16*, S7–S26. [CrossRef]
4. Erlandsson, M.; Byfors, K.; Lundin, J.S. *Byggsektorns Historiska Klimatpåverkan Och en Projektion för Nära Noll*; IVL Rapport C 277; IVL Svenska Miljöinstitutet: Stockholm, Sweden, 2018.
5. SCB Utsläpp av Växthusgaser Från Svensk Ekonomi Oförändrade. 2018. Available online: https://www.scb.se/hitta-statistik/statistik-efter-amne/miljo/miljoekonomi-och-hallbar-utveckling/miljorakenskaper/pong/statistiknyhet/miljorakenskaper--utslapp-till-luft-fjarde-kvartalet-2018/ (accessed on 25 February 2020).
6. Naturvårdsverket. *Fördjupad Analys av den Svenska Klimatomställningen 2019. Industrin i Fokus*; Naturvårdsverket: Stockholm, Sweden, 2019; Volume 6911, ISBN 9789162069117.
7 Naturvårdsverket. Boverket. *Klimatscenarier för Bygg och Fastighetssektorn. Förslag på Metod för Bättre Beslutsunderlag*; Naturvårdsverket: Stockholm, Sweden, 2019.

8.   Erlandsson, M. *Hur når Bygg- och Fastighetssektorn Klimatmålen 2045? Expertmöte för Utvärdering av Föreslagen Modell för Validering och Inspel inför Kommande Scenarioanalys*; IVL Svenska Miljöinstitutet: Stockholm, Sweden, 2020.

9.   Liljenström, C.; Toller, S.; Åkerman, J.; Björklund, A. Annual climate impact and primary energy use of swedish transport infrastructure. *Eur. J. Transp. Infrastruct. Res.* **2019**. [CrossRef]

10.   Steinbach, N.; Palm, V.; Cederberg, C.; Finnveden, G.; Persson, L.; Persson, M.; Berglund, M.; Björk, I.; Fauré, E.; Trimmer, C. *Miljöpåverkan Från Svensk Konsumtion-nya Indikatorer för Uppföljning Slutrapport för Forskningsprojektet PRINCE*; Naturvårdsverket: Stockholm, Sweden, 2018; ISBN 9789162068424.

11.   Erlandsson, M. *Modell för Bedömning av Svenska Byggnaders Klimatpåverkan*; Naturvårdsverket: Stockholm, Sweden, 2019.

12.   Säynäjoki, A.; Heinonen, J.; Junnonen, J.M.; Junnila, S. Input–output and process LCAs in the building sector: Are the results compatible with each other? *Carbon Manag.* **2017**, *8*, 155–166. [CrossRef]

13.   Crawford, R.H.; Bontinck, P.A.; Stephan, A.; Wiedmann, T.; Yu, M. Hybrid life cycle inventory methods—A review. *J. Clean. Prod.* **2018**, *172*, 1273–1288. [CrossRef]

14.   Crawford, R.H. *Life Cycle Assessment in the Built Environment*; Taylor & Francis: Oxfordshire, UK, 2011; ISBN 9780203868171.

15.   Soares, N.; Bastos, J.; Pereira, L.D.; Soares, A.; Amaral, A.R.; Asadi, E.; Rodrigues, E.; Lamas, F.B.; Monteiro, H.; Lopes, M.A.R.; et al. A review on current advances in the energy and environmental performance of buildings towards a more sustainable built environment. *Renew. Sustain. Energy Rev.* **2017**, *77*, 845–860. [CrossRef]

16.   Matthews, H.S.; Hendrickson, C.T.; Weber, C.L. The importance of carbon footprint estimation boundaries. *Environ. Sci. Technol.* **2008**, *42*, 5839–5842. [CrossRef]

17.   Nässén, J.; Holmberg, J.; Wadeskog, A.; Nyman, M. Direct and indirect energy use and carbon emissions in the production phase of buildings: An input-output analysis. *Energy* **2007**, *32*, 1593–1602. [CrossRef]

18.   Lenzen, M.; Treloar, G. Embodied energy in buildings: Wood versus concrete—Reply to Börjesson and Gustavsson. *Energy Policy* **2002**, *30*, 249–255. [CrossRef]

19.   Lenzen, M. Errors in conventional and input-output-based life-cycle inventories. *J. Ind. Ecol.* **2000**. [CrossRef]

20.   Peters, G.P.; Hertwich, E. Structural analysis of international trade: Environmental impacts of Norway. *Econ. Syst. Res.* **2006**. [CrossRef]

21.   Schuerch, R.; Kaenzig, J.; Jungbluth, N.; Nathani, C. 45th Discussion forum on LCA-environmentally extended input-output analysis and LCA, September 15, 2011, Berne, Switzerland. *Int. J. Life Cycle Assess.* **2012**, *17*, 840–844. [CrossRef]

22.   Fischedick, M.; Marzinkowski, J.; Winzer, P.; Weigel, M. Techno-economic evaluation of innovative steel production technologies. *J. Clean. Prod.* **2014**, *84*, 563–580. [CrossRef]

23.   Eurofer. *A Steel Roadmap for a Low Carbon Europe 2050*; European Confederation of Iron & Steel Industries: Brussels, Belgium, 2013.

24.   Wörtler, M.; Dahlmann, P.; Schuler, F.; Bodo Lüngen, H.; Voigt, N.; Ghenda, J.-T.; Schmidt, T. *Steel's Contribution to a Low-Carbon Europe 2050: Technical and Economic Analysis of the Sector's CO_2 Abatement Potential*; European Confederation of Iron & Steel Industries: Brussels, Belgium, 2013.

25.   Arens, M.; Worrell, E.; Eichhammer, W.; Hasanbeigi, A.; Zhang, Q. Pathways to a low-carbon iron and steel industry in the medium-term—The case of Germany. *J. Clean. Prod.* **2017**, *163*. [CrossRef]

26.   Jernkontoret. *Klimatfärdplan för en Fossilfri och Konkurrenskraftig Stålindustri i Sverige*; Jernkontoret: Stockholm, Sweden, 2018.

27.   IEA; CSI. *Technology Roadmap: Low-Carbon Transition in the Cement Industry.* 2018, Volume 66. Available online: https://www.iea.org/reports/technology-roadmap-low-carbon-transition-in-the-cement-industry (accessed on 10 June 2020).

28.   CEMBUREAU. *The Role of Cement in the 2050 Low Carbon Economy*; The European Cement Association: Brussels, Belgium, 2013.

29.   Favier, A.; De Wolf, C.; Scrivener, K.; Habert, G. *A Sustainable Future for the European Cement and Concrete Industry: Technology Assessment for Full Decarbonisation of the Industry by 2050*; ETH Zürich: Zürich, Switzerland, 2018; Volume 96. [CrossRef]

30.   IEA. *The Future of Trucks*; Oecd/International Energy Agency: Paris, France, 2017.

31.   Skinner, I.; van Essen, H.; Smokers, H.; Hill, N. *Towards the Decarbonisation of EU's Transport Sector by 2050*; European Commission: Brussels, Belgium, 2010.

32. Bondemark, A.; Jonsson, L. *Fossilfrihet för Arbetsmaskiner—En Rapport av WSP för Statens Energimyndighet*; World Scientific: Singapore, 2017.

33. Swedish Transport Administration. *Arbetsmaskiners Klimatpåverkan och hur den kan Minska—Ett Underlag till 2050-Arbetet*; Trafikverket: Borlänge, Sweden, 2012.

34. Gerres, T.; Chaves Ávila, J.P.; Llamas, P.L.; San Román, T.G. A review of cross-sector decarbonisation potentials in the European energy intensive industry. *J. Clean. Prod.* **2019**, *210*, 585–601. [CrossRef]

35. Bataille, C.; Åhman, M.; Neuhoff, K.; Nilsson, L.J.; Fischedick, M.; Lechtenböhmer, S.; Solano-Rodriquez, B.; Denis-Ryan, A.; Steiber, S.; Waisman, H.; et al. A review of technology and policy deep decarbonization pathway options for making energy-intensive industry production consistent with the Paris agreement. *J. Clean. Prod.* **2018**. [CrossRef]

36. Wyns, T.; Axelson, M. *The Final Frontier—Decarbonising Europe's Energy Intensive Industries*; Institute for European Studies: Bruxelles, Belgium, 2016; p. 64. [CrossRef]

37. Material Economics. *Industrial Transformation 2050—Pathways to Net-Zero Emissions from EU Heavy Industry*; Material Economics: Stockholm, Sweden, 2019.

38. Energy Transition Commission. *Mission Possible—Reaching Net Zero Carbon Emissions from Harder-to-Abate Sectors by Mid-Century*; Energy Transitions Commission: London, UK, 2018.

39. European Commission. In-Depth Analysis in Support of the Commission Communication com (2018) 773 A Clean Planet for all A European Long-Term Strategic Vision for a Prosperous, Modern, Competitive and Table of Contents. 2018. Available online: https://ec.europa.eu/knowledge4policy/publication/depth-analysis-support-com2018-773-clean-planet-all-european-strategic-long-term-vision_en (accessed on 10 June 2020).

40. Rootzén, J.; Johnsson, F. CO_2 emissions abatement in the Nordic carbon-intensive industry—An end-game in sight? *Energy* **2015**. [CrossRef]

41. Schneider, C.; Saurat, M.; Tönjes, A.; Zander, D.; Hanke, T.; Soukup, O.; Barthel, C.; Viebahn, P. *Decarbonisation Pathways for Key Economic Sectors*; Wuppertal Institute for Climate, Environment and Energy: Wuppertal, Germany, 2020.

42. Grønn Byggallianse; Norsk Eiendom. *The Property Sector's Roadmap Towards 2050*; Norwegian Green Building Council: Oslo, Norway, 2016.

43. World Green Building Council. *Bringing Embodied Carbon Upfront*; World Green Building Council: London, UK, 2019.

44. Allwood, J.M.; Cullen, J.M. *Sustainable Materials with Both Open Eyes*; UIT Cambridge: Cambridge, UK, 2012; ISBN 9781906860059.

45. Green Construction Board Low Carbon Routemap for the UK Built Environment. 2013. Available online: https://www.researchgate.net/publication/289245351_Green_Construction_Board_Low_Carbon_Routemap_for_the_Built_Environment_-_2015_Routemap_Progress_Technical_Report (accessed on 13 January 2018).

46. Giesekam, J.; Barrett, J.; Taylor, P.; Owen, A. The greenhouse gas emissions and mitigation options for materials used in UK construction. *Energy Build.* **2014**, *78*, 202–214. [CrossRef]

47. Allwood, J.M.; Dunant, C.F.; Lupton, R.C.; Cleaver, C.J.; Serrenho, A.C.H.; Azevedo, J.M.C.; Horton, P.M.; Clare, C.; Low, H.; Horrocks, I.; et al. Absolute Zero. Delivering the UK's Climate Change Commitment with Incremental Changes to Today's Technologies. 2019. Available online: https://researchportal.bath.ac.uk/en/publications/absolute-zero-delivering-the-uks-climate-change-commitment-with-i (accessed on 1 February 2019).

48. Fossilfritt Sverige. *Roadmaps for Fossil Free Competitiveness—A Summary of Roadmaps from Swedish Business Sectors*; Fossilfritt Sverige: Stockholm, Sweden, 2018.

49. Fossilfritt Sverige. *Färdplan för Fossilfri Konkurrenskraft Bygg- och Anläggningssektorn*; Fossilfritt Sverige: Stockholm, Sweden, 2018.

50. SWECO. *Klimatneutral Konkurrenskraft—Kvantifiering av Åtgärder i Klimatfärdplaner*; Sweco: Stockholm, Sweden, 2019.

51. Kungliga IngenjörsVetenskaps Akademien. *Så Klarar Svensk Industri Klimatmålen En Delrapport från IVA-Projektet Vägval för Klimatet*, Kungliga IngenjörsVetenskaps Akademien: Stockholm, Sweden, 2019.

52. Sveriges Byggindustrier; IVA. *Klimatpåverkan Från Byggprocessen*; Sveriges Byggindustrier, IVA: Stockholm, Sweden, 2014.

53. Trafikverket. *Miljökonsekvensbeskrivning av Förslag till Nationell Plan för Transportsystemet 2018–2029*; Trafikverket: Borlänge, Sweden, 2017.

54. Karlsson, I.; Rootzén, J.; Johnsson, F. Reaching net-zero carbon emissions in construction supply chains–Analysis of a Swedish road construction project. *Renew. Sustain. Energy Rev.* **2020**, *120*. [CrossRef]

55. Moncaster, A.M.; Rasmussen, F.N.; Malmqvist, T.; Houlihan Wiberg, A.; Birgisdottir, H. Widening understanding of low embodied impact buildings: Results and recommendations from 80 multi-national quantitative and qualitative case studies. *J. Clean. Prod.* **2019**, *235*, 378–393. [CrossRef]

56. Schwartz, Y.; Raslan, R.; Mumovic, D. The life cycle carbon footprint of refurbished and new buildings—A systematic review of case studies. *Renew. Sustain. Energy Rev.* **2018**, *81*, 231–241. [CrossRef]

57. Chastas, P.; Theodosiou, T.; Kontoleon, K.J.; Bikas, D. Normalising and assessing carbon emissions in the building sector: A review on the embodied CO_2 emissions of residential buildings. *Build. Environ.* **2018**, *130*, 212–226. [CrossRef]

58. Amer, M.; Daim, T.U.; Jetter, A. A review of scenario planning. *Futures* **2013**, *46*, 23–40. [CrossRef]

59. Energimyndigheten. *Scenarier över Sveriges Energisystem 2016*; Energimyndigheten: Eskilstuna, Sweden, 2017.

60. European Standards EN 15978:2011. *Sustainability of Construction Works—Assessment of Environmental Performance of Buildings—Calculation Method*; National Standards Authority of Ireland: Dublin, Ireland, 2011.

61. Swedish Transport Administration. *Förslag till Nationell Plan för Transportsystemet 2014–2025 Underlagsrapport—Miljökonsekvensbeskrivning*; Swedish Transport Administration: Borlänge, Sweden, 2013.

62. Kurkinen, E.; Norén, J.; Peñaloza, D.; Al-Ayish, N.; During, O. *Energi och Klimateffektiva Byggsystem: Miljövärdering av Olika Stomalternativ*; SP Technical Research Institute of Sweden: Borås, Sweden, 2017.

63. Svensk Betong. *Klimatförbättrad Betong*; Svensk Betong Service AB: Stockholm, Sweden, 2019.

64. Cementa Miljödata Slite. Available online: https://www.cementa.se/sv/miljodata-slite (accessed on 26 March 2019).

65. Ishak, S.A.; Hashim, H. Low carbon measures for cement plant—A review. *J. Clean. Prod.* **2015**, *103*, 260–274. [CrossRef]

66. Basbagill, J.; Flager, F.; Lepech, M.; Fischer, M. Application of life-cycle assessment to early stage building design for reduced embodied environmental impacts Application of life-cycle assessment to early stage building design for reduced embodied environmental impacts. *Build. Environ.* **2013**, *60*, 81–92. [CrossRef]

67. Chau, C.K.; Leung, T.M.; Ng, W.Y. A review on life cycle assessment, life cycle energy assessment and life cycle carbon emissions assessment on buildings. *Appl. Energy* **2015**, *143*, 395–413. [CrossRef]

68. Toller, S. *Rapport: Klimatkalkyl—Beräkning av Infrastrukturens Klimatpåverkan och Energianvändning i ett Livscykelperspektiv*; Modell 5.0 och 6.0; Trafikverket: Borlänge, Sweden, 2018.

69. Erlandsson, M. *Byggsektorns Miljöberäkningsverktyg BM1.0*; IVL Svenska Miljöinstitutet: Stockholm, Sweden, 2018.

70. Bianco, L.; Baracchini, G.; Cirilli, F. Sustainable Electric Arc Furnace Steel Production: GREENEAF. *BHM Berg-und Hüttenmännische Monatshefte* **2013**, *158*, 17–23. [CrossRef]

71. Otto, A.; Robinius, M.; Grube, T.; Schiebahn, S.; Praktiknjo, A.; Stolten, D. Power-to-Steel: Reducing CO_2 through the Integration of Renewable Energy and Hydrogen into the German Steel Industry. *Energies* **2017**, *10*, 451. [CrossRef]

72. Xylia, M.; Silveira, S.; Duerinck, J.; Meinke-Hubeny, F. Weighing regional scrap availability in global pathways for steel production processes. *Energy Effic.* **2018**. [CrossRef]

73. Gunarathne, D.S.; Mellin, P.; Yang, W.; Pettersson, M.; Ljunggren, R. Performance of an effectively integrated biomass multi-stage gasification system and a steel industry heat treatment furnace. *Appl. Energy* **2016**, *170*, 353–361. [CrossRef]

74. Zabalza Bribián, I.; Valero Capilla, A.; Aranda Usón, A. Life cycle assessment of building materials: Comparative analysis of energy and environmental impacts and evaluation of the eco-efficiency improvement potential. *Build. Environ.* **2011**, *46*, 1133–1140. [CrossRef]

75. Andersson, M.; Barkander, J.; Kono, J.; Ostermeyer, Y. Abatement cost of embodied emissions of a residential building in Sweden. *Energy Build.* **2018**. [CrossRef]

76. Mousa, E.; Wang, C.; Riesbeck, J.; Larsson, M. Biomass applications in iron and steel industry: An overview of challenges and opportunities. *Renew. Sustain. Energy Rev.* **2016**, *65*, 1247–1266. [CrossRef]

77. European General Galvanizers Association. *Batch Hot Dip Galvanizing of Steel Products to EN ISO 1461*; European Average; European General Galvanizers Association: London, UK, 2016.

78. Industrial Galvanizers. *Specifiers Manual*; Industrial Galvanizers: Brisbane, Australia, 2013.
79. Lasvaux, S.; Habert, G.; Peuportier, B.; Chevalier, J. Comparison of generic and product-specific Life Cycle Assessment databases: Application to construction materials used in building LCA studies. *Int. J. Life Cycle Assess.* **2015**, 1473–1490. [CrossRef]
80. Swedish Transport Administration. *Gröna Koncept Inom Asfaltsbeläggningar—Kunskapsöversikt*; Swedish Transport Administration (Trafikverket): Borlänge, Sweden, 2015.
81. Lehne, J.; Preston, F. *Chatham House Report: Making Concrete Change Innovation in Low-carbon Cement and Concrete*; Chatham House Report, Energy Enivronment and Resources Department: London, UK, 2018.
82. Martinsson, K.; Jacobson, T.; Lundberg, R. *Calculation of Energy and Carbon Dioxide on Asphalt Pavements*; Swedish Transport Administration: Orebro, Sweden, 2017.
83. Hill, C.; Norton, A.; Dibdiakova, J. A comparison of the environmental impacts of different categories of insulation materials. *Energy Build.* **2018**, *162*, 12–20. [CrossRef]
84. Pargana, N.; Pinheiro, M.D.; Silvestre, J.D.; De Brito, J. Comparative environmental life cycle assessment of thermal insulation materials of buildings. *Energy Build.* **2014**, *82*, 466–481. [CrossRef]
85. Gustavsson, L.; Joelsson, A.; Sathre, R. Life cycle primary energy use and carbon emission of an eight-storey wood-framed apartment building. *Energy Build.* **2010**, *42*, 230–242. [CrossRef]
86. Lushnikova, N.; Dvorkin, L. *Sustainability of Gypsum Products as a Construction Material*, 2nd ed.; Elsevier Ltd.: Amsterdam, The Netherlands, 2016; ISBN 9780081003701.
87. Quintana, A.; Alba, J.; del Rey, R.; Guillén-Guillamón, I. Comparative Life Cycle Assessment of gypsum plasterboard and a new kind of bio-based epoxy composite containing different natural fibers. *J. Clean. Prod.* **2018**, *185*, 408–420. [CrossRef]
88. Fischedick, M.; Roy, J.; Abdel-Aziz, A.; Acquaye, A.; Allwood, J.M.; Ceron, J.; Geng, Y.; Kheshgi, H.; Lanza, A.; Perczyk, D.; et al. Industry. In *Climate Change 2014: Mitigation of Climate Change. IPCC Working Group III Contribution to AR5*; Cambridge University Press: Cambridge, UK, 2014; pp. 739–810.
89. Sandberg, E.; Toffolo, A.; Krook-Riekkola, A. A bottom-up study of biomass and electricity use in a fossil free Swedish industry. *Energy* **2019**. [CrossRef]
90. Lechtenböhmer, S.; Nilsson, L.J.; Åhman, M.; Schneider, C. Decarbonising the energy intensive basic materials industry through electrification—Implications for future EU electricity demand. *Energy* **2016**, *115*, 1623–1631. [CrossRef]
91. Lolli, N.; Hestnes, A.G. The influence of different electricity-to-emissions conversion factors on the choice of insulation materials. *Energy Build.* **2014**, *85*, 362–373. [CrossRef]
92. Erlandsson, M.; Malmqvist, T.; Francart, N.; Kellner, J. *Minskad Klimatpåverkan från Nybyggda Flerbostadshus—Underlagsrapport*; Sveriges Byggindustrier: Stockholm, Sweden, 2018.
93. Liljenström, C.; Malmqvist, T.; Erlandsson, M.; Fredén, J.; Adolfsson, I.; Larsson, G.; Brogren, M. *Byggandets Klimatpåverkan Livscykelberäkning av Klimatpåverkan och Energianvändning för ett Nyproducerat Energieffektivt Flerbostadshus i Betong*; IVL Svenska Miljöinstitutet: Stockholm, Sweden, 2015.
94. Barandica, J.M.; Fernández-Sánchez, G.; Berzosa, Á.; Delgado, J.A.; Acosta, F.J. Applying life cycle thinking to reduce greenhouse gas emissions from road projects. *J. Clean. Prod.* **2013**, *57*, 79–91. [CrossRef]
95. Swedish Energy Agency. *Drivmedel 2017 Redovisning av Uppgifter Enligt Drivmedelslagen och Hållbarhetslagen*; Swedish Energy Agency: Eskilstuna, Sweden, 2018.
96. Shanks, W.; Dunant, C.F.; Drewniok, M.P.; Lupton, R.C.; Serrenho, A.; Allwood, J.M. How much cement can we do without? Lessons from cement material flows in the UK. *Resour. Conserv. Recycl.* **2019**, *141*, 441–454. [CrossRef]
97. Scrivener, K.L.; John, V.M.; Gartner, E.M. *Eco-Efficient Cements: Potential, Economically Viable Solutions for a Low-CO2 Cement Based Industry*; Elsevier: Amsterdam, The Netherlands, 2016.
98. Cementa; Fossilfritt Sverige. *Färdplan Cement för ett Klimatneutralt Betongbyggande*; Fossilfritt Sverige: Stockholm, Sweden, 2018.
99. ERMCO. *Ready-Mixed Concrete Industry Statistics. Year 2013*; ERMCO—European Ready Mixed Concrete Organization: Bruxelles, Belgium, 2014.
100. ERMCO. *European Ready Mix Concrete Organization (ERMCO) Statistics 2016*; ERMCO—European Ready Mixed Concrete Organization: Bruxelles, Belgium, 2017.
101. Hildebrandt, J.; Hagemann, N.; Thrän, D. The contribution of wood-based construction materials for leveraging a low carbon building sector in europe. *Sustain. Cities Soc.* **2017**, *34*. [CrossRef]

102. Hafner, A.; Schäfer, S. Comparative LCA study of different timber and mineral buildings and calculation method for substitution factors on building level. *J. Clean. Prod.* **2018**, *167*, 630–642. [CrossRef]

103. Moncaster, A.M.; Birgisdottir, H.; Malmqvist, T.; Nygaard Rasmussen, F.; Houlihan Wiberg, A.; Soulti, E. Embodied carbon measurement, mitigation and management within Europe, drawing on a cross-case analysis of 60 building case studies. *Embodied Carbon Build. Meas. Manag. Mitig.* **2018**, 443–462. [CrossRef]

104. Birgisdottir, H.; Moncaster, A.; Wiberg, A.H.; Chae, C.; Yokoyama, K.; Balouktsi, M.; Seo, S.; Oka, T.; Lützkendorf, T.; Malmqvist, T. IEA EBC annex 57 'evaluation of embodied energy and CO_{2eq} for building construction'. *Energy Build.* **2017**, *154*, 72–80. [CrossRef]

105. Bahramian, M.; Yetilmezsoy, K. Life cycle assessment of the building industry: An overview of two decades of research (1995–2018). *Energy Build.* **2020**, *219*. [CrossRef]

106. Wilhelmsson, B.; Kolberg, C.; Larsson, J.; Eriksson, J.; Eriksson, M. *CemZero—Feasibility Study*; Vattenfall: Solna, Sweden, 2018.

107. Lindgren, Å.; Simonsson, P.; Olofsson, J.; Uppenberg, S.; Ekström, D.; Liljenroth, U.; Magnusson, J.; Söderqvist, J.; Wilhelmsson, B.; Lindvall, A.; et al. Klimatoptimerat Byggande av Betongbroar—Råd och Vägledning. 2017. Available online: https://vpp.sbuf.se/Public/Documents/ProjectDocuments/5091a3fe-9f6c-4f98-b1e2-c2416df0aa42/FinalReport/SBUF%2013207%20Slutrapport%20Klimatoptimerat%20byggande%20av%20betongbroar.pdf (accessed on 10 June 2020).

108. Celsa Steel Service AS. *Environmental Product Declaration, Steel Reinforcement Products for Concrete*; EPD International: Stockholm, Sweden, 2012.

109. Allwood, J.M. Transitions to material efficiency in the UK steel economy. *Philos. Trans. A Math. Phys. Eng. Sci.* **2013**, *371*, 20110577. [CrossRef]

110. Fleiter, T.; Herbst, A.; Rehfeldt, M.; Arens, M. *Industrial Innovation: Pathways to Deep Decarbonisation of Industry. Part 2: Scenario Analysis and Pathways to Deep Decarbonisation*; Fraunhofer-Institut für System-und Innovationsforschung ISI: Karlsruhe, Germany, 2019.

111. HYBRIT. *Slutrapport HYBRIT—Hydrogen Breakthrough Ironmaking Technology—Genomförbarhetsstudie*; Hybrit: Luleå, Sweden, 2018.

112. Vogl, V.; Åhman, M.; Nilsson, L.J. Assessment of hydrogen direct reduction for fossil-free steelmaking. *J. Clean. Prod.* **2018**, *203*. [CrossRef]

113. Norgate, T.; Haque, N.; Somerville, M.; Jahanshahi, S. Biomass as a source of renewable carbon for iron and steelmaking. *ISIJ Int.* **2012**, *52*, 1472–1481. [CrossRef]

114. Wei, R.; Zhang, L.; Cang, D.; Li, J.; Li, X.; Xu, C.C. Current status and potential of biomass utilization in ferrous metallurgical industry. *Renew. Sustain. Energy Rev.* **2017**, *68*, 511–524. [CrossRef]

115. HYBRIT. SSAB, LKAB and Vattenfall to Build a Globally-Unique Pilot Plant for Fossil-Free Steel—Press Release. Available online: https://www.ssab.com/company/newsroom/media-archive/2018/02/01/06/31/ssab-lkab-and-vattenfall-to-build-a-globallyunique-pilot-plant-for-fossilfree-steel (accessed on 13 March 2018).

116. Energy Transition Commission. *How to Decarbonize Energy Systems through Electrification–An Analysis of Electrification Opportunities in Transport, Buildings and Industry*; Energy Transitions Commission: London, UK, 2017.

117. Wesseling, J.H.; Lechtenböhmer, S.; Åhman, M.; Nilsson, L.J.; Worrell, E.; Coenen, L. The transition of energy intensive processing industries towards deep decarbonization: Characteristics and implications for future research. *Renew. Sustain. Energy Rev.* **2017**, *79*, 1303–1313. [CrossRef]

118. Fick, G.; Mirgaux, O.; Neau, P.; Patisson, F. Using biomass for pig iron production: A technical, environmental and economical assessment. *Waste Biomass Valorization* **2014**, *5*, 43–55. [CrossRef]

119. Suopajärvi, H.; Kemppainen, A.; Haapakangas, J.; Fabritius, T. Extensive review of the opportunities to use biomass-based fuels in iron and steelmaking processes. *J. Clean. Prod.* **2017**, *148*, 709–734. [CrossRef]

120. Mandova, H.; Leduc, S.; Wang, C.; Wetterlund, E.; Patrizio, P.; Gale, W.; Kraxner, F. Possibilities for CO2emission reduction using biomass in European integrated steel plants. *Biomass Bioenergy* **2018**, *115*, 231–243. [CrossRef]

121. Anheden, M.; Uhlir, L. *Roadmap 2015 to 2025 Biofuels for Low-carbon Steel Industry—A Report by RISE*; RISE: Gothenburg, Sweden, 2015.

122. Suopajärvi, H.; Umeki, K.; Mousa, E.; Hedayati, A.; Romar, H.; Kemppainen, A.; Wang, C.; Phounglamcheik, A.; Tuomikoski, S.; Norberg, N.; et al. Use of biomass in integrated steelmaking—Status quo, future needs and comparison to other low-CO_2 steel production technologies. *Appl. Energy* **2018**, *213*, 384–407. [CrossRef]

123. Van der Stel, J.; Louwerse, G.; Sert, D.; Hirsch, A.; Eklund, N.; Pettersson, M. Top gas recycling blast furnace developments for 'green' and sustainable ironmaking. *Ironmak. Steelmak.* **2013**, *40*, 483–489. [CrossRef]

124. Abdul Quader, M.; Ahmed, S.; Dawal, S.Z.; Nukman, Y. Present needs, recent progress and future trends of energy-efficient Ultra-Low Carbon Dioxide (CO_2) Steelmaking (ULCOS) program. *Renew. Sustain. Energy Rev.* **2016**, *55*, 537–549. [CrossRef]

125. Biermann, M.; Ali, H.; Sundqvist, M.; Larsson, M.; Normann, F.; Johnsson, F. Excess heat-driven carbon capture at an integrated steel mill—Considerations for capture cost optimization. *Int. J. Greenh. Gas Control* **2019**. [CrossRef]

126. Biermann, M.; Normann, F.; Johnsson, F.; Skagestad, R. Partial Carbon Capture by Absorption Cycle for Reduced Specific Capture Cost. *Ind. Eng. Chem. Res.* **2018**. [CrossRef]

127. Dylewski, R.; Adamczyk, J. Life cycle assessment (LCA) of building thermal insulation materials. *Eco-Efficient Constr. Build. Mater. Life Cycle Assess. Eco-Labelling Case Stud.* **2013**, 267–286. [CrossRef]

128. Schiavoni, S.; D'Alessandro, F.; Bianchi, F.; Asdrubali, F. Insulation materials for the building sector: A review and comparative analysis. *Renew. Sustain. Energy Rev.* **2016**, *62*, 988–1011. [CrossRef]

129. Isover. *Planet, People, Prosperity—Our Commitment to Sustainable Construction*; Isover: Ottawa, ON, Canada, 2009. Available online: https://www.isover.com/sites/isover.com/files/assets/documents/lca_brochure.pdf (accessed on 14 August 2019).

130. Pedreño-Rojas, M.A.; Fořt, J.; Černý, R.; Rubio-de-Hita, P. Life cycle assessment of natural and recycled gypsum production in the Spanish context. *J. Clean. Prod.* **2020**, *253*. [CrossRef]

131. Hill, N.; Brannigan, C.; Wynn, D.; Milnes, R.; van Essen, H.; den Boer, E.; van Grinsven, A.; Ligthart, T.; van Gijlswijk, R. EU Transport GHG: Routes to 2050 II. The Role of GHG Emissions from Infrastructure Construction, Vehicle Manufacturing, and ELVs in Overall Transport Sector Emissions. 2012. Available online: http://eutransportghg2050.eu/cms/assets/Uploads/Reports/EU-Transport-GHG-2050-II-Task-2-draftfinal1Mar12.pdf (accessed on 10 June 2020).

132. Thives, L.P.; Ghisi, E. Asphalt mixtures emission and energy consumption: A review. *Renew. Sustain. Energy Rev.* **2017**, *72*, 473–484. [CrossRef]

133. WRAP. *Cutting Embodied Carbon in Construction Projects*; WRAP: Banbury, UK, 2013.

134. Kungliga IngenjörsVetenskaps Akademien. *Så klarar Sveriges Transporter Klimatmålen—En Delrapport från IVA-Projektet Vägval för Klimatet*; Kungliga IngenjörsVetenskaps Akademien: Stockholm, Sweden, 2019.

135. Jakobsen, J.; Roussanaly, S.; Anantharaman, R. A techno-economic case study of CO_2 capture, transport and storage chain from a cement plant in Norway. *J. Clean. Prod.* **2017**, *144*, 523–539. [CrossRef]

136. Chan, Y.; Petithuguenin, L.; Fleiter, T.; Herbst, A.; Arens, M.; Stevenson, P. *Industrial Innovation: Pathways to Deep Decarbonisation of Industry. Part 1: Technology Analysis*; ICF: Fairfax, VA, USA, 2019.

137. European Commission. *A Clean Planet for all A European Strategic Long-term Vision for a Prosperous, Modern, Competitive and Climate Neutral Economy*; European Commission: Brussels, Belgium, 2018.

138. Swedish Energy Agency. *Scenarier över Sveriges Energisystem 2018*; Swedish Energy Agency: Eskilstuna, Sweden, 2019.

139. Boverket. *Behov av nya Bostäder 2018–2025*; Boverket: Karlskrona, Sweden, 2018; ISBN 9789175635729.

140. Näringsdepartementet. *Fastställelse av Nationell Trafikslagsövergripande Plan för Transportinfrastrukturen för Perioden 2018–2029*; Ministry of Enterprise and Innovation: Stockholm, Sweden, 2018.

141. Swedbank. *Makroanalys—Höghastighetståg kan Bidra till Tillväxt, Jobb och Bättre Miljö, Men Statens Finanser Försämras*; Swedbank: Stockholm, Sweden, 2018.

142. Liljelund, L.-E.; Lundstedt, M.; Nilsson, B.O.; Nordlöf, G.; Olofsson, M.; Norr, R. *Resurseffektivitet—Färdvägar mot 2050*; Kungliga IngenjörsVetenskaps Akademien: Stockholm, Sweden, 2015.

143. Energimyndigheten. *Industrins processrelaterade utsläpp av växthusgaser och hur de kan minskas. ER 2018:24.*; Swedish Energy Agency (Energimyndigheten): Eskilstuna, Sweden, 2018.

144. Köhler, N. *Importen av Byggvaror Ökar*; Byggindustrin: Stockholm, Sweden, 2011.

145. Transport och Logisitik. Sjöfarten kan minska antalet riskfyllda godstransporter" TransportochLogistik.se: 2019. Available online: https://www.transportochlogistik.se/index.php/20190803/6007/sjofarten-kan-minska-antalet-riskfyllda-godstransporter (accessed on 9 March 2020).

146. SCB. Priserna för Nyproducerade Bostadshus Lägre än Föregående år. Available online: https://www.scb.se/hitta-statistik/statistik-efter-amne/boende-byggande-och-bebyggelse/byggnadskostnader/priser-for-nyproducerade-bostader/pong/statistiknyhet/priser-for-nyproducerade-bostader-2018/ (accessed on 9 March 2020).

147. Naturvårdsverket. *Klimatscenarier för Bygg—och Fastighetssektorn*; Naturvårdsverket: Stockholm, Sweden, 2018.

148. Zhang, C.; Hu, M.; Yang, X.; Amati, A.; Tukker, A. Life cycle greenhouse gas emission and cost analysis of prefabricated concrete building façade elements. *J. Ind. Ecol.* **2020**, 1–15. [CrossRef]

149. Rootzén, J.; Johnsson, F. Managing the costs of CO_2 abatement in the cement industry. *Clim. Policy* **2017**. [CrossRef]

150. Svensk Betong. *Betongindikatorn–Resultat 2011–2018*; Svensk Betong Service AB: Stockholm, Sweden, 2018.

151. Almeida, M.; Ferreira, M.; Barbosa, R. Relevance of embodied energy and carbon emissions on assessing cost effectiveness in building renovation-Contribution from the analysis of case studies in six European countries. *Buildings* **2018**, *8*, 103. [CrossRef]

152. Piccardo, C.; Dodoo, A.; Gustavsson, L.; Tettey, U.Y.A. Energy and carbon balance of materials used in a building envelope renovation. *IOP Conf. Ser. Earth Environ. Sci.* **2019**, *225*. [CrossRef]

153. Jernkontoret Utrikeshandel. Available online: https://www.jernkontoret.se/sv/stalindustrin/branschfakta-och-statistik/utrikeshandel/ (accessed on 27 February 2020).

154. Moynihan, M.C.; Allwood, J.M. The flow of steel into the construction sector. *Resour. Conserv. Recycl.* **2012**, *68*, 88–95. [CrossRef]

155. Ekvall, T.; Fråne, A.; Hallgren, F.; Holmgren, K. Material pinch analysis: A pilot study on global steel flows. *Metall. Res. Technol.* **2014**, *111*, 359–367. [CrossRef]

156. Energimyndigheten. *Energiläget 2019—En Översikt*; Swedish Energy Agency (Energimyndigheten): Eskilstuna, Sweden, 2019.

157. Svensk Författningssamling. *Förordning (2018: 195) om Reduktion av Växthusgasutsläpp Genom Inblandning av Biodrivmedel i Bensin och Dieselbränslen*; Svensk Författningssamling: Stockholm, Sweden, 2018.

158. Material Economics. *The Circular Economy: A Powerful Force for Climate Mitigation*; Material Economics: Stockholm, Sweden, 2018. Available online: https://materialeconomics.com/publications/the-circular-economy-a-powerful-force-for-climate-mitigation-1 (accessed on 4 November 2019).

159. Bonde, I.; Kuylenstierna, J.; Bäckstrand, K.; Eckerberg, K.; Kåberger, T.; Löfgren, Å.; Rummukainen, M.; Sörlin, S. Klimatpolitiska Rådets Rapport 2020. Available online: https://www.klimatpolitiskaradet.se/en/rapport-2020/ (accessed on 27 March 2020).

160. Vogt-Schilb, A.; Hallegatte, S. Marginal abatement cost curves and the optimal timing of mitigation measures. *Energy Policy* **2014**, *66*, 645–653. [CrossRef]

161. Neuhoff, K.; Chiappinelli, O.; Gerres, T.; Haussner, M.; Ismer, R.; May, N.; Pirlot, A.; Richstein, J. *Building Blocks for a Climate- Neutral European Industrial Sector*; Climate Strategies: London, UK, 2019.

162. Rootzén, J.; Johnsson, F. *A Transformation Fund for Financing High-cost Measures for Deep Emission Cuts in the Construction Industry*, Submitted for publication in Climate Policy. 2019.

163. Wyns, T.; Khandekar, G.; Axelson, M.; Sartor, O.; Neuhoff, K. *Towards an Industrial Strategy for a Climate Neutral Europe*; European Climate Foundation: Brussels, Belgium, 2019.

164. Swedish Energy Agency. *Hinder för Klimatomställning i Processindustrin*; Swedish Energy Agency: Eskilstuna, Sweden, 2019; Volume 49.

165. Swedish Energy Agency. *Riksintresse Vindbruk 2013*; Swedish Energy Agency: Eskilstuna, Sweden, 2013.

166. Energimyndigheten; Naturvårdsverket. *Strategi för Hållbar Vindkraftsutbyggnad—Miljömålsrådsåtgärd 2018*; Naturvårdsverket: Stockholm, Sweden, 2018.

167. CEMA; CECE. *CECE and CEMA: Optimising our Industry to Reduce Emissions*; CECE: Kuala Lumpur, Malaysia, 2011.

168. Preston Aragonès, M.; Serafimova, T. *Zero Emission Construction Sites: The Possibilities and Barriers of Electric Construction Machinery—Bellona Europa*; Bellona Publishing House: Warsaw, Poland, 2018.

169. Davis, S.J.; Lewis, N.S.; Shaner, M.; Aggarwal, S.; Arent, D.; Azevedo, I.L.; Benson, S.M.; Bradley, T.; Brouwer, J.; Chiang, Y.M.; et al. Net-zero emissions energy systems. *Science* **2018**, *360*, 6396. [CrossRef] [PubMed]

170. Giesekam, J.; Barrett, J.R.; Taylor, P. Construction sector views on low carbon building materials. *Build. Res. Inf.* **2016**, *44*, 423–444. [CrossRef]

171. Kadefors, A.; Uppenberg, S.; Olsson, J.A.; Balian, D. *Procurement Requirements for Carbon Reduction in Infrastructure Construction Projects*; Formas: Stockholm, Sweden, 2019.

172. Åhman, M. Perspective: Unlocking the "Hard to Abate" Sectors. 2020. Available online: https://www.researchgate.net/publication/340837433_Perspective_Unlocking_the_Hard_to_Abate_Sectors (accessed on 10 June 2020).

173. Bataille, C.G.F. Physical and policy pathways to net-zero emissions industry. *Wiley Interdiscip. Rev. Clim. Chang.* **2020**, *11*, 1–20. [CrossRef]

174. Abergel, T.; Dean, B.; Dulac, J. *GLOBAL STATUS REPORT 2017 Towards a Zero-emission, Efficient, and Resilient Buildings and Construction Sector*; United Nations Environment Programme: Nairobi, Kenya, 2017; ISBN 9789280736861.

175. IRP; UN Environment. *The Weight of Cities: Resource Requirements of Future Urbanization—A Report by the International Resource Panel*; United Nations Environment Programme: Nairobi, Kenya, 2019.

176. UNEP. *City-Level Decoupling: Urban Resource Flows and the Governance of Infrastructure Transitions. A Report of the Working Group on Cities of the International Resource Panel*; United Nations Environment Programme: Nairobi, Kenya, 2013.

177. Gibbs, M.J.; Soyka, P.; Coneely, D.; Kruger, D. *Global CO_2 Emissions from Cement Production–Good Practice Guidance and Uncertainty Management in National Greenhouse Gas Inventories CO_2*; CICERO Center for International Climate Research: Oslo, Norway, 2006; Volume 10.

178. Leeson, D.; Mac Dowell, N.; Shah, N.; Petit, C.; Fennell, P.S. A Techno-economic analysis and systematic review of carbon capture and storage (CCS) applied to the iron and steel, cement, oil refining and pulp and paper industries, as well as other high purity sources. *Int. J. Greenh. Gas Control* **2017**, *61*, 71–84. [CrossRef]

179. Vatopoulos, K.; Tzimas, E. Assessment of CO2 capture technologies in cement manufacturing process. *J. Clean. Prod.* **2012**, *32*, 251–261. [CrossRef]

180. Kuramochi, T.; Ramírez, A.; Turkenburg, W.; Faaij, A. Comparative assessment of CO2 capture technologies for carbon-intensive industrial processes. *Prog. Energy Combust. Sci.* **2012**, *38*, 87–112. [CrossRef]

181. Cormos, C.C.; Cormos, A.M.; Petrescu, L. *Assessing the CO2Emissions Reduction from Cement Industry by Carbon Capture Technologies: Conceptual Design, Process Integration and Techno-economic and Environmental Analysis*; Elsevier Masson SAS: Issy-les-Moulineaux, France, 2017; Volume 40, ISBN 9780444639653.

182. Limbachiya, M.; Bostanci, S.C.; Kew, H. Suitability of BS EN 197-1 CEM II and CEM V cement for production of low carbon concrete. *Comput. Chem. Eng.* **2014**, *71*, 397–405. [CrossRef]

183. Zhou, D.; Wang, R.; Tyrer, M.; Wong, H.; Cheeseman, C. Sustainable infrastructure development through use of calcined excavated waste clay as a supplementary cementitious material. *J. Clean. Prod.* **2017**, *168*, 1180–1192. [CrossRef]

184. Scrivener, K.L.; John, V.M.; Gartner, E.M. Eco-efficient cements: Potential economically viable solutions for a low-CO_2 cement-based materials industry. *Cem. Concr. Res.* **2018**, *114*, 2–26. [CrossRef]

185. HYBRIT. *Summary of Findings from HYBRIT Pre-Feasibility Study 2016–2017*; Hybrit: Stockholm, Sweden, 2018. Available online: https://www.wko.at/service/aussenwirtschaft/hybrit-brochure.pdf (accessed on 13 March 2018).

186. Worrell, E.; Blinde, P.; Neelis, M.; Blomen, E.; Masanet, E. *Energy Efficiency Improvement and Cost Saving Opportunities for the U.S. Iron and Steel Industry–An ENERGY STAR® Guide for Energy and Plant Managers–Iron_Steel_Guide.pdf*; 2020. Available online: https://www.osti.gov/servlets/purl/1026806 (accessed on 6 June 2020).

187. Bastante-Ceca, M.J.; Cerezo-Narváez, A.; Piñero-Vilela, J.-M.; Pastor-Fernández, A. Determination of the Insulation Solution that Leads to Lower CO2 Emissions during the Construction Phase of a Building. *Energies* **2019**, *12*, 2400. [CrossRef]

188. IEA. *CO_2 Emissions from Fuel Combustion—Highlights*; International Energy Agency: Paris, France, 2017; Volume 53.

189. Naturvårdsverket. *Vägledning i Klimatklivet-Beräkna Utsläppsminskning*; Naturvårdsverket: Stockholm, Sweden, 2018; Volume 1.
190. Benhelal, E.; Zahedi, G.; Shamsaei, E.; Bahadori, A. Global strategies and potentials to curb CO_2 emissions in cement industry. *J. Clean. Prod.* **2013**, *51*, 142–161. [CrossRef]
191. Aranda Usón, A.; López-Sabirón, A.M.; Ferreira, G.; Llera Sastresa, E. Uses of alternative fuels and raw materials in the cement industry as sustainable waste management options. *Renew. Sustain. Energy Rev.* **2013**. [CrossRef]
192. Moro, A.; Lonza, L. Electricity carbon intensity in European Member States: Impacts on GHG emissions of electric vehicles. *Transp. Res. Part D Transp. Environ.* **2018**, *64*, 5–14. [CrossRef]
193. European Energy Agency CO_2 Emissions Intensity Electricity Generation. Available online: https://www.eea.europa.eu/ds_resolveuid/3f6dc9e9e92b45b9b829152c4e0e7ade (accessed on 1 May 2020).
194. Statens Energimyndighet. *Fyra Framtider—Energisystemet Efter 2020*. 2016. Available online: https://www.energimyndigheten.se/globalassets/klimat--miljo/fyra-framtider/fyra-framtider-for-skarmlasning.pdf (accessed on 12 September 2018).
195. EEA. *Trends and Projections in Europe 2018—Tracking Progress towards Europe's Climate and Energy Targets*; European Environment Agency: Copenhagen, Denmark, 2018.
196. Junnila, S.; Horvath, A.; Guggemos, A.A. Life-cycle assessment of office buildings in Europe and the United States. *J. Infrastruct. Syst.* **2006**, *12*, 10–17. [CrossRef]
197. Wallhagen, M.; Glaumann, M.; Malmqvist, T. Basic building life cycle calculations to decrease contribution to climate change—Case study on an office building in Sweden. *Build. Environ.* **2011**, *46*, 1863–1871. [CrossRef]
198. Ylmén, P.; Peñaloza, D.; Mjörnell, K. Life cycle assessment of an office building based on site-specific data. *Energies* **2019**, *12*, 2588. [CrossRef]
199. Luo, Z.; Yang, L.; Liu, J. Embodied carbon emissions of office building: A case study of China's 78 office buildings. *Build. Environ.* **2016**, *95*, 365–371. [CrossRef]
200. Rodrigues, V.; Martins, A.A.; Nunes, M.I.; Quintas, A.; Mata, T.M.; Caetano, N.S. LCA of constructing an industrial building: Focus on embodied carbon and energy. *Energy Procedia* **2018**, *153*, 420–425. [CrossRef]
201. Bonamente, E.; Cotana, F. Carbon and energy footprints of prefabricated industrial buildings: A systematic life cycle assessment analysis. *Energies* **2015**, *8*, 12685–12701. [CrossRef]
202. Bonamente, E.; Merico, M.C.; Rinaldi, S.; Pignatta, G.; Pisello, A.L.; Cotana, F.; Nicolini, A. Environmental impact of industrial prefabricated buildings: Carbon and Energy Footprint analysis based on an LCA approach. *Energy Procedia* **2014**, *61*, 2841–2844. [CrossRef]
203. Erlandsson, M.; Malmqvist, T. Olika byggsystem av betong och trä där mix av material inklusive stål ger klimatfördelar. *Bygg och Tek.* **2018**, *7*, 25–29.
204. Larsson, M.; Erlandsson, M.; Malmqvist, T.; Kellner, J. *Climate Impact of Constructing an Apartment Building with Exterior Walls and Frame of Cross-laminated Timber—the Strandparken Residential Towers*; IVL Swedish Environmental Research Institute AB: Stockholm, Sweden, 2017.
205. Monahan, J.; Powell, J.C. An embodied carbon and energy analysis of modern methods of construction in housing: A case study using a lifecycle assessment framework. *Energy Build.* **2011**, *43*, 179–188. [CrossRef]
206. Petrovic, B.; Myhren, J.A.; Zhang, X.; Wallhagen, M.; Eriksson, O. Life cycle assessment of a wooden single-family house in Sweden. *Appl. Energy* **2019**, *251*, 113253. [CrossRef]
207. Hall, A.; Högberg, A.; Ingelhag, G.; Ljungstedt, H.; Perzon, M.; Stalheim, N.J. Hoppet—The first fossil free preschool. *IOP Conf. Ser. Earth Environ. Sci.* **2019**, *323*. [CrossRef]
208. Tulevech, S.M.; Hage, D.J.; Jorgensen, S.K.; Guensler, C.L.; Himmler, R.; Gheewala, S.H. Life cycle assessment: A multi-scenario case study of a low-energy industrial building in Thailand. *Energy Build.* **2018**, *168*, 191–200. [CrossRef]

Article

Multi-Agent Cooperation Based Reduced-Dimension Q(λ) Learning for Optimal Carbon-Energy Combined-Flow

Huazhen Cao [1], Chong Gao [1], Xuan He [1], Yang Li [1] and Tao Yu [2,*]

[1] Power Grid Planning Center of Guangdong Power Grid Co., Ltd., Guangzhou 510640, China; 13922120994@139.com (H.C.); gao__chong@163.com (C.G.); hexuan1216@163.com (X.H.); liyang2010vip@163.com (Y.L.)

[2] School of Electric Power, South China University of Technology, Guangzhou 510640, China

* Correspondence: taoyu1@scut.edu.cn; Tel.: +86-20-2223-6205

Received: 17 June 2020; Accepted: 31 August 2020; Published: 14 September 2020

Abstract: This paper builds an optimal carbon-energy combined-flow (OCECF) model to optimize the carbon emission and energy losses of power grids simultaneously. A novel multi-agent cooperative reduced-dimension Q(λ) (MCR-Q(λ)) is proposed for solving the model. Firstly, on the basis of the traditional single-objective Q(λ) algorithm, the solution space is reduced effectively to shrink the size of Q-value matrices. Then, based on the concept of ant cooperative cooperation, multi-agents are used to update the Q-value matrices iteratively, which can significantly improve the updating rate. The simulation in the IEEE 118-bus system indicates that the proposed technique can decrease the convergence speed by hundreds of times as compared with conventional Q(λ), keeping high global stability, which is very suitable for dynamic OCECF in a large and complex power grid compared with other algorithms.

Keywords: multi-agent cooperation; reduced-dimension Q(λ); optimal carbon-energy combined-flow

---

1. Introduction

With the increasing impact of the greenhouse effect on the environment, low-carbon economy has gradually become the key development direction of various energy consumption industries. As the largest CO_2 emitter, the electric power industry will play an important role in low-carbon economic development [1]. All kinds of energy-consuming enterprises have also commenced on focusing on the control of carbon emissions, especially in the power industry, which makes up approximately 40% of CO_2 emissions in the whole world [2]. Generally speaking, low-carbon power involves four sectors: generation, transmission, distribution and consumption. Therefore, how to reduce the carbon emissions of transmission and distribution sectors in the power grid industry has turned into an instant issue to be solved [3,4].

Up to now, numerous scholars have carried out research on all aspects of low-carbon power, including optimal power flow (OPF) [5–7], economic emission dispatching [8,9], low-carbon power system dispatch [10], unit commitment [11,12], carbon storage and capture [13,14] and other issues. However, the previous studies mainly focused on the carbon emissions of the generation side, with a lack of research on how to reduce the carbon emissions of the power network (i.e., the transmission and distribution sides). Therefore, the optimal carbon-energy combined-flow (OCECF) model, which can reflect the energy flow and carbon flow distribution of the power grid, is further established in this paper. Basically, the OCECF is on the basis of the conventional reactive power optimization model, which should not only attempt to minimize the power loss and voltage deviation, but also aim to minimize the carbon emission of the power network while satisfying the various operating constraints of power systems.

Obviously, the OCECF is a complicated nonlinear planning problem considering the carbon flow losses of power grids, which can be solved by traditional optimization strategies including nonlinear planning [15], the Newton method [16] and the interior point method [17]. However, due to the strong nonlinearity of power systems, the discontinuity of the objective function and constraint conditions, as well as the existence of multiple local optimal solutions, usually hinder the effectiveness or applications of the classical optimization methods. On the other hand, meta-heuristic algorithms including the genetic algorithm (GA) [18], particle swarm optimization (PSO) [19,20], grouped grey wolf optimizer (GWO) [21] and the memetic salp swarm algorithm (MSSA) [22] have relatively low dependence on specific models, and can obtain relatively satisfactory results when solving such problems. However, due to the low convergence stability of the algorithm, these algorithms may only converge to a local optimal solution. Thus, the conventional $Q(\lambda)$ reinforcement learning algorithm with better convergence robustness and stability is proposed in [23]. Nevertheless, because of the search ergodicity of the single agent $Q(\lambda)$ algorithm, its convergence is relatively long for large-scale system optimization due to the low learning efficiency, while the "dimension disaster" problem with the increasing number of variables can also occur. Moreover, the on-line optimization requirement of the OCECF is also difficult to be met.

Therefore, the author of ant colony optimization (ACO) introduces the concept of ant colony in the classical Q-learning algorithm and puts forward the multiagent Ant-Q algorithm with a faster optimization speed [24]. Based on this, a new multi-agent cooperation-based reduced-dimension $Q(\lambda)$ (MCR-$Q(\lambda)$) learning is proposed for OCECE in this paper, which mainly contains the following contributions:

(i) Most of existing low-carbon power studies did not consider the carbon emissions of the power network due to the energy flow and carbon flow from the generation side to the load side, which cannot satisfy the low-carbon requirement from the viewpoint of the power network. In contrast, the presented OCECF can further reduce the carbon emissions of the power network, which can improve the benefit of the power grid company in a carbon trading market.

(ii) The proposed MCR-$Q(\lambda)$ can effectively shorten the dimension of the solution space of the Q algorithm to solve the OCECF problem by introducing the eligibility trace (λ) returns mechanism [23]. Besides, it also can accelerate the convergence rate and avoid trapping into a low-quality optimum for OCECE via multi-agent cooperation.

The framework of this paper mainly includes: firstly, Section 2 which concludes the related work; Section 3 presents the establishment of the OCECF mathematical model; then, the principle of MCR-$Q(\lambda)$ learning is described in Section 4; Section 5 gives the concrete steps of solving the OCECF problem; Section 6 undertakes simulation studies on the IEEE 118 node system to verify the convergence and stability of MCR-$Q(\lambda)$ learning. Finally, the conclusion of the whole paper is presented in Section 7.

2. Related Work

2.1. Low-Carbon Power

To achieve a low-carbon operation of a power system, extensive studies were devoted to addressing the environmental economic dispatch (EED). In EED, the minimization of emissions [25] is generally designed as one part of the objective function. To further improve the operation economy, the uncertainty of wind power was considered in [26,27], in which the power output of a wind turbine was evaluated based on a probability distribution function of the wind speed. Besides, a modified EED, by combining heat and power economic dispatch, was presented in [28], which can achieve an optimal operation for the heat and power system simultaneously. Furthermore, a coordinated operation of an integrated regional energy system with various energies (e.g., a CO_2-capture-based power) was proposed in [29], while the demand response was also introduced in EED. To further reduce carbon emissions, the CO_2 emission trading system was combined into the daily operation of an energy system. In [30], a decentralized economic dispatch was proposed by considering the carbon capture power plants with carbon emission trading. Moreover, the power uncertainty of wind and photovoltaic energy was fully taken into account in [31,32] based on carbon emission trading. For the purpose of clarifying the internal relation between energy consumption and carbon emissions from power grids, the concept of carbon emission flow is put forward for the first time in reference [33]. On this basis, the authors of [34–36] carried out a theoretical analysis and case verification on the carbon emission flow calculation and the carbon flow tracking of a power system, respectively.

2.2. Application of Meta-Heuristic Algorithms

In fact, the optimal low-carbon operation of a power system faces with various complex and difficult optimization problems, e.g., EED. Hence, various meta-heuristic algorithms have been employed for these optimization problems due to their strong searching ability and high application flexibility. In [25], an improved PSO combining the differential evolution algorithms was designed for EED. In [26], a so-called exchange market algorithm was used for EED due to its fast convergence and strong global searching ability. In [27], a population-based honey bee mating optimization with an online learning mechanism was presented. Inspired by the well-known tag-team game in India, the novel Kho-Kho optimization algorithm [28] with an excellent optimization performance was proposed for EED. To achieve a distributed optimization for real-time power dispatch, a novel adaptive distributed auction-based algorithm with a varying swap size was proposed in [37]. On the other hand, the reinforcement learning-based optimization attracted many investigations for optimal operations of power systems. In [23], a distributed multi-step Q(λ) learning was proposed for the complex OPF of a large-scale power system. To satisfy the requirement of multi-objective optimization, an approximate ideal multi-objective solution Q(λ) learning was presented in [36] via a design of multiple Q matrices for different objective functions.

3. OCECF Mathematical Model

3.1. Carbon-Energy Combined-Flow

The carbon-energy combined-flow (CECF) of the power grid is a comprehensive network flow [36], which combines the power flow of the power grid with the carbon emission flow attached to the power flow of the power grid. Among them, the energy flow is the actual network flow, and the carbon emission flow is the virtual network flow, which can be referred to as the carbon flow in the power system. Carbon flow is generated in the power generation, which represents the concept that the carbon emission is transferred from the generation side to the demand side. The energy flow transfers from the power supply end to the receiving end, but unlike the energy flow, only the power supply that produces carbon emissions at the power supply end can be called a carbon source, as shown in Figure 1. For a given carbon source, the carbon emission is equivalent to the product of the energy flow and the carbon emission rate of the corresponding power generation side [35].

Energy flow is the transmission of electric energy in the power grid. In the process of transmission, there will be power losses, commonly known as network losses, which are generally described as follows:

$$P_{\text{loss}} = \sum_{i,j \in N_{\text{L}}} g_{ij}\left[V_i^2 + V_j^2 - 2V_iV_j\cos\theta_{ij}\right] \tag{1}$$

where V_i and V_j are the voltage amplitudes of the interconnection node i and j, respectively; θ_{ij} means the voltage phase angle difference between node i and j; g_{ij} denotes the conductance between node i and j; N_{L} denotes the branch set of the power network.

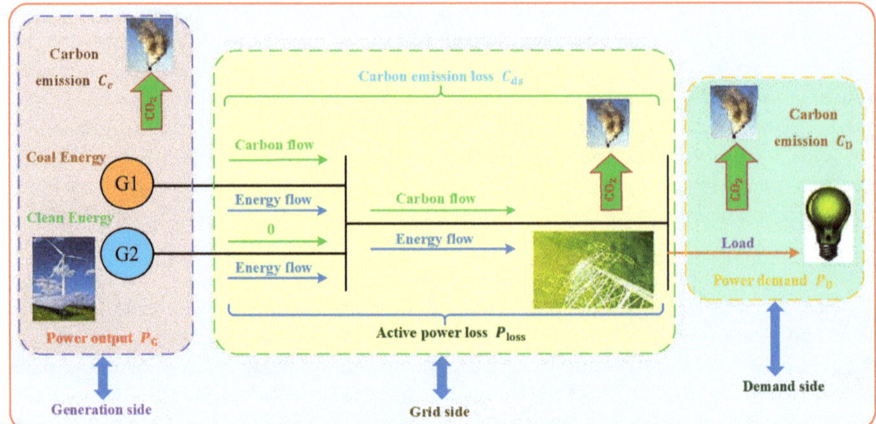

Figure 1. The carbon-energy combined-flow (CECF) structure in power systems.

In the process of power transmission, the energy flow should bear the corresponding amount of carbon flow losses. The tracking of the grid carbon emission flow is based on load flow tracking, and the source of network loss is traced in light of the proportional sharing rule [35]. The ratio of the wth generator to the whole active power injected at node j is

$$\beta_{wj} = \frac{a_{jw}^{(-1)} P_{sw}}{P'_{nj}} \tag{2}$$

where P_{sw} is the active output of the wth generator; P'_{nj} represents the whole active power injection of the j node in the equivalent lossless network; $a_{jw}^{(-1)}$ means the active power injection weight of the wth generator at node j, its specific derivation process can be found in [23].

The proportion of the wth generator outgoing line at node j is the same, and the line loss is decomposed according to the utilization share of the carbon source to the line. Hence, β_{wj} is the component ratio of the active power losses of the wth generator in line i–j. Here, the active power losses of line i–j can be expressed as follows:

$$\Delta P_{ij} = \sum_{w \in W} \left(\frac{a_{jw}^{(-1)} \Delta P_{ij}}{P'_{nj}} \right) P_{sw} \tag{3}$$

where W denotes the generator set.

Therefore, the total carbon flow losses of the power grid can be described by

$$C_{ds} = \sum_{i,j \in N_L} \sum_{w \in W} \left(\frac{a_{jw}^{(-1)} \Delta P_{ij}}{P'_{nj}} \right) P_{sw} \delta_{sw} \tag{4}$$

where δ_{sw} denotes the carbon emission rate of the wth generator.

3.2. OCECF Model

The OCECF model aims to reduce the network losses and carbon flow losses as much as possible according to satisfying the constraints of the power grid and maintaining the stability of the power system voltage. Therefore, the OCECF model is able to describe as follows [23,36]:

$$
\begin{cases}
\min \ \mu_1 f_1(x) + \mu_2 f_2(x) + (1 - \mu_1 - \mu_2)V_d \\[4pt]
s.t.\ P_{Gi} - P_{Di} - V_i \sum_{j \in N_i} V_j\big(g_{ij}\cos\theta_{ij} + b_{ij}\sin\theta_{ij}\big) = 0 \\[4pt]
Q_{Gi} - Q_{Di} - V_i \sum_{j \in N_i} V_j\big(g_{ij}\sin\theta_{ij} + b_{ij}\cos\theta_{ij}\big) = 0 \\[4pt]
P_{Gi}^{\min} \le P_{Gi} \le P_{Gi}^{\max} \ i \in N_G \\[4pt]
Q_{Gi}^{\min} \le Q_{Gi} \le Q_{Gi}^{\max} \ i \in N_G \\[4pt]
V_i^{\min} \le V_i \le V_i^{\max} \ i \in N_B \\[4pt]
Q_{Ci}^{\min} \le Q_{Ci} \le Q_{Ci}^{\max} \ i \in N_C \\[4pt]
k_{ti}^{\min} \le k_{ti} \le k_{ti}^{\max} \ i \in N_k \\[4pt]
|S_i| \le S_i^{\max} \ i \in N_L
\end{cases}
\tag{5}
$$

where nonlinear functions $f_1(x)$ and $f_2(x)$ are the components of carbon flow loss and active power loss; V_d is the voltage stability component; μ_1 and μ_2 are the weight coefficients, $\mu_1 \in [0, 1]$, $\mu_2 \in [0, 1]$, $\mu_1 + \mu_2 \le 1$; $x = [V, \theta, k_t, Q_C]^T$ corresponds to the voltage value of each node of the power grid V, the phase angle of each node θ and the on-load tap changer (OTLC) ratio k_t, reactive power compensation Q_C. The remaining variables can be referenced in the nomenclature and V_d can be described as [23]

$$
V_d = \sum_{j=1}^{n} \left| \frac{2V_j - V_{jmax} - V_{jmin}}{V_{jmax} - V_{jmin}} \right|
\tag{6}
$$

where n represents the number of load nodes; V_j is the node voltage of load node j; and V_{jmax} and V_{jmin} denote the maximal and minimal voltage ranges of load node j, respectively.

4. MCR-Q(λ) Learning

4.1. Q(λ) Learning

Multi-step backtrack Q(λ) learning is a conventional algorithm of RL, in which Q-learning combines the idea multi-step TD(λ) returns [38] and introduces the eligibility trace, such that the convergence speed of the algorithm can be improved to a certain extent. The eligibility trace can be described as [38]

$$
e_k(s,a) = \begin{cases}
\gamma \lambda e_{k-1}(s,a) + 1, & \text{if } (s,a) = (s_k, a_k) \\
\gamma \lambda e_{k-1}(s,a), & \text{otherwise}
\end{cases}
\tag{7}
$$

where $e_k(s,a)$ stands for the eligibility trace under a state-action pair (s, a) corresponding to the kth iteration; (s_k, a_k) denotes the actual state-action pair of the kth iteration; γ means the discount factor; and λ represents the trace-decay factor.

The eligibility trace (λ) uses the "backward estimation" mechanism to approximate the optimal value function matrix Q^*, and sets Q_k as the kth iterative value of the estimated value Q^*, thus the value function of the algorithm can be updated iteratively as follows [39]:

$$
\rho_k = R(s_k, s_{k+1}, a_k) + \gamma Q_k(s_{k+1}, a_g) - Q_k(s_k, a_k)
\tag{8}
$$

$$
\delta_k = R(s_k, s_{k+1}, a_k) + \gamma Q_k(s_{k+1}, a_g) - Q_k(s_k, a_g)
\tag{9}
$$

$$
Q_{k+1}(s,a) = Q_k(s,a) + \alpha \delta_k e_k(s,a)
\tag{10}
$$

$$
Q_{k+1}(s_k, a_k) = Q_{k+1}(s_k, a_k) + \alpha \rho_k
\tag{11}
$$

where α is the learning factor, $R(s_k, s_{k+1}, a_k)$ is the reward function value of the kth iterative time environment from state s_k to s_{k+1} through the selected action a_k; and a_g is the greedy action strategy, which also represents the action corresponding to the highest Q-value in the current state, which can be written by [39]

$$a_g = \underset{a \in A}{\mathrm{argmax}} Q_k(s_{k+1}, a) \tag{12}$$

where A represents the action set, which is also the alternative action set for each variable.

4.2. MCR-Q(λ) Learning

4.2.1. Reduced-Dimension of Solution Space

As shown in Figure 2, the traditional single-objective Q(λ) algorithm does not decompose the action space of all the variables. Assume that the ith variable x_i has m_i alternative solutions, the number of action set elements $|A| = m_1 m_2 \cdots m_n$, when the number of variables n is large, the alternative action combination will increase accordingly, which leads to a slow convergence and difficulties in the iterative calculation. Up to now, the most usual way to work out this "dimension disaster" issue is hierarchical reinforcement learning (HRL) [40]. However, it is difficult to determine the hierarchical design and connection, which usually leads to the convergence of the algorithm to the local optimal solution.

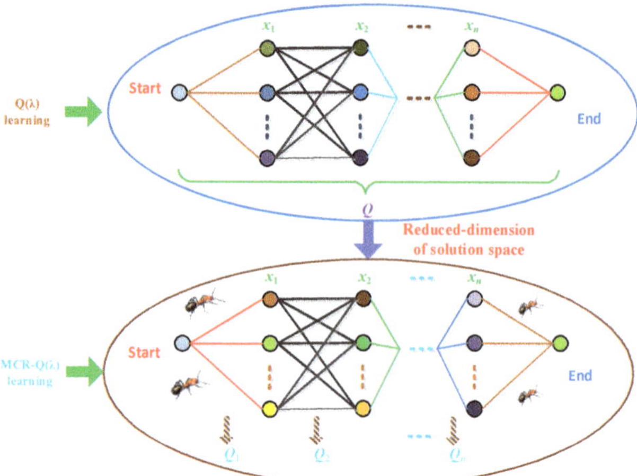

Figure 2. Difference between Q(λ) and MCR-Q(λ).

Under the framework of the proposed MCR-Q(λ) learning algorithm, each variable has a corresponding value function Q_i matrix, and the action set is respectively divided into $(A_1, A_2, \cdots, A_n)$ with $|A_i| = m_i$. In the iterative optimization of each Q matrix, the difficulty of optimization is greatly reduced due to the action space being obviously smaller. Meanwhile, the action space of each variable is the state space of the next variable, which enhances the internal relationship between variables, as can be illustrated in Figure 2. The state space of the first variable is divided according to the load scenario.

4.2.2. Multi-Agent Cooperative Search

In the iterative optimization of Q(λ) learning, which only employs a single agent for exploration and exploitation, the Q matrix is less efficient at updating just one element per iteration. On the contrary, in MCR-Q(λ) learning, there are multiple agents for exploration and exploitation at the same time, in which multiple elements of the Q matrix can be updated at each iteration, and the update speed of the Q matrix is greatly improved. Here, the value function of MCR-Q(λ) learning can be updated iteratively as follows [23]:

$$\rho_k^{ij} = R^{ij}\big(s_k^{ij}, s_{k+1}^{ij}, a_k^{ij}\big) + \gamma Q_k^i\big(s_{k+1}^{ij}, a_g^i\big) - Q_k^i\big(s_k^{ij}, a_g^i\big) \tag{13}$$

$$\delta_k^{ij} = R^{ij}\left(s_k^{ij}, s_{k+1}^{ij}, a_k^{ij}\right) + \gamma Q_k^i\left(s_{k+1}^{ij}, a_g^i\right) - Q_k^i\left(s_k^{ij}, a_g^i\right) \tag{14}$$

$$Q_{k+1}^i\left(s^i, a^i\right) = Q_k^i\left(s^i, a^i\right) + \alpha \delta_k^{ij} e_k^i\left(s^i, a^i\right) \tag{15}$$

$$Q_{k+1}^i\left(s_k^{ij}, a_k^{ij}\right) = Q_{k+1}^i(s_k, a_k) + \alpha \rho_k^{ij} \tag{16}$$

where the superscript i represents the ith variable or the ith Q-value matrix; the superscript j represents the jth objective; $e_k^i\left(s^i, a^i\right)$ and a_g^i are similar to Equations (7) and (12), respectively.

As with the Ant-Q algorithm, MCR-Q(λ) does not calculate the global reward function after each individual selects all the variables, i.e., from the start to the end, as shown in Figure 2. The reward function value can be calculated as follows [24]:

$$R^{ij}\left(s_k^{ij}, s_{k+1}^{ij}, a_k^{ij}\right) = \begin{cases} \frac{W}{L_{Best}}, & \text{if } \left(s_k^{ij}, a_k^{ij}\right) \in SA_{Best} \\ 0, & \text{otherwise} \end{cases} \tag{17}$$

where L_{Best} represents the function value of an individual (i.e., the best individual) that has the lowest value of the objective function value at the kth iteration; W is a positive constant; SA_{Best} denotes the state-action pair set of the optimal individual executed at the kth iteration.

4.2.3. Action Selections

As all individuals are exploring and learning, they are faced with action selections. When the individual j prepares to determine the variable x_i, its action selection is based on the following equation [41]:

$$a_{k+1}^{ij} = \begin{cases} \underset{a^i \in A_i}{\arg\max} Q_{k+1}^i\left(s_{k+1}^{ij}, a^i\right), & \text{if } q \leq q_0 \\ a_S, & \text{otherwise} \end{cases} \tag{18}$$

where q is a random number; q_0 is a positive constant for determining the probability of a pseudo-random selection; a_s denotes the action determined by the pseudo-random selection. In this paper, the rotary selection method is adopted to determine the action to be selected according to the P_k^i distribution of the action probability matrix, and the probability matrix is calculated as follows:

$$P_{k+1}^i\left(s_{k+1}^{ij}, a_{k+1}^i\right) = \frac{Q_{k+1}^i\left(s_{k+1}^{ij}, a_{k+1}^i\right)}{\sum_{a^i \in A_i} Q_{k+1}^i\left(s_{k+1}^{ij}, a^i\right)} \tag{19}$$

When an individual finds the best value of the objective function, the probability of its state-action for the corresponding action will be increased, which will attract other individuals to perform the same action. When the algorithm converges, all individuals will perform the same state-action pair when selecting all variables from the start to the end.

5. OCECF Based on MCR-Q(λ) Learning

5.1. Design of State and Action

As mentioned above, the action space of each variable is designed to be the state space of the next variable, in which the state space of the first variable is designed to be the state set of the environment (i.e., the power grid). For OCECF, the power grid load scenario can be designed as the state of the first variable, where a load scenario is divided at every 15 min and the scenarios with similar loads are set to the same state, e.g., the power grid load scenarios with different loads at 11:00 a.m. and 11:15 a.m. can be regarded as two different states.

In addition, OCECF mainly optimizes the carbon emissions on the power grid side, and the variables in the model are mainly divided into two categories: (a) reactive power compensation device and (b) the OTLC ratio. Thus, the action set corresponding to each variable is a discrete optional action of the reactive power compensation quantity or transformer changer ratio.

5.2. Design of Reward Function

As shown in Equation (17), L_{Best} represents the optimal objective function value of all individuals. According to the OCECF model described by Equation (5), the inequality constraint is brought in by the objective function, and then the objective function value obtained by the individual j becomes [41]

$$L^j = \mu_1 f_1(x^j) + \mu_2 f_2(x^j) + (1 - \mu_1 - \mu_2)V_d^j + N^j \tag{20}$$

$$L_{Best} = \min_{j \in J} L^j \tag{21}$$

where N^j denotes the number of unsatisfied inequality constraints calculated by the power flow after the individual j determines the variable, and J is the number of groups.

5.3. Parameter Setting

In MCR-Q(λ) learning, six parameters γ, λ, α, q_0, J and W, have great influence on the effect of the algorithm [36]. After a large number of simulation tests using trial-and-error, all the parameters can be set as indicated in Table 1.

5.4. Algorithm Flow of the OCECF

Generally speaking, the algorithm flow of OCECF based on MCR-Q(λ) learning is shown in Algorithm 1.

Algorithm 1 Flow of MCR-Q(λ) Learning for OCECF

1:    Initialization: functions Q^I, action probability P^i, eligibility trace matrices e^i, and $i = 1, 2, \cdots, n$;
2:    Input power flow calculation result;
3:    Calculate fitness values of all individuals;
4:    Set $k: = 0$;
5:    **WHILE** $k < k_{max}$;
6:      **FOR** $i = 1$ to n
7:      According to Equations (18) and (19), individual j selects the corresponding action a_k^i of each variable in turn and records the next state;
8:      Calculate power flow for all variables x determined by individuals;
9:      **END FOR**
10:    According to Equations (1) and (4)–(6) respectively calculate the linear loss P_{loss}, the carbon loss C_{ds}, the number of constraints N of dissatisfaction inequality, and the voltage stable component V_d;
11:    Calculate the reward function R^{ij} from Equations (17)–(21);
12:    Update the Q-value functions by Equations (13)–(16);
13:   **END WHILE**
14:   Output: optimal variable x and corresponding optimal function value.

Table 1. Parameter setting of MCR-Q(λ) learning.

Parameters	Range	Value
γ	$0 < \gamma < 1$	0.1
λ	$0 < \lambda < 1$	0.5
α	$0 < \alpha < 1$	0.1
q_0	$0 < q_0 < 1$	0.8
J	$J > 1$	20
W	$W > 0$	1

6. Case Studies

For purpose of testing the optimization performance of MCR-Q(λ) learning, the simulation results of Q(λ) learning, Q learning [41], quantum genetic algorithm (QGA) [42], GA [43], PSO [44], ant colony system (ACS) [45], group search optimizer (GSO) [46] and artificial bee colony (ABC) [47] were also introduced for comparison. Note that the weight coefficient in Equation (5) can be adjusted according to the preference on different components of the objective function. In the simulation analysis, since three components of the objective function in Equation (5) have the same preferences, and the weight coefficient in Equation (5) is set to be 1/3, both the testing IEEE 118-bus system and IEEE 300-bus system are referenced from the tool called MATPOWER [48], in which the detailed parameters can be found in [49]. Besides, it assumes that both the wind and solar energy outputs can be accurately acquired by using effective forecasting techniques, e.g., the deep long-short-term memory recurrent neural network [50]. Among them, the algorithms are simulated and tested in Matlab 2016b by a personal computer with an Intel(R) Core TM i5-4210 CPU at 2.6 GHz with 8 GB of RAM.

6.1. Case Study of IEEE 118-Bus System

6.1.1. Simulation Model

According to different generator types, the carbon emission rate δ_{sw} of each unit in the IEEE 118-bus system is summarized in Table 2. Besides, this paper adopts the same benchmark model of IEEE 118-bus system in all case studies, related detail parameters can be referenced in [36].

Moreover, the system load of the IEEE 118-bus system is mainly divided into five scenarios, as shown in Table 3. Particularly, the scenarios from 1 to 5 represent the system with different load demands, where the load demand gradually increases from scenarios 1 to 5 for all the presented nodes in Table 3. As mentioned above, Tables 2 and 3 are obtained under the same benchmark model of IEEE 118-bus system [36].

In fact, reactive power compensation can be designed for the nodes with generators or load demand to provide adequate reactive power, while the OLTC ratio can be selected for the line with two different voltage nodes. According to this rule, the reactive power compensation of nodes 45, 79, and 105, and the OLTC ratio of lines 8–5, 26–25, 30–17, 63–59, and 64–61 are respectively selected as controllable variables, which are defined in sequence as $(x_1, x_2, x_3, x_4, x_5, x_6, x_7, x_8)$, with

(1) The reactive power compensation is divided into five configurations as {−40%, −20%, 0%, 20%, 40%} with its reference value;

(2) The OLTC ratio is divided into three grades, which are {0.98, 1.00, 1.02}.

Hence, the optimization variables of the IEEE 118-bus system can be found in Table 4, where the variables can be divided into two types, i.e., the reactive power compensation and OLTC ratio; the "no. of bus" represents the location of each variable in the power network; the "action space" denotes the set of the alternative control actions for each variable; and the "variable number" is the number of all the optimization variables.

Table 2. Carbon emission rate of the IEEE 118-bus system.

Generator Node	Generator Type	δ_{sw} (kg/kW·h)	Generator Node	Generator Type	δ_{sw} (kg/kW·h)
1	Gas	0.5	65	Hydro	0
4	Hydro	0	66	Wind	0
6	Coal	1.06	69	Gas	0.5
8	Coal	1.01	70	Hydro	0
10	Coal	0.95	72	Coal	1.06
12	Coal	1.5	73	Coal	1.01
15	Coal	0.7	74	Coal	0.95
18	Gas	0.5	76	Coal	1.5
19	Hydro	0	77	Coal	0.7
24	Hydro	0	80	Hydro	0
25	Coal	1.01	85	Hydro	0
26	Coal	0.95	87	Gas	0
27	Coal	1.5	89	Wind	0
31	Wind	0	90	Gas	1.01
32	Coal	1.06	91	Coal	0.95
34	Coal	1.01	92	Coal	1.5
36	Coal	0.95	99	Coal	0
40	Coal	1.5	100	Hydro	0
42	Coal	0.7	103	Hydro	0
46	Hydro	0	104	Gas	1.06
49	Hydro	0	105	Coal	1.01
54	Gas	0.5	107	Coal	0.95
55	Photovoltaic	0	110	Coal	1.5
56	Coal	1.01	111	Coal	0.7
59	Coal	0.95	112	Coal	0
61	Coal	1.5	113	Hydro	0
62	Hydro	0	116	Hydro	0

Table 3. Load statistical conditions employed in five scenarios.

Scenarios	Active Power (MW)				
	Node 54	Node 59	Node 80	Node 90	Node 116
1	91	221	105	131	148
2	102	249	118	147	166
3	113	277	131	163	184
4	124	305	144	179	202
5	135	333	157	192	220

Table 4. Optimization variables of the IEEE 118-bus system.

Variable Type	Number of Bus	Action Space	Variable Number
Reactive power compensation	45, 79, 105	{−40%, −20%, 0%, 20%, 40%}	3
OLTC ratio	8–5, 26–25, 30–17, 63–59, 64–61	{0.98, 1.00, 1.02}	5

6.1.2. Convergence Analysis

Figure 3 illustrates the convergence process of the Q-value deviation between Q(λ) learning and MCR-Q(λ) learning under scenario 1, where the Q-value deviation is defined as the 2-norm of matrix $(Q_{k+1} - Q_k)$, that is, $\|Q_{k+1} - Q_k\|_2$. As obtained from Figure 3a, since the Q matrix of single-objective Q(λ) learning is large and the updating speed is slow, the algorithm can converge to the optimal Q^* matrix through a variety of trial-and-error explorations, while the convergence time is about 530s. In contrast, after reducing the dimension of the solution space of MCR-Q(λ) learning, the Q^i matrix corresponding to each variable is very small, and 20 objectives are updated at the same time. The optimization speed is more than 100 times of that of Q(λ) learning, which can converge after about 3.5 s, as shown in Figure 3b. Moreover, it can be obtained from the convergence of the objective function values in Figure 4 that the optimization speed of MCR-Q(λ) learning is much faster, and both algorithms can converge to the global optimal solution.

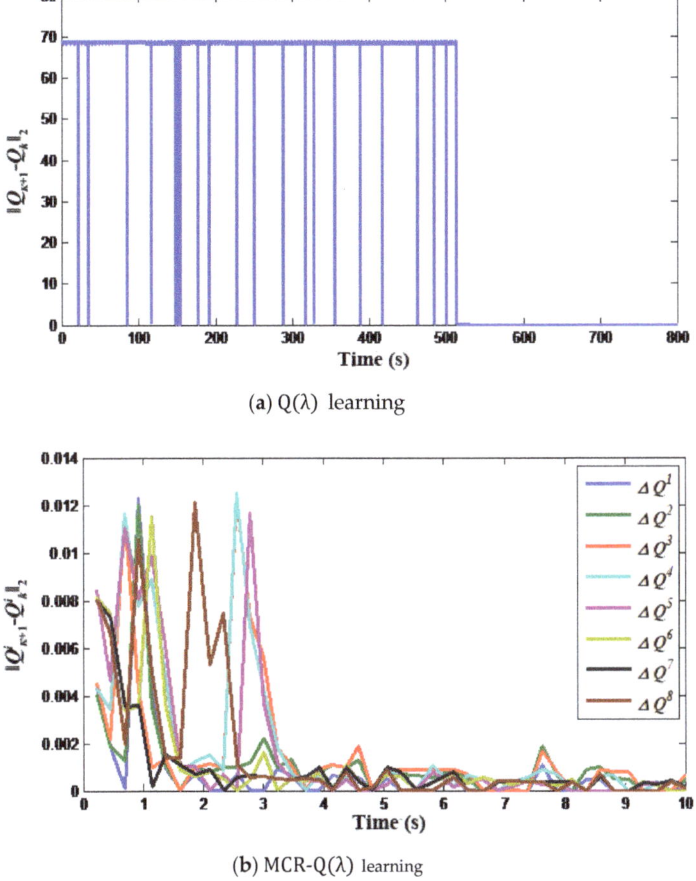

(a) Q(λ) learning

(b) MCR-Q(λ) learning

Figure 3. Q-value difference convergence.

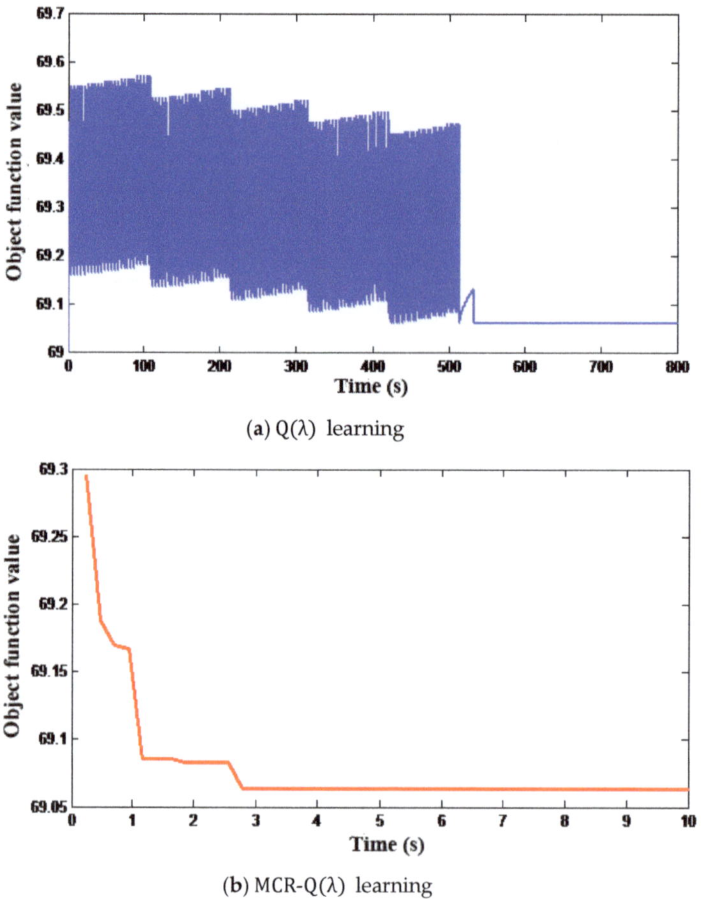

(**a**) Q(λ) learning

(**b**) MCR-Q(λ) learning

Figure 4. Convergence process of the objective function value.

When MCR-Q(λ) learning converges, the value function matrix Q^i and probability matrix P^i corresponding to all variables will prefer a state-action pair, and all individuals will tend to be consistent in selecting the action, as demonstrated in Figure 5.

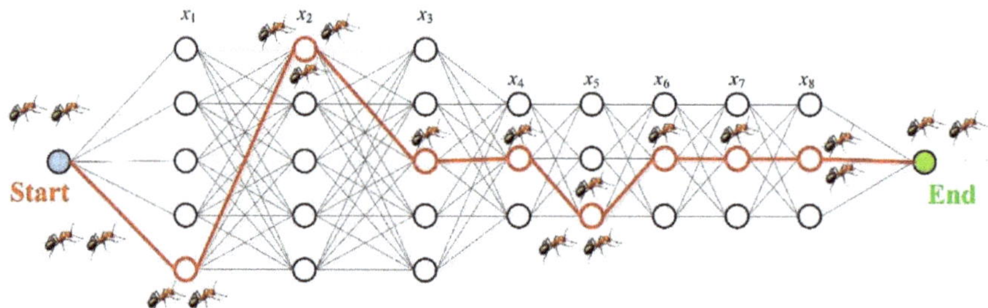

Figure 5. Convergent results of state-action pairs by MCR-Q(λ) learning.

6.1.3. Comparative Analysis of Simulation Results

For the purpose of evaluating the optimization capability of MCR-Q(λ) learning, this section applies all the algorithms to solve the OCECF model for 10 repetitions. For each method, the objective function value is directly taken to evaluate the quality of a solution during the searching process, which is the most crucial index to evaluate the optimization performance.

Table 5 indicates the average convergence results of 10 repetitions for the different algorithms, and it can be found that:

(a) The optimal solution obtained by Q learning and Q(λ) learning is the best, but the optimization time is also the longest, which also shows the strong ergodicity of RL;

(b) The convergence objective value of MCR-Q learning and MCR-Q(λ) learning is the closest to Q learning and Q(λ) learning, and the convergence time is the shortest, while the convergence speed is about 100 times that of single-objective Q learning and Q(λ) learning;

(c) RL improves the algorithmic speed by up to 37.13% with the introduction of the eligibility trace (λ) returns mechanism;

(d) With the increase in the load scenario, the line losses and carbon losses of the power grid will also increase correspondingly. However, since the power system has a sufficient reactive power supply, its voltage stability component just changes slightly.

Table 5. Average results of different algorithms on the IEEE 118-bus system in 10 runs.

Scenarios	Indexes	ABC	GSO	ACS	PSO	GA	QGA	Q	Q(λ)	MCR-Q	MCR-Q(λ)
	Time (s)	55.08	13.30	13.68	31.44	17.14	20.53	660.00	608.00	5.75	5.27
	C_{ds} (t/h)	50.71	50.71	50.71	50.71	50.77	50.71	50.71	50.71	50.71	50.71
1	P_{loss} (MW)	128.85	128.85	128.85	128.85	128.91	128.85	128.85	128.85	128.85	128.85
	V_d	27.65	27.63	27.63	27.64	27.86	27.65	27.63	27.63	27.63	27.64
	Objective	69.07	69.07	69.06	69.07	69.18	69.07	69.06	69.06	69.06	69.06
	Time (s)	65.73	15.83	8.93	29.72	16.44	16.52	646.00	450.00	4.14	3.43
	C_{ds} (t/h)	52.69	52.69	52.69	52.69	52.73	52.70	52.69	52.69	52.69	52.69
2	P_{loss} (MW)	130.24	130.23	130.23	130.23	130.28	130.24	130.23	130.23	130.23	130.23
	V_d	27.58	27.56	27.56	27.57	27.70	27.58	27.56	27.56	27.57	27.57
	Objective	70.17	70.16	70.16	70.17	70.23	70.17	70.16	70.16	70.17	70.16
	Time (s)	36.75	12.66	23.69	49.40	15.57	12.35	671.00	445.00	4.92	3.09
	C_{ds} (t/h)	54.92	54.92	54.92	54.92	54.95	54.92	54.92	54.92	54.92	54.92
3	P_{loss} (MW)	132.50	132.50	132.49	132.49	132.53	132.49	132.49	132.49	132.49	132.49
	V_d	27.52	27.52	27.52	27.53	27.74	27.52	27.52	27.52	27.53	27.52
	Objective	71.65	71.65	71.64	71.65	71.74	71.64	71.64	71.64	71.64	71.64
	Time (s)	44.11	16.65	10.16	52.77	15.93	14.33	663.00	447.00	4.70	4.30
	C_{ds} (t/h)	57.48	57.48	57.48	57.48	57.52	57.48	57.48	57.48	57.48	57.48
4	P_{loss} (MW)	135.66	135.66	135.66	135.66	135.72	135.66	135.66	135.66	135.66	135.66
	V_d	27.49	27.48	27.48	27.48	27.85	27.48	27.48	27.48	27.48	27.48
	Objective	73.54	73.54	73.54	73.54	73.70	73.54	73.54	73.54	73.54	73.54
	Time (s)	26.43	18.41	7.67	42.65	14.27	12.92	658.00	441.00	6.37	5.01
	C_{ds} (t/h)	60.36	60.36	60.36	60.36	60.40	60.36	60.36	60.36	60.36	60.36
5	P_{loss} (MW)	139.73	139.73	139.73	139.72	139.76	139.73	139.72	139.72	139.73	139.72
	V_d	27.45	27.45	27.45	27.45	27.74	27.45	27.45	27.45	27.45	27.45
	Objective	75.84	75.85	75.85	75.84	75.97	75.84	75.84	75.84	75.84	75.84

Figure 6 gives the results comparison between different methods, where each value is the average of the sum value of five scenarios in 10 runs. It is obvious that the result obtained by GA is the worst among all the methods due to its premature convergence. On the other hand, the proposed MCR-Q(λ) learning only has a slight improvement on each index compared with the other methods, but it also can obtain the lowest total carbon flow loss and objective function. It verifies that the proposed method can effectively satisfy the low-carbon requirement from the viewpoint of power networks.

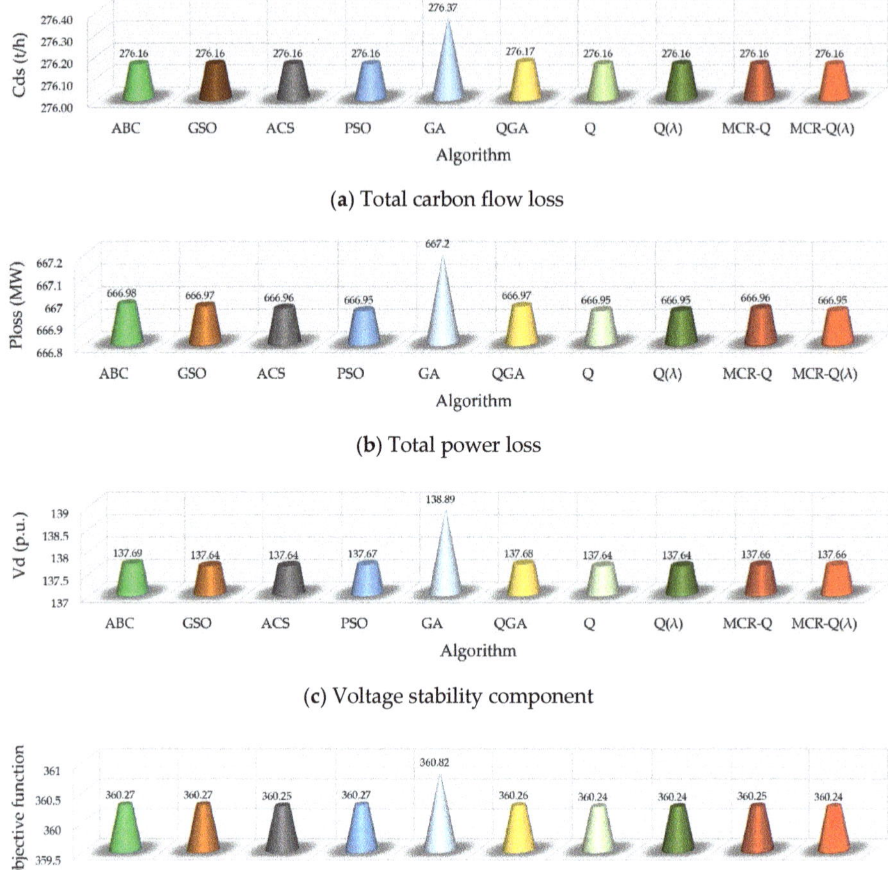

(**a**) Total carbon flow loss

(**b**) Total power loss

(**c**) Voltage stability component

(**d**) Objective function

Figure 6. Comparison of results obtained by different methods in the IEEE 118-bus system.

Lastly, Table 6 gives the statistic convergence results of 10 repetitions for the different algorithms, and it can be found that:

(a) The Q learning and Q(λ) learning have the highest convergence stability and can converge to the global optimal solution every time;

(b) The statistical variance and standard deviation of MCR-Q(λ) learning are the closest to Q learning and Q(λ) learning, which have a relatively high convergence stability;

(c) Except RL, other algorithms are more likely to trap at a local optimum because of the parameter setting and the lack of learning ability.

Table 6. Distribution statistics of the objective function under different algorithms in the IEEE 118-bus system in 10 runs.

Scenarios	Criteria	ABC	GSO	ACS	PSO	GA	QGA	Q	Q(λ)	MCR-Q	MCR-Q(λ)
1	Best	69.06	69.06	69.06	69.06	69.06	69.06	69.06	69.06	69.06	69.06
	Worst	69.09	69.08	69.07	69.09	69.36	69.11	69.06	69.06	69.06	69.07
	Variance	1.2×10^{-4}	2.7×10^{-5}	5.9×10^{-6}	5.5×10^{-5}	8.4×10^{-3}	2.2×10^{-4}	0	0	0	1.6×10^{-6}
	Standard deviation	1.1×10^{-2}	5.2×10^{-3}	2.4×10^{-3}	7.4×10^{-3}	9.1×10^{-2}	1.5×10^{-2}	0	0	0	1.3×10^{-3}
2	Best	70.16	70.16	70.16	70.16	70.16	70.16	70.16	70.16	70.16	70.16
	Worst	70.22	70.17	70.16	70.19	70.33	70.20	70.16	70.16	70.20	70.17
	Variance	3.0×10^{-4}	5.7×10^{-6}	0	5.5×10^{-5}	4.1×10^{-3}	2.3×10^{-4}	0	0	1.5×10^{-4}	2.8×10^{-6}
	Standard deviation	1.7×10^{-2}	2.4×10^{-3}	0	7.4×10^{-3}	6.4×10^{-2}	1.5×10^{-2}	0	0	1.2×10^{-2}	1.7×10^{-3}
3	Best	71.64	71.64	71.64	71.64	71.64	71.64	71.64	71.64	71.64	71.64
	Worst	71.66	71.65	71.66	71.69	71.96	71.65	71.64	71.64	71.65	71.64
	Variance	2.4×10^{-5}	1.3×10^{-5}	2.3×10^{-5}	2.7×10^{-4}	1.0×10^{-2}	5.6×10^{-6}	0	0	8.2×10^{-6}	0
	Standard deviation	5.3×10^{-3}	3.6×10^{-3}	4.8×10^{-3}	1.6×10^{-2}	1.0×10^{-1}	2.4×10^{-3}	0	0	2.9×10^{-3}	0
4	Best	73.54	73.54	73.54	73.54	73.54	73.54	73.54	73.54	73.54	73.54
	Worst	73.57	73.55	73.55	73.54	73.87	73.54	73.54	73.54	73.54	73.54
	Variance	7.6×10^{-5}	5.7×10^{-6}	2.3×10^{-5}	0	1.0×10^{-2}	0	0	0	0	0
	Standard deviation	8.7×10^{-3}	2.4×10^{-3}	4.8×10^{-3}	0	1.0×10^{-1}	0	0	0	0	0
5	Best	75.84	75.84	75.84	75.84	75.84	75.84	75.84	75.84	75.84	75.84
	Worst	75.85	75.85	75.86	75.84	76.12	75.85	75.84	75.84	75.85	75.84
	Variance	5.7×10^{-6}	1.3×10^{-5}	2.6×10^{-5}	0	8.7×10^{-3}	1.6×10^{-6}	0	0	5.7×10^{-6}	0
	Standard deviation	2.4×10^{-3}	3.6×10^{-3}	5.1×10^{-3}	0	9.3×10^{-2}	1.3×10^{-3}	0	0	2.4×10^{-3}	0

6.2. Case Study of the IEEE 300-Bus System

6.2.1. Simulation Model

According to different generator types, the carbon emission rate δ_{sw} of each unit in the IEEE 300-bus system is summarized in Table 7. Besides, 96 different load scenarios are designed to simulate different optimization tasks in a day for the IEEE 300-bus system, as shown in Figure 7. Moreover, the optimization variables are given in Table 8.

Table 7. Carbon emission rate of the IEEE 300-bus system.

Generator Node	Generator Type	δ_{sw} (kg/kWh)	Generator Node	Generator Type	δ_{sw} (kg/kWh)	Generator Node	Generator Type	δ_{sw} (kg/kWh)
8	Hydro	0	171	Hydro	0	7002	Hydro	0
10	Photovoltaics	0	176	Hydro	0	7003	Coal	1.06
20	Coal	1.01	177	Hydro	0	7011	Coal	1.5
63	Coal	0.95	185	Coal	1.01	7012	Coal	0.7
76	Coal	1.5	186	Coal	0.95	7017	Photovoltaics	0
84	Coal	0.7	187	Coal	1.5	7023	Gas	0.5
91	Coal	0.95	190	Hydro	0	7024	Hydro	0
92	Coal	1.5	191	Hydro	0	7039	Wind	0
98	Coal	0.7	198	Hydro	0	7044	Coal	1.5
108	Hydro	0	213	Hydro	0	7049	Coal	0.7
119	Gas	0.5	220	Wind	0	7055	Hydro	0
124	Coal	1.06	221	Gas	0.5	7057	Wind	0
125	Coal	1.01	222	Coal	1.06	7061	Coal	1.06
138	Hydro	0	227	Coal	1.01	7062	Coal	1.01
141	Hydro	0	230	Coal	0.95	7071	Coal	1.01
143	Coal	1.06	233	Coal	1.5	7130	Hydro	0
146	Coal	1.01	236	Coal	0.7	7139	Hydro	0
147	Coal	0.95	238	Coal	0.95	7166	Coal	0.7
149	Coal	1.5	239	Hydro	0	9002	Gas	0.5
152	Hydro	0	241	Hydro	0	9051	Coal	1.06
153	Photovoltaics	0	242	Coal	0.95	9053	Coal	1.01
156	Coal	1.06	243	Coal	1.5	9054	Hydro	0
170	Coal	0.95	7001	Coal	0.95	9055	Photovoltaics	0

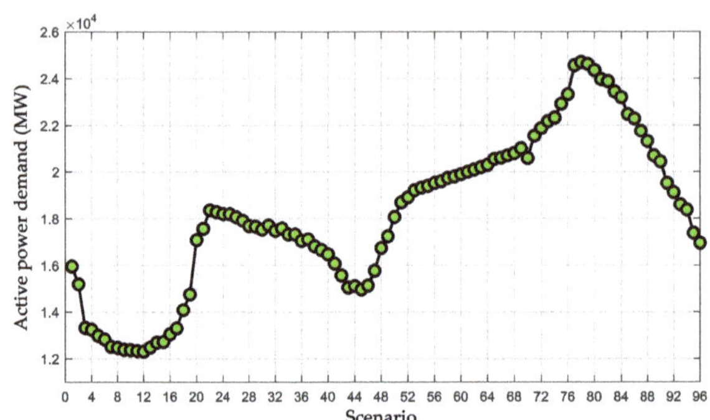

Figure 7. The load scenarios of the IEEE 300-bus system.

Table 8. Optimization variables of the IEEE 300-bus system.

Variable Type	Number of Bus	Action Space	Variable Number
Reactive power compensation	117, 120, 154, 164, 166, 173, 190, 231, 238, 240, 248	{−40%, −20%, 0%, 20%, 40%}	11
OLTC ratio	9021–9022, 9002–9024, 9023–9025, 9023–9026, 9007–9071, 9007–9072, 9003–9031, 9003–9032, 9003–9033, 9004–9041, 9004–9042, 9004–9043, 9003–9034, 9003–9035, 9003–9036, 9003–9037, 9003–9038, 213–214, 222–237, 227–231, 241–237, 45–46, 73–74, 81–88, 85–99, 86–102, 122–157, 142–175, 145–180, 200–248, 211–212, 223–224, 196–2040, 7003–3, 7003–61, 7166–166, 7024–24, 7001–1, 7130–130, 7011–11, 7023–23, 7049–49, 7139–139, 7012–12	{0.98, 1.00, 1.02}	44

6.2.2. Comparative Analysis of Simulation Results

For the purpose of evaluating the optimization capability of MCR-Q(λ) learning, this section applies all the algorithms to solve the OCECF model for 10 runs. Since the number of optimization variables of the IEEE 300-bus system dramatically increases, the conventional Q and Q(λ) algorithms cannot implement an optimization due to the dimension disaster. Figure 8 provides the results comparison between different methods, where each value is the average of the sum value of a day in 10 runs. It can be found that the proposed MCR-Q(λ) learning significantly outperforms other methods on the total carbon flow loss, total power loss, voltage stability component and the objective function. Hence, the MCR-Q(λ) learning-based OCECF can achieve a low-carbon operation for the power network. Particularly, these values obtained by MCR-Q(λ) learning are 2.0%, 3.4%, 45.9% and 10.3% lower than that obtained by GSO. It verifies that the optimization performance of MCR-Q(λ) is much better than other conventional meta-heuristic algorithms as the system scale increases.

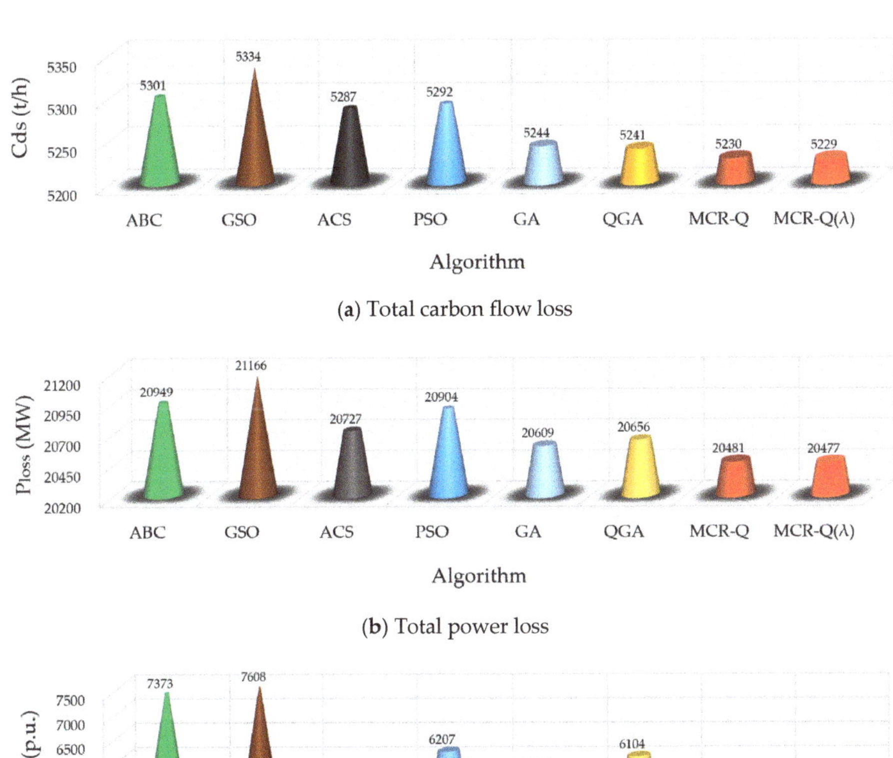

(**a**) Total carbon flow loss

(**b**) Total power loss

(**c**) Voltage stability component

Figure 8. *Cont.*

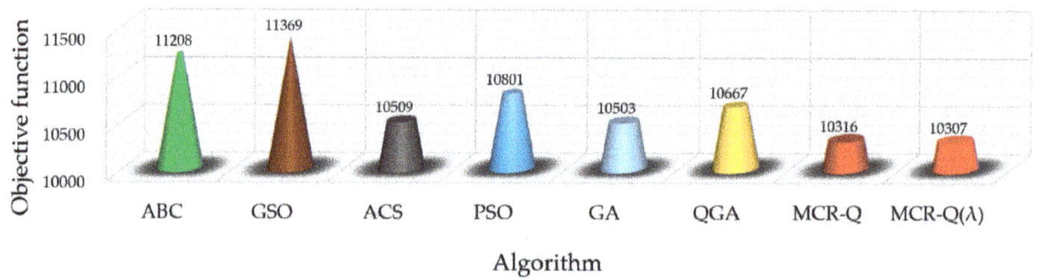

(**d**) Objective function

Figure 8. Comparison of results obtained by different methods in the IEEE 300-bus system.

Besides, Table 9 gives the distribution statistics of the objective function under different algorithms in the IEEE 300-bus system, where each value is the sum value of the objective function of a day in 10 runs; the best, worst, variance and standard deviation (Std. Dev.) are calculated to evaluate the convergence stability [51]. It can be seen from Table 9 that the convergence stability of MCR-Q(λ) learning is the highest among all the methods with the smallest variance and standard deviation of the objective function.

Table 9. Distribution statistics of the objective function under different algorithms in the IEEE 300-bus system in 10 runs.

Criteria	ABC	GSO	ACS	PSO	GA	QGA	MCR-Q	MCR-Q(λ)
Best	11,182.97	11,328.38	10,505.58	10,795.73	10,495.03	10,658.28	10,312.84	10,305.54
Worst	11,229.61	11,404.35	10,513.40	10,812.54	10,509.03	10,675.34	10,320.86	10,308.30
Variance	246.73	541.79	10.25	45.62	23.96	32.57	10.12	1.20
Standard deviation	15.71	23.28	3.20	6.75	4.89	5.71	3.18	1.09

7. Conclusions

This paper builds an OCECF model to optimize the carbon emission and energy losses of power grids simultaneously and proposes a new MCR-Q(λ) learning to solve this problem, which has the following four contributions/novelties:

(1) The OCECF model carefully considers the distribution of carbon flow in the power grid, which effectively resolves the carbon emission optimization at the power grid side;
(2) MCR-Q(λ) learning is proposed for the first time, which largely reduces the dimension of the solution space, and significantly accelerates the updating rate of the Q-value matrix via multi-agent cooperative exploration learning, such that the optimization speed can be considerably accelerated;
(3) Compared with Q(λ) learning, the convergence rate of MCR-Q(λ) learning can be increased by about 100 times, while a higher global convergence stability is guaranteed. Hence, it is very suitable for resolving dynamic OCECF in a large and complex power grid compared with other algorithms;
(4) Like ACO, MCR-Q(λ) learning is also suitable for solving various complex optimization problems.

To further improve the operation benefit of power grids, future works can focus on the carbon trading system-based optimal power flow and the Pareto-based multi-objective learning methods, while a decentralized optimization will be studied for high operation privacy and reliability.

Author Contributions: H.C. established the model, implemented the simulation and wrote this article; C.G. guided and revised the paper and refined the language; X.H. collected references; Y.L. guided the research; T.Y. assisted in writing algorithms. All authors have read and agreed to the published version of the manuscript.

Funding: This research was funded by Technical Projects of China Southern Power Grid grant number [GDK JXM20173256].

Conflicts of Interest: The authors declare no conflict of interest.

Nomenclature

P_{Gi}, Q_{Gi}	active and reactive power generation of the ith node
P_{Di}, Q_{Di}	active and reactive power demand of the ith node
V_i, V_j	voltage magnitude of the ith and jth node
b_{ij}	susceptance of line i–j
S_i	apparent power flow of the ith transmission line
N_i	node set
N_L	set of branches of the power network
N_G	set of units
N_H	set of hydro units
N_B	set of PQ nodes
N_C	set of compensation equipment
N_K	set of on-load transformers
k_t	on-load tap changer ratio
Q_c	reactive power compensation
θ	phase angle of each node
V_d	component of voltage stability
$V_{j\min}, V_{j\max}$	minimum and maximum voltage limit of load node j
μ_1, μ_2	weight coefficients
W	generator set
(s_k, a_k)	actual state-action pair of the kth iteration
δ_k, ρ_k	estimates of Q-function errors
$R(s_k, s_{k+1}, a_k)$	reward function value of the kth iterative time environment from state s_k to s_{k+1} through a selected action a_k
a_g	greedy action strategy
A	action set
L_{Best}	function value of an individual (i.e., the best individual) that has the least value of the target function value at the kth iteration
SA_{Best}	state-action pair set of the best individual executed at the kth iteration
γ	discount factor
λ	trace-decay factor
α	learning factor
J	number of groups

Abbreviations

OCECF	optimal carbon energy combined flow
OTLC	on-load tap changer
MCR-Q(λ)	multi-agent cooperative reduced-dimension Q(λ)
HRL	hierarchical reinforcement learning
EED	environmental economic dispatch

References

1. Yang, B.; Jiang, L.; Wang, L.; Yao, W.; Wu, Q.H. Nonlinear maximum power point tracking control and modal analysis of DFIG based wind turbine. *Int. J. Electr. Power Energy Syst.* **2016**, *74*, 429–436. [CrossRef]
2. Yang, Y.D.; Song, A.J.; Liu, H.; Qin, Z.J.; Deng, J.; Qi, J.J. Parallel computing of multi-contingency optimal power flow with transient stability constraints. *Prot. Control Mod. Power Syst.* **2018**, *3*, 204–213. [CrossRef]
3. Yang, B.; Yu, T.; Zhang, X.S.; Li, H.F.; Shu, H.C.; Sang, Y.Y.; Jiang, L. Dynamic leader based collective intelligence for maximum power point tracking of PV systems affected by partial shading condition. *Energy Convers. Manag.* **2019**, *179*, 286–303. [CrossRef]

4. Yang, B.; Wang, J.B.; Zhang, X.S.; Yu, T.; Yao, W.; Shu, H.C.; Zeng, F.; Sun, L.M. Comprehensive overview of meta-heuristic algorithm applications on pv cell parameter identification. *Energy Convers. Manag.* **2020**, *208*, 112595. [CrossRef]

5. Yang, B.; Yu, T.; Shu, H.C.; Zhang, Y.M.; Chen, J.; Sang, Y.Y.; Jiang, L. Passivity-based sliding-mode control design for optimal power extraction of a PMSG based variable speed wind turbine. *Renew. Energy* **2018**, *119*, 577–589. [CrossRef]

6. Badal, F.R.; Das, P.; Sarker, S.K.; Das, S.K. A survey on control issues in renewable energy integration and microgrid. *Prot. Control Mod. Power Syst.* **2019**, *4*, 87–113. [CrossRef]

7. Li, Y.; Li, Y. Two-step many-objective optimal power flow based on knee point-driven evolutionary algorithm. *Processes* **2018**, *6*, 250. [CrossRef]

8. Li, G.Y.; Li, G.D.; Zhou, M. Comprehensive evaluation model of wind power accommodation ability based on macroscopic and microscopic indicators. *Prot. Control Mod. Power Syst.* **2019**, *4*, 215–226. [CrossRef]

9. Yang, B.; Yu, T.; Shu, H.C.; Dong, J.; Jiang, L. Robust sliding-mode control of wind energy conversion systems for optimal power extraction via nonlinear perturbation observers. *Appl. Energy* **2018**, *210*, 711–723. [CrossRef]

10. Ji, Z.; Kang, C.; Chen, Q.; Xia, Q.; Jiang, C.; Chen, Z. Low-carbon power system dispatch incorporating carbon capture power plants. *IEEE Trans. Power Syst.* **2013**, *28*, 4615–4623. [CrossRef]

11. Kuo, C.C.; Lee, C.Y.; Sheim, Y.C. Unit commitment with energy dispatch using a computationally effifient encoding structure. *Energy Convers Manag.* **2011**, *52*, 1575–1582. [CrossRef]

12. Ji, B.; Yuan, X.; Li, X.; Huang, Y.; Li, W. Application of quantum-inspired binary gravitational search algorithm for thermal unit commitment with wind power integration. *Energy Convers Manag.* **2014**, *87*, 589–598. [CrossRef]

13. Wall, T.; Stanger, R.; Santos, S. Demonstrations of coal-fired oxy-fuel technology for carbon capture and storage and issues with commercial deployment. *Int. J. Greenh. Gas Control* **2011**, *5*, S5–S15. [CrossRef]

14. Coninck, H.D.; Benson, S.M. Carbon Dioxide capture and storage: Issues and prospects. *Annu. Rev. Environ. Resour.* **2014**, *39*, 243–270. [CrossRef]

15. Chen, J.; Yao, W.; Zhang, C.K.; Ren, Y.; Jiang, L. Design of robust MPPT controller for grid-connected PMSG-Based wind turbine via perturbation observation based nonlinear adaptive control. *Renew. Energy* **2019**, *134*, 478–495. [CrossRef]

16. Giras, T.C.; Talukdar, S.N. Quasi-newton method for optimal power flows. *Int. J. Electr. Power Energy Syst.* **1981**, *3*, 59–64. [CrossRef]

17. Zhang, X.S.; Xu, Z.; Yu, T.; Yang, B.; Wang, H. Optimal mileage based AGC dispatch of a GenCo. *IEEE Trans. Power Syst.* **2020**, *35*, 2516–2526. [CrossRef]

18. Azzam, M.; Mousa, A.A. Using genetic algorithm and TOPSIS technique for multi-objective reactive power compensation. *Electr. Power Syst. Res.* **2010**, *80*, 675–681. [CrossRef]

19. Juan, L.I.; Yang, L.; Liu, J.L.; Yang, D.L.; Zhang, C. Multi-objective reactive power optimization based on adaptive chaos particle swarm optimization algorithm. *Power Syst. Prot. Control* **2011**, *39*, 26–31.

20. Han, P.P.; Fan, G.J.; Sun, W.Z.; Shi, B.L.; Zhang, X.A. Research on identification of LVRT characteristics of photovoltaic inverters based on data testing and PSO algorithm. *Processes* **2019**, *7*, 250. [CrossRef]

21. Yang, B.; Zhang, X.S.; Yu, T.; Shu, H.C.; Fang, Z.H. Grouped grey wolf optimizer for maximum power point tracking of doubly-fed induction generator based wind turbine. *Energy Convers. Manag.* **2017**, *133*, 427–443. [CrossRef]

22. Yang, B.; Zhong, L.E.; Yu, T.; Li, H.F.; Zhang, X.S.; Shu, H.C.; Sang, Y.Y.; Jiang, L. Novel bio-inspired memetic salp swarm algorithm and application to MPPT for PV systems considering partial shading condition. *J. Clean. Prod.* **2019**, *215*, 1203–1222. [CrossRef]

23. Yu, T.; Liu, J.; Chan, K.W.; Wang, J.J. Distributed multi-step $Q(\lambda)$ learning for optimal power flow of large-scale power grids. *Int. J. Electr. Power Energy Syst.* **2012**, *42*, 614–620. [CrossRef]

24. Liana, M.; Roberto, S. The Ant-Q algorithm applied to the nuclear reload problem. *Ann. Nucl. Energy* **2002**, *29*, 1455–1470.

25. Zhao, X.-G.; Liang, L.; Meng, J.; Zhou, Y. An improved quantum particle swarm optimization algorithm for environmental economic dispatch. *Expert Syst. Appl.* **2020**, *152*, 113370.

26. Hagh, M.T.; Kalajahi, S.M.S.; Ghorbani, N. Solution to economic emission dispatch problem including wind farms using exchange market algorithm method. *Appl. Soft Comput. J.* **2020**, *88*, 106044. [CrossRef]

27. Ghasemi, A.; Gheydi, M.; Golkar, M.J.; Eslami, M. Modeling of wind/environment/economic dispatch in power system and solving via an online learning meta-heuristic method. *Appl. Soft Comput.* **2016**, *43*, 454–468. [CrossRef]

28. Srivastava, A.; Das, D.K. A new Kho-Kho optimization algorithm: An application to solve combined emission economic dispatch and combined heat and power economic dispatch problem. *Eng. Appl. Artif. Intell.* **2020**, *94*, 103763. [CrossRef]

29. He, L.; Lu, Z.; Geng, L.; Zhang, J.; Li, X.; Guo, X. Environmental economic dispatch of integrated regional energy system considering integrated demand response. *Int. J. Electr. Power Energy Syst.* **2020**, *116*, 105525. [CrossRef]

30. Zhang, R.; Yan, K.; Li, G.; Jiang, T.; Li, X.; Chen, H. Privacy-preserving decentralized power system economic dispatch considering carbon capture power plants and carbon emission trading scheme via over-relaxed ADMM. *Int. J. Electr. Power Energy Syst.* **2020**, *121*, 106094. [CrossRef]

31. Jin, J.; Zhou, P.; Li, C.; Guo, X.; Zhang, M. Low-carbon power dispatch with wind power based on carbon trading mechanism. *Energy* **2019**, *170*, 250–260. [CrossRef]

32. Tan, Q.; Ding, Y.; Ye, Q.; Mei, S.; Zhang, Y.; Wei, Y. Optimization and evaluation of a dispatch model for an integrated wind-photovoltaic-thermal power system based on dynamic carbon emissions trading. *Appl. Energy* **2019**, *253*, 113598. [CrossRef]

33. Kang, C.; Zhou, T.; Chen, Q. Carbon emission flow in network. *Sci. Rep.* **2012**, *2*, 479. [CrossRef] [PubMed]

34. Zhou, T.; Kang, C.; Qianyao, X.U.; Chen, Q. Preliminary Investigation on a Method for carbon emission flow calculation of power system. *Autom. Electr. Power Syst.* **2012**, *36*, 44–49.

35. Li, B.; Song, Y.; Hu, Z. Carbon flow tracing method for assessment of demand side carbon emissions obligation. *IEEE Trans. Sustain. Energy* **2013**, *4*, 1100–1107. [CrossRef]

36. Zhang, X.S.; Yu, T.; Yang, B.; Zheng, L.M.; Huang, L.N. Approximate ideal multi-objective solution Q(λ) learning for optimal carbon-energy combined-flow in multi-energy power systems. *Energy Convers. Manag.* **2015**, *106*, 543–556. [CrossRef]

37. Zhang, X.; Tan, T.; Zhou, B.; Yu, T.; Yang, B.; Huang, X. Adaptive distributed auction-based algorithm for optimal mileage based AGC dispatch with high participation of renewable energy. *Int. J. Electr. Power Energy Syst.* **2021**, *124*, 106371. [CrossRef]

38. Sutton, R.S. Learning to predict by the methods of temporal differences. *Mach. Learn.* **1988**, *3*, 9–44. [CrossRef]

39. Tao, Y.U.; Wang, Y.M.; Zhen, W.G.; Wenjia, Y.E.; Liu, Q.J. Multi-step backtrack Q-learning based dynamic optimal algorithm for auto generation control order dispatch. *Control Theory Appl.* **2011**, *28*, 58–64.

40. Ghavamzadeh, M.; Mahadevan, S. Hierarchical average reward reinforcement learning. *J. Mach. Learn. Res.* **2017**, *8*, 2629–2669.

41. Cao, H.Z.; Yu, T.; Zhang, X.S.; Yang, B.; Wu, Y.X. Reactive power optimization of large-scale power system: A transfer bees optimizer application. *Processes* **2019**, *7*, 321. [CrossRef]

42. Xiong, Y.; Chen, H.H.; Miao, F.Y.; Wang, X.F. A quantum genetic algorithm to solve combinatorial optimization problem. *Acta Electron. Sin.* **2004**, *32*, 1855–1858.

43. Kumari, M.S.; Maheswarapu, S. Enhanced genetic algorithm based computation technique for multi-objective optimal power flow solution. *Int. J. Electr. Power Energy Syst.* **2010**, *32*, 736–742. [CrossRef]

44. Hazra, J.; Sinha, A.K. A multi-objective optimal power flow using particle swarm optimization. *Eur. Trans. Electr. Power* **2011**, *21*, 1028–1045. [CrossRef]

45. Han, Y.; Shi, P. An improved ant colony algorithm for fuzzy clustering in image segmentation. *Neurocomputing* **2007**, *70*, 665–671. [CrossRef]

46. Basu, M. Group search optimization for solution of different optimal power flow problems. *Electr. Mach. Power Syst.* **2016**, *44*, 10. [CrossRef]

47. Karaboga, D.; Basturk, B. On the performance of artificial bee colony (ABC) algorithm. *Appl. Soft Comput.* **2008**, *8*, 687–697. [CrossRef]

48. Zimmerman, R.D.; Murillo-Sánchez, C.E.; Thomas, R.J. Matpower: Steady-state operations, planning and analysis tool for power systems research and education. *IEEE Trans. Power Syst.* **2011**, *26*, 12–19. [CrossRef]

49. MATPOWER—Free, Open Source Tools for Electric Power System Simulation and Optimization. Available online: https://matpower.org/ (accessed on 21 May 2020).

50. Mahmoud, K.; Abdel-Nasser, M.; Mustafa, E.; Ali, Z.M. Improved salp-swarm optimizer and accurate forecasting model for dynamic economic dispatch in sustainable power systems. *Sustainability* **2020**, *12*, 576. [CrossRef]
51. Zhang, X.S.; Yu, T.; Yang, B.; Cheng, L.F. Accelerating bio-inspired optimizer with transfer reinforcement learning for reactive power optimization. *Knowl. Based Syst.* **2017**, *116*, 26–38. [CrossRef]

Article

Identification and Categorization of Factors Affecting the Adoption of Energy Efficiency Measures within Compressed Air Systems

Andrea Trianni [1], Davide Accordini [2,*] and Enrico Cagno [2]

[1]   Faculty of Engineering and IT, University of Technology Sydney, Ultimo, NSW 2007, Australia;
    Andrea.Trianni@uts.edu.au

[2]   Department of Management, Economics & Industrial Engineering, Politecnico di Milano, 20133 Milano, Italy;
    enrico.cagno@polimi.it

*   Correspondence: davide.accordini@polimi.it; Tel.: +39-348-148-0926

Received: 11 August 2020; Accepted: 22 September 2020; Published: 1 October 2020

Abstract: Understanding the factors driving the implementation of energy efficiency measures in compressed air systems is crucial to improve industrial energy efficiency, given their low implementation rate. Starting from a thorough review of the literature, it is thus clear the need to support companies in the decision-making process by offering an innovative framework encompassing the most relevant factors to be considered when adopting energy efficiency measures in compressed air systems, inclusive of the impacts on the production resources and the operations of a company. The framework, designed following the perspective of the industrial decision-makers, has been validated, both theoretically and empirically, and preliminarily applied to a heterogeneous cluster of manufacturing industries. Results show that, beside operational, energetic, and economic factors, in particular contextual factors such as complexity, compatibility, and observability may highlight critical features of energy efficiency measures whose absence may change the outcome of a decision-making process. Further, greater awareness and knowledge over the important factors given by the implementation of the framework could play an important role in fostering the implementation of energy efficiency measures in compressed air systems. The paper concludes with further research avenues to further promote energy efficiency and sustainability oriented practices in the industrial sector.

Keywords: energy efficiency; compressed air systems; energy efficiency measures; nonenergy benefits; assessment factors

---

1. Introduction

Industrial energy efficiency is widely recognized as crucial means to mitigate the growing final energy consumption (by more than 25% in the 2018–2040 time span [1]), given that industry is responsible for 35% of global total final energy use [2]. Energy efficiency can also lead to other benefits, such as enhanced security of the energy production systems and a healthier and more comfortable environment [3], plus strategic advantages connected to a less volatile energy market [4], especially in countries strongly dependent on energy imports [5,6]. As discussed by [7,8], previous research has mainly focused on sector-specific energy efficiency measures (EEMs). However, the extreme heterogeneity of the industrial sectors calls for a different approach aimed at promoting specific cross-cutting technologies. Among others, the Compressed Air System (CAS) looks particularly interesting, being widely diffused as ancillary technology within many industrial processes [9] due to its cleanness, practicality, and ease of use [10]. Usually, industrial compressed air (CA) is generated by using electricity as energy source and can account for about 10% of the total electricity bill in some contexts [10]. By taking a life-cycle costs perspective on CAS, the largest portion of costs is covered by operating costs (almost 80% [11]). Therefore, improved energy efficiency in

CAS by implementing EEMs (both implying both technological and behavioral changes [12]) should be abundantly cost-effective, and lead to other benefits, such as reduced scrap rates, greater capacity utilization, enhanced safety, and many others [13].

Nonetheless, despite the huge potentials for energy efficiency gains (up to 20% [11,14]) and continuous development in the field [15], EEMs are not diffused as expected, leading to the so-called energy efficiency gap [16,17], particularly critical for small and medium-sized enterprises (SMEs), which everywhere represent the vast majority of companies and are responsible for the largest share of consumption [18,19]. Previous research noted that SMEs particularly suffer from a lack of internal competences as well as standard procedures hindering EEMs adoption [20–22]. This is also confirmed by studies on barriers to energy efficiency [17,20,23], which only partially refers to costs, rather pointing the attention on the lack of awareness and specific knowledge [22–24] as well as unperfect information and irrational behavior [25], therefore suggesting that it is of primary importance to highlight the single factors driving the decision-making process over EEMs. The literature has so far identified assessment factors for EEMs (e.g., [26]); however, they are referred to other technologies other than CAS. Since different technologies are characterized by different EEMs [27], different factors should be analyzed as well.

Classifications of interventions in CAS have been proposed by literature [11,28,29]; nevertheless, a mere technical EEM description does not sufficiently pinpoint some relevant factors, such as specific implications at the operational level that, beyond energy and monetary savings, are crucial for wise decision-making, representing a major research gap. Therefore, starting from an overview of CAS (Section 2) and literature review in Section 3, we offered a novel framework encompassing the most important factors for decision-making over industrial CAS EEMs (Section 4). The framework, which includes the specific EEMs description, broadens the effects of their implementation beyond energy and economic considerations, offering a genuine and innovative contribution to the academic discussion over the impacts of EEMs on industrial operations. Further, the proposed framework also aims to effectively contribute to supporting decision-makers and policymakers in fostering the adoption of EEMs in CAS, as well as technology and service providers in tailoring their services. A validation and preliminary application of the framework was conducted in several manufacturing enterprises (Sections 5 and 6, respectively), giving valuable insights and opening further research avenues (Section 7).

2. EEMs in CAS: An Overview

Overall, CAS are usually characterized by reduced energy efficiency [10,30]. However, CAS energy efficiency can be improved through well-known EEMs, in terms of technologies and practices available in the market. Understanding the characteristics of CAS EEMs is of primary importance to shed light on the factors driving their adoption and foster their implementation.

A valuable source for the analysis of EEMs in CAS is represented by the US DOE Industrial Assessment Center (IAC) [31], which identified 16 EEMs labelled with an Assessment Recommendation Code (ARC). Such EEMs, as noted by previous literature [7,26,32], represent a broad range of activities to improve the energy efficiency of CAS, including (as summarized in Table 1):

- installation of new equipment (e.g., ARC 2,4226 "Use/purchase optimum sized compressors", 2,4224 "Upgrade control compressors", 2,4225 "Install common header on compressors");
- optimization of existing equipment (e.g., ARC 2,4231 "Reduce the pressure of compressed air to the minimum required", 2,4235 "Remove or close off unneeded compressed air lines");
- recovery of extant working conditions (e.g., ARC 2,4236 "Eliminate leaks in inert gas and compressed air lines/valves");
- replacement of compressed air medium (e.g., ARC 2,4232 "Eliminate or reduce the compressed air used for cooling, agitating liquids, moving products or drying", 2,4233 "Eliminate permanently the use of compressed air");
- energy recovery (e.g., ARC 2,2434 from either compressors or ARC 2,2435 from air dryers).

Moreover, efficiency in CAS may be reached following three directions: preventing energy losses, minimizing energy input, and recovering energy [33]. The IAC database covers the first two areas, however, the latter is partially lacking since the database only refers to the recovery of thermal energy. Hence, to cover the gap, an additional EEM related to the adoption of energy harvesting units was added to Table 1.

With respect to other literature addressing EEMs in CAS (e.g., Nehler [11]), the IAC has been preferred, given that Nehler [11] has clustered EEMs according to their physical local location to recognize the effect on the system and their interrelations, however leading to a significant overlapping, since multiple EEMs seem to target the same energy efficiency issue. Rather, IAC classification allows assessing EEMs with an industrial decision-maker perspective. In fact, as reported in Table 1, the implementation of those EEMs should consider several additional operational issues (e.g., accessibility, location, noise) and impacts on other production resources (e.g., labor through an impact on maintenance activities and/or safety) that are important for industrial decision-makers and other literature, industrial and scientific. Interestingly, the existence of such implications seems to show the need for academic literature to more thoroughly and systematically address the factors that should be considered when adopting an EEM in CAS.

Table 1. Industrial Assessment Center (IAC) classification of EEMs in CAS.

ARC Code	EEMs	Type of EEM	Description	Important Characteristics for the Adoption	References
2,4221	Install compressor air intakes in the coolest location	Installation of new equipment	Aspiring from the coolest location [34], may they be outside [35] or inside the plant [36], could provide multiple benefits, ranging from efficiency up to the regulation range, passing by avoidance of shutdowns, according to the type of compressor installed [37,38].	• The location may be difficult to access with a consequent negative impact on maintenance practices [37,38]; • continuous air monitoring required (external installation) [36]; • the installation of an additional ventilation system may be required (internal installation) [36].	[34–43]
2,4222	Install adequate dryers on air lines to eliminate blowdown	Installation of new equipment	Applications of compressed air or wear requirements of the components need a certain level of air dryness [44], usually guaranteed by refrigerated dryers, coupled with a moisture separator and condensate traps.	• Cycling or noncycling refrigerated dryers are usually adopted, characterized by different implementation and operation costs [45–47]; • periodic maintenance required [30].	[30,42,44–52]
2,4224	Upgrade controls on compressors	Installation of new equipment	The control system ensures high efficiency by matching the supplied compressed air to meet the demand, ensuring that the minimum required pressure is maintained. Control can be achieved for a single unit or the entire system to optimize the operations [29].	• Different control systems exist, with the optimal one depending on the specific application (e.g., see [29,45,53]); • a reduction in the required number of compressors may be achieved through a central control system [29]; • if a monitoring system is installed with the central control system, benefits in terms of maintenance and unscheduled downtimes may be obtained.	[29,34,39,42,44,45, 47–59]

Table 1. *Cont.*

ARC Code	EEMs	Type of EEM	Description	Important Characteristics for the Adoption	References
2,4225	Install common header on compressors	Installation of new equipment	The closed-loop configuration represents the best air distribution system layout, saving up to 12% of power requirements [42,48,57]. Moreover, the installation of a common header enables compressors to work together, taking advantage of load sharing.	• Higher bore improves air storage capacity, which enables operations with a higher output of compressors and avoidance of unexpected switching on or off [42,52]; • installation must be performed by CA experts [60]; • there may be accessibility issues.	[30,42,44,48–52,57, 60,61]
2,4226	Use/purchase optimum sized compressors	Installation of new equipment	Use a compressor able to handle the demand of the system at any time with efficient operation, since oversizing is one of the major problems in the supply side of compressed air systems [48].	• High efficiency units must be preferred [11,50,57,62,63]; • noise may be reduced; • space requirements may be reduced; • installation must be performed by CA experts.	[11,29,34,35,39,42, 45,48–50,55,57,62– 64]
2,4227	Use compressor air filter	Installation of new equipment	A filtering system may be necessary to provide air of the right quality, designed considering (i) extraction efficiency, (ii) air flow rate, and (iii) dust capacity.	• Noise may be reduced; • useful life of compressors may be increased, and unplanned downtimes reduced; • filters should be inspected and replaced regularly [51,65]; • simple installation and maintenance practices.	[39,42,44,45,47,48, 50,51,57,65–69]
2,4231	Reduce the pressure of compressed air to the minimum required	Optimization of existing equipment	Pressure should be minimized according to the requirements of end-users [30,51,70], proceeding then backward in the identification of losses [29,71].	• The number of working compressors [68] may be reduced; • end-use pressure should be reached avoiding losses rather than increasing the generated pressure [42,72,73].	[29,30,34,40,42,47, 51,55,58,69–74]
2,4232	Eliminate or reduce the compressed air used for cooling, agitating liquids, moving products or drying	Replacement of compressed air medium	CA is a simple and readily available form of energy, but it is often used inappropriately; many operations in a plant, such as agitating liquids, moving product, aspirating, atomizing, padding, can be accomplished more economically through alternative technologies [29].	• The alternatives to CA are vast, ranging from blowers to air amplification high-performance nozzles [29,39,75], each of them characterized by different features; • blowers, for instance, require more space but are easy to implement [76] and are much more efficient for high volume low-pressure applications [50,77].	[29,30,39,40,45,47– 51,70,75–79]
2,4233	Eliminate permanently the use of compressed air	Replacement of compressed air medium	When the wrong use of CA is discovered, it should be converted to other types of equipment (e.g., electric driven equipment for vacuum pump [29,79])	• Specific characteristics depend on the alternative solution chosen.	[29,30,39,40,45,47– 51,70,75–79]
2,4234	Cool compressor air intake with heat exchanger	Installation of new equipment	Lowering the inlet temperature may provide multiple benefits to CAS (see ARC 2,4221). Beside moving the compressor air intake, it is possible to obtain a cooling effect of inlet air using a heat exchanger [37,38].	• Heat exchangers are easier to install with respect to a change in the compressor air intake, but they require more space [37,38].	[37,38,57]

Table 1. *Cont.*

ARC Code	EEMs	Type of EEM	Description	Important Characteristics for the Adoption	References
2,4235	Remove or close off unneeded compressed air lines	Optimization of existing equipment	Compressed air lines should be removed in case of permanent disuse or temporarily closed, e.g., through shut-off valves, when they remain idle for a certain time during the production cycle [50,80,81].	• The disconnection may reduce noise, enhance safety, and save space once occupied by the equipment itself [29,82]; • there may be issues in the accessibility of pipes with consequent hidden costs.	[29,42,50,80–82]
2,4236	Eliminate leaks in inert gas and compressed air lines/valves	Recovery of extant working conditions	Leaks are the major single sources of consumption in compressed air systems [35,70]. They can be reduced following operational good practices [49,83] and performing maintenance activities, beside introducing a leak management program [29].	• Leak reduction may enhance the equipment lifetime reducing the pressure of operating time [29]; • experienced personnel are required to design and carry out the activity [84]; • accessibility may represent an issue.	[29,30,35,39,42,44, 47,49,56,57,59,70, 83–85]
2,4237	Substitute compressed air cooling with water or air cooling	Replacement of compressed air medium	Cooling air at the compressor outlet enables the blowdown collection and the avoidance of heat exchangers in the points of use; different cooling system exists, with the optimal fit depending on the specific case (e.g., see [86,87]).	• Maintenance, operating costs, and installation costs depend on the specific choice; • water usage costs and water waste management costs should be considered when dealing with a cooling system where water is the main medium [88]; • noise may be reduced after the replacement [88].	[39,86–89]
2,4238	Do not use compressed air for personal cooling	Replacement of compressed air medium	Personnel cooling describes the self-application, made by operators, of compressed air for ventilation purposes. An efficient and secure alternative is provided by electrical fans [29].	• Enhance personnel safety since the flow of compressed air can inject particles into the human skin [29].	[29]
2,2434	Recover heat from air compressor	Energy recovery	Up to 93% of the electrical energy used by an industrial air compressor is converted into heat, which can be mostly recovered with a properly designed heat recovery unit [27,42,90]	• Maintenance efforts are higher due to the requirements of the added equipment [39,91]; • equipment lifetime may be improved [36].	[27,29,34,36,39,40, 42,44,45,51,57,90, 91]
2,2435	Recover heat from compressed air dryers	Energy recovery	As for air compressors, heat can be recovered from dryers. This intervention is one of the most convenient concerning energy efficiency, since the source of energy is often waste [34].	• Maintenance efforts are higher due to the requirements of the added equipment [39,91]; • equipment lifetime may be improved [36].	[27,29,34,36,39,40, 42,44,45,51,57,90, 91]
/	Install energy harvesting units	Energy recovery	Energy can be recovered from the wasted pressurized air when discharged in the environment or from the presence of moving masses (kinetic energy) [92,93]. It can be transformed into electricity [93] or directly used to power other devices [94].	• Maintenance requirements increase [92]; • energy-saving circuits might be difficult to implement and might affect system performance [33,95].	[33,92–98]

3. Literature Review, Critiques, and Needs

Section 2 highlighted several EEMs characteristics helpful to identify technical and operative factors that should be assessed when dealing with the adoption of EEMs in CAS. Similarly, assessment factors have been discussed by previous academic literature. A breakthrough contribution is represented by the study by Fleiter et al. [26], who developed a framework based on 12 factors grouped into three categories, namely relative advantage, technical context, and information context. Interestingly, the factors considered refer to the profitability side of the EEMs, but point also toward their complexity, with thus some links to research by Rogers focused on the adoption of innovation into industry [99]. The relative advantage and the complexity indeed represent the only factors, among the ones considered by Rogers [99], which are statistically related to the adoption of interventions, together with the compatibility of an innovation [100], considered however as a "rather broad and subjective characteristic that is heavily dependent on the potential adopter", thus neglected in the analysis by Fleiter et al. [26]. Roberts and Ball [101], referring more generally to sustainability practices (thus with a broader focus than energy efficiency), encompassed most of the aforementioned considerations, defining a framework that also pointed out the importance of including the time dimension in the analysis, which was not included by Fleiter et al. [26]. Similarly, factors for the characterization of EEM were considered by Trianni et al. [7], who maintained the profitability dimension but also the description of the complexity of an EEMs, as suggested by Fleiter et al. [26], through factors such as the activity type, the ease of implementation, and the likelihood of success/acceptance. Noteworthy, both Roberts and Ball [101] and Trianni et al. [7] made a further step preliminarily suggesting to include among the assessment factors also the nonenergy benefits (NEBs), i.e., all the benefits coming from the adoption of an EEM beyond the energy savings, as defined by Mills and Rosenfeld [102], but not explicitly.

However, NEBs represent the positive impacts that EEMs have on the operations and the other production resources. They were considered mainly as additional benefits to stimulate the implementation of industrial energy efficiency, since their value may exceed that of the energy savings [7,103]. However, recent research has pointed out that there may be also negative implications stemming from the adoption (e.g., [103,104]), which should likewise be included in the assessment also as a necessary acknowledgement to gain credibility with the industrial sector [102]. In a nutshell, regardless of being positive or negative, NEBs describe impacts stemming from the EEMs adoption and, as such, they should be assessed during the decision-making process to make a sound decision.

Literature identified NEBs stemming from the adoption of a variety of technologies and EEMs, referring them to a set of categories according to their nature and targeted area (e.g., relative advantage, technical context, information context [26]; complexity, compatibility, observability [99,100]; waste, emission, operation and maintenance, production, working environment, and other [105,106]). In this regard, Table 2 shows the most significant contributions (NEBs encompassed by literature are indicated with an "X"; the green background helps to graphically highlight the areas most frequently covered by the past studies). Unfortunately, the majority of literature over NEBs does look to specific technologies not including CAS (e.g., [107–109]), or considers CAS together with other technologies [11]. To the best of our knowledge, only very few studies were conducted targeting CAS specifically. Gordon et al. [49] first attempted to analyze NEBs referring to CAS exclusively, listing a variety of NEBs, ranging from maintenance and insurance and labor costs to improved system performance and workers' safety conditions. More recently, Nehler et al. [27] highlighted a simple list of 34 specific NEBs for CAS, ranked according to their importance as perceived by users and experts, with the top positions occupied by organizational related factors (e.g., commitment from top management; people with real ambition), energy-related factors (cost-reductions resulting from lowered energy use; energy management system; the threat of rising energy prices), and strategic factors (long-term energy strategy). Doyle and Cosgrove [110] further delved into this issue by identifying the benefits stemming from one EEM, i.e., compressed air leaks repair, in terms of reduction of the required working units and the consequent drop in the plant room temperature, which in turn improve the efficiency of CAS. Interestingly, Table 2 shows that, despite referring specifically to CAS, these studies consider about the same NEBs already defined by Worrell et al. [105]. The only exception is represented by the improvements in system performance, which address improved pressure levels, consistency of pressure, and the ability to address spikes in usage [49], which are indeed specific of the technology. On the other hand, if many manuals deal with CAS technology (e.g., [29,39,111]) they refer solely to technical aspects, such as the impact on parameters like pressure or temperature, which are critical for the adoption of the technology, nonetheless representing a limited perspective, not even naming the wider concepts of assessment factor nor NEBs.

Table 2. Factors used in literature to describe EEMs.

Categories	Factors	CAS (Specific)						CAS (Among Other Technologies)								Other Technologies/Innovations			
		[49]	[110]	[27]	[105]	[112]	[106]	[7]	[107]	[113]	[108]	[109]	[114]	[26]	[115]	[100]	[99]	[102]	[116]
Relative advantage/economic	IRR													X					
	Pay-back	X						X				X		X		X (relative advantage)	X (relative advantage)		
	Increased sales										X								
	Initial expenditure/implementation cost	X						X						X					
	Interest cost on capital investment	X																	
Technical context	Distance to core process							X						X					
	Type of modification													X					
	Scope of impact													X					
	Lifetime (of the measure)													X					
	Complexity													X		X	X		
	Compatibility															X	X		
Informational context	Transaction cost													X					
	Knowledge for planning and implementation													X					
	Diffusion progress													X					
	Sectoral applicability													X					
	Trialability															X	X		
	Observability															X	X		
	Communicability															X			
	Divisibility															X			
	Social approval															X			
Implementation related	Saving strategy							X											
	Activity type							X											
	Ease of implementation							X											
	Likelihood of success/acceptance							X											
	Corporate involvement							X											
	Check-up frequency							X											
Waste	Use of waste fuel, heat, gas								X										
	Reduced product waste			X	X	X	X				X		X		X				
	Reduced wastewater			X	X	X	X	X (waste)	X				X		X				
	Reduced hazardous waste (and hazardous water)			X	X	X	X		X						X			X (waste)	
	Waste disposal cost										X	X							
	Material reduction (raw material)			X	X	X			X		X		X		X				

Table 2. *Cont.*

Categories	Factors	CAS (Specific)						CAS (Among Other Technologies)									Other Technologies/Innovations		
		[49]	[110]	[27]	[105]	[112]	[106]	[7]	[107]	[113]	[108]	[109]	[114]	[26]	[115]	[100]	[99]	[102]	[116]
Emissions	Reduced dust emission (ashes)			X	X	X	X			X									
	Reduced CO, CO2, NOX, SOX emissions			X	X	X	X	X (emission)	X (emission)	X (emission)		X							X (emission)
	Reduced cost of environmental compliance (fines included)											X							
Operations and maintenance	Reduced need for engineering control				X	X													
	Better control/improved process control												X					X	
	Lowered cooling requirements			X	X	X	X		X			X			X				
	Increased facility reliability				X	X													
	Reduced wear and tear on equipment				X	X						X							
	Increased lifetime			X					X	X	X		X		X				
	Reduced labor requirements (cost, savings)	X		X	X	X	X	X (operations and maintenance)		X			X		X			X	
	Reduced maintenance (maintenance cost)	X		X			X		X	X			X		X				
	Reduced water consumption						X												
	Lower cost of treatment chemicals						X												
	Reduced operating time									X									
	Reduced purchases of ancillary materials						X												
	Reduced nonenergy operational cost											X	X						
	Improved ease of system operations	X																	
	Productivity							X	X	X	X								
Production	Increased product output/yield			X	X	X	X		X			X	X	X	X				
	Improved equipment performance	X			X	X													
	Shorter process cycle time			X	X	X	X												
	Improved product quality/decrease scrap			X	X	X	X		X	X	X	X			X				X (production)
	Increased system capacity											X							
	Reduced cost of production disruption								X										
	Increased reliability in production	X		X	X	X	X			X					X				

Table 2. *Cont.*

Categories	Factors	CAS (Specific)						CAS (Among Other Technologies)									Other Technologies/Innovations		
		[49]	[110]	[27]	[105]	[112]	[106]	[7]	[107]	[113]	[108]	[109]	[114]	[26]	[115]	[100]	[99]	[102]	[116]
Working environment	Reduced need for PPE/increased safety/reduced illness or injuries	X		X	X	X	X			X	X		X						
	Decreased personnel needs										X								
	Improved lighting				X	X							X						
	Improved noise level				X	X	X	X (working environment)	X (working environment)	X			X					X (working environment)	X (working environment)
	Improved temperature control/reduced temperature		X	X	X	X						X	X						
	Better aesthetics												X						
	Comfort												X						
	Reduced glare, eyestrain												X						
	Improved air quality			X	X	X	X						X		X				
Other	Decreased liability				X	X													
	Improved public image			X	X	X			X						X				X
	Delay/reduce capital expenditure	X		X	X	X	X								X				
	Achieved rebate/incentives						X												
	Reduced/eliminated demand charges						X												
	Reduced/eliminated rental equipment cost						X												
	Additional space				X	X													
	Improved workers morale (satisfaction)			X	X	X			X			X			X				
	Direct and indirect economic benefits (downsizing)																	X	
	Reduced currency risk								X										
	Reduced number of devices required	X	X																
	Reduced insurance cost from fewer compressors	X																	

By analyzing the literature, and in particular the area surrounded by the red line in Table 2, the main literary gap is clearly represented by the lack of study encompassing for the entire range of factors that should be considered by decision-makers during the assessment of EEMs, especially when dealing with CAS. Referring to a single technology is necessary since different technologies require different EEMs, which might provide different NEBs [27] and be characterized by different assessment factors. Moreover, without this specificity, the work might lose the practical interest by decision-makers because it is too general to describe the broadest set of possible industrial contexts where to consider the adoption of EEMs on CAS. Furthermore, it is clear how most studies dealing with assessment factors on CAS, regardless from the addressed technology, do not address the context in which the technology is called to operate, therefore missing a (potentially) crucial element for a complete decision-making. Moreover, it should be noted that most studies are focused on NEBs from the service phase of the equipment, whilst both the drawbacks stemming from the adoption and the implementation phase itself of the EEM have been rarely considered in the analysis [117].

4. A Novel Framework of Factors for Decision-Making Over CAS EEMs

The framework, designed to provide a holistic perspective for decision-making purposes, has been created by tailoring factors and the broader categories to the specific features of CAS EEMs. The factors, which should be relevant to the adoption of EEMs and, if possible, should avoid overlaps, derive from either a thorough review of the industrial literature about the technology behind single EEMs (Table 1) or from the scientific literature on EEMs characteristic. This dual perspective guarantees the completeness of the analysis, being therefore inclusive of the impacts on the operations and the other productive resources of a company. This completeness was maintained during the following synthesis process, which made it possible to obtain a synthetic framework thanks to the grouping of factors into categories and subcategories. Furthermore, the grouping process was carried out in such a way that the framework obtained corresponds to the perspective adopted by decision-makers regarding the adoption of EEMs to CAS. As summarized in Table 3, 22 factors were identified and organized in three categories, respectively: (i) operative factors, (ii) economic-energetic factors, and (iii) contextual factors, which in turn were divided into three further subcategories, i.e., (i) complexity, (ii) compatibility, and (iii) observability.

4.1. Operational Factors

The need for compressed air is primarily defined by end-users' requirements in terms of:

- *air flow rate* [29];
- *pressure level* [29];
- *air temperature* [39].

The CAS performance and efficiency do not rely exclusively on such primary factors. Yet, primary factors are strictly interconnected to several secondary ones: among these, we can find heat and thermal capacity, linked to the air temperature, power, work, but also volume, density and mass flow rate of air, directly connected to its pressure and flow rate.

4.2. Economic and Energetic Factors

Pay-back time. Pay-back time has been widely recognized as an easy yet indicative factor supporting industrial decision-makers with limited resources [7,118].

Initial expenditure. Regardless of the type of investment [119], the initial expenditure is a crucial factor and may represent a major hurdle hindering EEMs adoption, especially among SMEs, due to their limited capital availability [120,121].

Energy savings. The amount of saved energy is a critical indicator of savings stemming from the adoption of an EEM [7] and it refers to monetary quantification of the physical energy source (either primary or secondary).

4.3. Contextual Factors

Other than considering operative and economic-energetic factors, CAS EEMs can be characterized by many factors strongly dependent on the specific industrial context for which they are considered. We took inspiration from the study conducted by Rogers [99] who broadly reviewed the characteristics of innovation in general. Since the adoption of EEMs into a specific context can represent a process innovation, those characteristics were transferred and adapted to CAS EEMs, as detailed in the following.

4.3.1. Complexity Factor

Complexity describes the difficulty one might encounter when adopting an EEM, inversely proportional to the adoption rate of the measure itself [99]. Understanding in which cases the adoption is revealed to be complex is a fundamental passage to characterize it. Literature on innovation refers to the radicalness as an index of complexity, since it is correlated to the degree of change required for the adopters [122]. This is a rather vague definition for the specific study and a potential source of misunderstanding [26,123]. Hence, we decomposed the complexity into factors whose definitions are specifically intended for the analysis of EEMs.

Activity type distinguishes if an EEM constitutes a simple refurbishment or recovery of the existing functions, an optimization in the use of an existing technology, a retrofitting of the equipment or a new energy-efficient equipment installation [7]. Indeed, a simple retrofit is easier than a new investment in equipment [124].

Expertise required refers to the range of skills required for the correct implementation of an EEM. Since different levels of expertise are required for each EEM and considering their variety, the skill range can be wide enough to be hard for firms in finding technology experts, especially for SMEs, where CAS is used almost exclusively as a service [125].

Independency from other components/EEMs refers to the influence of the implementation of an EEM on the existing system, to underline the nature of the impact [26,100,126]. The possible impacts can influence CAS equipment working conditions, other systems or can generate cause–effect relationships with other EEMs, with the magnitude of the influence being inversely proportional to the easiness of understanding the consequences of the installation and predicting the total savings.

Change in maintenance effort. The variation of maintenance requirements as a consequence of the adoption of EEMs has been often considered an important factor by previous literature [102,105,106].

Accessibility. Difficulties in accessing equipment may require higher efforts from personnel or a greater amount of technological resources to carry out operations; this can be even harder for CAS, in which the distribution system is usually difficult to access. Moreover, accessibility may also refer to space unavailability for maintenance procedures when technology add-on measures are installed.

4.3.2. Compatibility Factors

Compatibility explains to which degree EEMs can be adapted to the existing system. According to Rogers [99], it can be referred, among others, to the compatibility with previously introduced ideas, that can be translated into technological compatibility, as suggested by Tornatzky and Klein [100], or to layout features or operating conditions that difficultly fits in the existing system. Nonetheless, despite being relevant for the adoption, compatibility and related factors have not been adequately considered in EEMs literature, being strongly dependent on the adopters' contextual characteristics [26].

Technological compatibility analyzes the technological constraints related to EEMs, pointing out the conditions where their implementation is suggested or should be avoided, highlighting a strict connection to the specific context. Indeed, in several cases, more technologies concur for the adoption of the specific EEM, and the best choice depends on their matching with the existing system, as well as their suitability [127]. Without technological compatibility, the EEMs expected performance may not be guaranteed, with also possible lack of trust for future interventions [128].

Presence of difference pressure loads outlines the existence of different pressure levels at the end-use which may be a source of high inefficiencies and incompatibilities in the system [129]. This may be due to (i) the widespread availability of lamination valves that, although can be easily installed, are meant to disperse the pressure generated; (ii) the generation of a high-pressure point, which is recommended only when a considerable amount of air is required at that pressure.

Adaptability to different conditions may be referred to demand needs as well as to different ambient conditions, which can influence, e.g., the air conditions at the compressor intake (e.g., see IAC ARC 2,4221). It represents a critical factor considering the flexibility of use usually required for CAS [29].

Synergy with other activities. During the EEM implementation, synergies among different EEMs may occur, leading to potential benefits coming from the coordination of multiple activities (e.g., similar interventions that are suggested contemporarily, taking advantage of the same downtime of the equipment [130]). Nonetheless, synergies may also be negative for EEMs adoption [131].

Distance to the electric service. The distance of the point of use to the electric service can be a reason for the low adoption rates of EEMs requiring the technology substitution from compressed air-driven to electric driven devices [132].

Presence of thermal loads. The quality level of the fluid delivered by the heat exchangers from heat recovery units represents the major problems for the low diffusion of this solution throughout CAS. Although the EEM can be theoretically installed for each compressor type (both packaged or not), [29,36], its profitability depends on the fluid quantity and temperature. If the compressor load is variable, heat may be delivered discontinuously in time, potentially representing an issue for the end-use application [36].

4.3.3. Observability Factors

Observability, when referred to innovations, relates to their visibility and the communicability of their effects to others [99]. Concerning CAS EEMs, observability can be translated into focus towards the sensible changes detected in both the CAS and the working environment once the EEM is implemented.

Safety. Since difficulties may arise when handling compressed air for high fluid pressure and high-speed rotating parts, safety requirements are tight, aiming at reducing the accident rates [133].

Air quality. Pollution in an indoor environment is one of the more underestimated problems within a production facility. Paying attention to air quality monitoring and improvement is on the one hand related to enhanced health and performance of operators [106,113]; on the other hand, to improved operating conditions for all the parts in contact with the fluid, thanks to lower values of solid and liquid contaminants.

Wear and tear variation of the equipment is widely considered in scientific literature, mostly with a positive meaning [105]. The same factor can be perceived, in turn, as influencing the lifetime of the equipment [103,113]. For the specific case of CAS, a reduction of wear and tear of the equipment may be obtained because of the lower stress impressed by the fluid, attained with the reduction of pressure or through enhanced control capabilities.

Noise coming from the equipment may affect the working environment and possibly the performance of the operators [102,103,105]. Nonetheless, the quantification of noise variation stemming from the implementation of a CAS EEM can be extremely difficult, being related to several parameters such as e.g., cost of absenteeism, accidents, and variation in workers productivity, that are extremely complex and with impacts measurable almost exclusively in the long-term.

Artificial demand. Air flow demand increases at higher pressure, especially when air is open blown to the atmosphere; hence, the sizing of the system based on the maximum pressure creates an over-pressurization that minimizes efficiency [134]. This further demand, defined as artificial demand, is considered one of the major causes of inefficiencies in compressed air systems. On the other hand, each time an EEM entails a reduction of the CAS pressure level or the reduction of its unregulated use, this affects positively the amount of air being delivered, representing a further benefit of the adoption.

Table 3. Categories, subcategories, and factors of the new framework.

Categories	Subcategories	Factors	References
Operational factors		Pressure	[29,49]
		Temperature	[39]
		Flow rate	[29]
Economic-energetic factors		Pay-back time	[7,26,118]
		Initial expenditure	[7,26,119–121]
		Energy savings	[7,135]
Contextual factors	Complexity	Activity type	[7,26,124,136]
		Expertise required	
		Independency from other components/EEMs	[26,100,107,126]
		Change in maintenance effort	[102,105,106,137]
		Accessibility	/
	Compatibility	Technological	[99,127,128]
		Presence of different pressure load	[129]
		Adaptability to different conditions	[29]
		Synergy with other activities	[130,131]
		Distance to the electric service	[132]
		Presence of thermal load	[36]
	Observability	Safety	[102,103,105,106,133]
		Air quality	[103,105–107,113]
		Wear and tear	[103,105,106,113]
		Noise	[72,102,103,105]
		Artificial demand	[29,134]

5. Validation of the Framework

The validation of the model, intended to reach the analytical generalization as defined by Yin [138], is performed following two separate steps: theoretical and empirical. The theoretical validation is based on the assessment of the factors that compose the model and their capacity to describe the selected EEMs through the analysis of literature contributions, both scientific and industrial, as discussed in Section 5.1. On the other hand, the empirical validation, structured according to the case study methodology following Yin [138] and Voss et al. [139], is required to validate with industrial decision-makers the framework and its composing elements, basing the analysis on a set of predetermined indicators (Section 5.2). For the purpose of the present study, i.e., understanding the main factors that rule the adoption rate of EEMs in CAS and their influence on the decision-making process, multiple case study is the most appropriate research methodology. Discrete experiments that serve as replications, contrasts, and extension to the emerging theory [138] are considered so that each of the case-studies gives a contribution to the theory development beside emphasizing the rich real-world context in which the phenomena will occur [140]. The combined approach for validation, successfully undertaken by previous research on similar topics ([7,141]), provides better generalizability of results, avoiding relying uniquely on the data obtained from a limited number of investigations.

5.1. Theoretical Validation

The theoretical validation is used (i) to verify the ability of the developed framework in characterizing the EEMs addressing CAS and (ii) to provide a qualitative evaluation of factors, which could result in interesting insights for decision-makers. The process involves a revision of the EEMs highlighted in Section 2 and it is accomplished thanks to a thorough review of the literature performed following the perspective imposed by the factors considered in the model. The results of the theoretical validation are reported in Table 4. In a nutshell, the framework proved to be able to fully describe EEMs in CAS, also supported by the inclusion of a qualitative evaluation of interventions, intended however to provide general guidelines rather than absolute and specific insights.

Table 4. Theoretical validation of the framework.

Description	Ref. Description	Economic Energetic Factors			Operative Factors			Contextual Factors — Complexity						Compatibility					Observability					Ref. Factors
		Investment Cost [a]	Payback Time [b]	Energy Savings [a]	Pressure [c]	Temperature [c]	Fluid flow Rate [c]	Activity Type [d]	Expertise Required [e]	Independency from Other Components/EEMs [f]	Change in Maintenance Effort [g]	Accessibility [h]	Technological [i]	Presence of Different Pressure Loads [j]	Adaptability to Different Conditions [k]	Synergy with Other Activities [l]	Distance from Electric Service [m]	Presence of Different Thermal Loads [j]	Safety [n]	Noise [n]	Air Quality [n]	Wear and Tear [n]	Artificial Demand [n]	
Install compressor air intakes in coolest location (ARC 2.4221)	[34–43]	L	S	L		X	X	R	H/L	L	N/A	I	H	0	H	H (N/A)	0	0	0	+	I	+	0	[7,37,72,133,142–144]
Install adequate dryers on air lines to eliminate blowdown (ARC 2.4222)	[30,42,44–52]	H	M	N/A		X	X	R	M	L	T	N/A	H	0	0	H; -D	-	0	0	-	+	+	0	[7,29,72,142,145–148]
Upgrade controls on compressors (ARC 2.4224)	[29,34,39,42,44,45,47–59]	M	S/M	M/H	X	X	X	R/N	H/L	L	T	N/A	H	0	+	H (N/A)	0	0	-	+	0	+	+	[7,29,72,133,142,149]
Install common header on compressors (ARC 2.4225)	[30,42,44,48–52,57,60,61]	H	M	N/A	X		X	N	M	L	+	-	H	-	+	H; -D	0	0	0	0	0	+	0	[7,29,129,142,150]
Use/purchase optimum sized compressor (ARC 2.4226)	[11,29,34,35,39,42,45,48–50,55,57,62–64]	H	M/L	M/H	X	X	X	N	H	L	+	I	H	H	+	H; (+D)	0	0	0	I	0	I	0	[7,29,64,72,129,133,134,142,151]
Use compressor air filter (ARC 2.4227)	[6,9,11,12,14,16,18,19,26,35–39]	L	S	L	X			O	L	H	-	I	L	0	0	M (N/A)	0	0	+	+	+	+	0	[7,133,142,143]
Reduce the pressure of compressed air to the minimum required (ARC 2.4231)	[29,30,34,40,42,47,51,55,58,69–74]	L	S	L	X		X	O	M	L +	-	-	0	H	0	H (N/A)	0	0	+	+	I	+	+	[7,29,72,142,150,152]

Table 4. *Cont.*

Description	Ref. Description	Economic Energetic Factors			Operative Factors				Contextual Factors — Complexity					Contextual Factors — Compatibility					Contextual Factors — Observability					Ref. Factors
		Investment Cost [a]	Payback Time [b]	Energy Savings [a]	Pressure [c]	Temperature [c]	Fluid flow Rate [c]	Activity Type [d]	Expertise Required [e]	Independency from Other Components/EEMs [f]	Change in Maintenance Effort [g]	Accessibility [h]	Technological [i]	Presence of Different Pressure Loads [j]	Adaptability to Different Conditions [k]	Synergy with Other Activities [l]	Distance from Electric Service [m]	Presence of Different Thermal Loads [j]	Safety [n]	Noise [n]	Air Quality [n]	Wear and Tear [n]	Artificial Demand [n]	
Eliminate or reduce the compressed air used for cooling, agitating liquids, moving products, or drying (ARC 2.4232)	[29,30,39,40,45,47–51,70,75–79]	M	S	H	X		X	O	H	L	T	I	H	T	-	L; -D	-	0	+	+	I	+	+	[7,29,142,153–155]
Eliminate permanently the use of compressed air (ARC 2.4233)	[29,30,39,40,45,47–51,70,75–79]	M	S	H			X	O	L	H	0	-	0	0	0	H (N/A)	0	0	0	+	0	0	+	[7,29,142,152,155]
Cool compressor air intake with heat exchanger (ARC 2.4234)	[37,38,57]	M	M	M		X	X	N	L	H	N/A	-	0	0	0	H; +M	0	0	0	0	0	0	0	[7,133,142–144,156–158]
Remove or close off unneeded compressed air line (ARC 2.4235)	[29,42,50,80–82]	M	S	N/A				O	L	H	+	-	0	0	0	H (N/A)	0	0	-	0	0	+	+	[7,29,142,152,155]
Eliminate leaks in inert gas and compressed air lines/valves (ARC 2.4236)	[29,30,35,39,42,44,47,49,56,57,59,70,83–85]	L	S	H	X		X	Rec	M	L	-	-	0	0	L	H; -D	0	0	+	+	0	+	+	[7,29,72,142]
Substitute compressed air cooling with water or air cooling (ARC 2.4237)	[39,86–89]	M	S	N/A		X	X	N	M	L	-	-	M	0	0	H; +M	0	0	0	I	0	0	0	[7,87,89,133,142,155,157–160]

Table 4. Cont.

Description	Ref. Description	Economic Energetic Factors			Operative Factors			Contextual Factors — Complexity						Contextual Factors — Compatibility					Contextual Factors — Observability					Ref. Factors
		Investment Cost [a]	Payback Time [b]	Energy Savings [a]	Pressure [c]	Temperature [c]	Fluid flow Rate [c]	Activity Type [d]	Expertise Required [e]	Independency from Other Components/EEMs [f]	Change in Maintenance Effort [g]	Accessibility [h]	Technological [i]	Presence of Different Pressure Loads [j]	Adaptability to Different Conditions [k]	Synergy with Other Activities [l]	Distance from Electric Service [m]	Presence of Different Thermal Loads [j]	Safety [n]	Noise [n]	Air Quality [n]	Wear and Tear [n]	Artificial Demand [n]	
Do not use compressed air for personal cooling (ARC 2,4238)	[29]	M	S	N/A			X	O	L	M	0	-	0	0	0	H (N/A)	-	0	+	-	+	0	+	[7,29,133,142]
Recover heat from air compressor (ARC 2,2434)	[27,29,34,36,39,40,42,44,45, 51,57,90,91]	M	M	H		X	X	R	H	L	-	-	M	+	-	M (N/A)	0	+	0	0	+	+	0	[7,29,36,39,72,89,142]
Recover heat from compressed air dryers (ARC 2,2435)	[27,29,34,36,39,40,42,44,45, 51,57,90,91]	M	M	H		X	X	R	H	L	-	-	M	+	-	M (N/A)	0	+	0	0	0	+	0	[7,29,36,39,72,89,142]
Install energy harvesting units	[33,92–98]	N/A	M	M			X	R	H	L/M -	+	-	H	T	-	N/A	0	0	0	0	0	0	0	[33,92–98]

[a] Low (L) if less than \$2.000; medium (M) if between \$2.000 and \$10.000; high (H) if higher than \$10.000; not available (N/A). [b] Short (S) if less than 1 year; medium (M) if between 1 and 2 years; long (L) if more than 2 years. [c] Impacted by the adoption of the EEM: (X); not impacted: ( ). [d] Retrofit (R); new installation (N); optimization (O); procedure of recovery (Rec). [e] Low (L) if the presence of maintenance personnel is enough; Medium (M) if engineering is required; High (H) if the support of a technology expert is needed. [f] Magnitude: Low (L); medium (M); high (H). Orientation: positive (+); negative (-). [g] Changes with technological change (T); maintenance effort is decreased (+); maintenance effort is increased (-); the factor is not influent for the EEM (0); not available (N/A). [h] Accessibility problems negatively influence the EEM adoption (-); the EEM may change accessibility to some point, but the influence is negative or positive depending on the context (I); not available (N/A). [i] High (H); medium (M); low (L); not influencing the EEM (0). [j] If the condition is verified, the factor may: highly influence the adoption (H); have a little influence (L); not influence at all (0). The influence may also depend on the technology (T). Orientation: the influence positively (+) or negatively (-) affects the adoption. [k] The factor may: highly influence the adoption (H); have a little influence (L); not influence at all (0). Orientation: the influence positively (+; ++) or negatively (-) affects the adoption. [l] Possibility of installation with other EEMs: low (L), medium (M), high (H). Orientation: positive (+) or negative (-) influence on the synergy with similar maintenance activities (M) or required shutdown of equipment (D); not available (N/A). [m] The factor may positively (+) or negatively (-) influence the adoption or may be not influencing at all (0). [n] The factor may positively (+) or negatively (-) influence the adoption. In some cases, the factor is influencing but not in a precise direction (I) or it does not influence at all (0).

5.2. Empirical Validation

We sampled firms across several sectors, limiting the analysis to SMEs, as discussed in the introduction [161]. In this exploratory phase, different industrial sectors are considered, since the usage of CA may vary according to the application, as well as its energy intensity. Five companies embodying the previously stated criteria were considered for the empirical validation (details provided in Table 5).

Table 5. Heterogeneity of the sample for the framework empirical validation.

Company	Sector	Dimensions (Employees)	Turnover [M€]	Energy Intensity (EI/NEI) [a]	Role of the Interviewee
V1	Plastic and packaging	150 ÷ 199	≤20	EI	Site manager
V2	Test and inspection of electric/ mechanical components	10 ÷ 49	≤2	EI	Maintenance responsible
V3	Machine design and construction	100 ÷ 149	≤10	EI	Quality and energy responsible
V4	Tires regeneration	10 ÷ 49	≤10	EI	Quality and energy responsible
V5	Food and beverage	100 ÷ 149	≤50	NEI	Quality and energy responsible

[a] The threshold between energy intensive and non-energy intensive companies is defined by the value of energy costs compared to the total turnover; in the present study such value is set at 2% [162].

The interviews followed a semi-structured format [156], to give higher flexibility and customization, being able to encompass a broader set of situations. In each case study, in the first part we collected various information regarding company profile, including sector, size, energy intensity and turnover, the role of the interviewees—ranging from the owner to the maintenance or energy manager—and their status and main responsibilities in the decision-making process over the adoption of CAS EEMs. Moreover, the perceived importance of energy and energy efficiency were investigated, together with the past EEMs implemented. Additionally, the CAS was analyzed to understand the applications and purposes of compressed air usage.

In the second part of the interview, respondents evaluated the proposed set of factors based on four performances, i.e., completeness, usefulness, clearness, and absence of overlapping, exploiting an even Likert scale from 1 (poor) to 4 (excellent) to avoid any neutral output. In particular, the validation process was divided into two separate steps: first, the foundations of the framework were assessed, i.e., its general structure, scope and perspective, as well as categories, subcategories, and factors considered as clusters in their own (top-level analysis). Second, the analysis delved into the investigation of the single elements of the framework, i.e., categories, subcategories, and factors (bottom-level analysis). The dual step process was designed to provide the interviewee with the general picture and only later moving into details, to avoid losing his attention releasing too much information in a single instance. The indicators used for the evaluation are displayed in Table 6, with detailed scores for the five companies reported in Appendix A.

The overall evaluation is extremely positive for each indicator, with no changes in the framework suggested:

- *usefulness:* the framework can provide useful insights to industrial decision-makers when dealing with the adoption of EEMs in CAS;
- *completeness:* all the critical factors are identified, especially those which are usually neglected due to a lack of awareness or specific knowledge about the technology;

- *clearness:* the factors are clearly defined and easy to understand for industrial decision-makers;
- *absence of overlapping:* the framework does not contain any unnecessary repetition.

Table 6. Parameters for the framework validation.

Framework		Completeness	Usefulness	Clearness	Absence of Overlapping
	Structure	X		X	
	Scope		X	X	
	Perspective		X		
Categories		X (cluster)	X	X	X
Subcategories		X (cluster)	X	X	X
Factors		X (cluster)	X	X	X

The importance of pointing out all the consequences stemming from the adoption is moreover stressed by the interviewee of company V4, suggesting that technology providers should also use the framework to highlight the consequences when proposing CAS EEMs. On the other hand, as noted by company V5, such increased knowledge might empower industrial decision-makers, since he recognized that usually service providers lean on a greater set of competences, thus limiting the company to implement suggested EEM, rather than proposing EEMs by themselves.

6. Application of the Model

Multiple case-study with semistructured interviews was selected as research methodology also for the empirical application of the framework into a second sample composed by 11 companies, sampled with the same rationale previously presented in Section 5 (details in Table 7). In order to apply the framework and test its effectiveness, considering the sample heterogeneity, we focused our analysis on the most recommended interventions, by considering the IAC database as reference (Table 8). Considering the timeline of the companies, EEMs are divided into:

(i)   past EEMs when recommended and backed up by an investment plan but never implemented;
(ii)   present EEMs if recommended and adopted, so the companies experienced the result; and
(iii) future EEMs if not yet recommended or only recently recommended, with no decision about their implementation undertaken.

In Box 1, we reported the application of the framework to a selected company (A5). In the following, we present the results of the application, displayed in Table 9. By looking at the implementation of the proposed framework, it appears clear how the operational factors are always considered during the assessment, with the only exception represented by the temperature, neglected in the assessment conducted by company A1 for the adoption of a controller, which nonetheless did not compromise the result. Referring to the economic-energetic factors, decision-makers stated how important they are for the correct assessment of EEMs, hence are usually the major set of factors considered in the decision-making process.

Nevertheless, the contextual factors pointed out on multiple occasions their capability to highlight critical features whose absence may change the adoption outcome. Particularly, the type of activity, providing information regarding the complexity of an EEM, was considered of primary importance in all the assessments, pointing out the huge perceived differences between the different nature of EEMs. The installation of a new device, or even a retrofit entailing the addition of new equipment, was indeed perceived as a complex operation by A1, which installed control systems and considered the movement of the compressors air intakes in a cooler place, or even by A5, which considered the replacement of the transportation system based on compressed air. On the other hand, completely different perceptions came from the companies which considered an optimization, e.g., companies A3, A6, A8, and A10, where the EEM relates to the repair of leaks. A2 stated how the type of activity was an important factor in his assessment, since the EEM, i.e., the reduction of the pressure level to the minimum required, is a simple optimization which does not imply any structural change in the system, hence requiring only a low level of involvement.

Table 7. Heterogeneity of the sample for the framework application.

Company	Sector	Dimensions (Employees)	Turnover (M€)	Energy Intensity (EI/NEI)	Role of the Interviewee
A1	Plastic and packaging	150 ÷ 199	≤20	EI	Site manager
A2	Test and inspection of electric/mechanical components	10 ÷ 49	≤2	EI	Maintenance responsible
A3	Machine design and construction	100 ÷ 149	≤10	EI	Quality and energy responsible
A4	Tires regeneration	10 ÷ 49	≤10	EI	Quality and energy responsible
A5	Food and beverage	100 ÷ 149	≤50	NEI	Quality and energy responsible
A6	Thermoforming of plastic and PVC materials	10 ÷ 49	≤10	N/A	Quality and energy responsible
A7	Microelectronic components	100 ÷ 149	≤20	EI	Site manager
A8	Plastic manufacture, thermoplastic, and plastic welding	10 ÷ 49	≤2	EI	Owner/site manager
A9	Manufacture and distribution of paints	10 ÷ 49	≤20	NEI	Site manager
A10	Food and beverage	10 ÷ 49	≤10	EI	Owner/site manager
A11	Food and beverage	10 ÷ 49	≤20	N/A	Site manager

Similarly, the expertise required to carry out the adoption is assessed as one of the main factors to be taken into consideration by decision-makers, especially for complex EEMs or in case of lack of knowledge, e.g., for the EEM considered by A9, which would imply the elimination of the compressed air used for dense phase transport but would be completely outsourced because of lack of internal competences. The expertise required guides A2 on the choice of simply consulting the compressor technical manual or contacting a technology expert for the adoption of the planned EEM. Further, in the case of A7, one of the main reasons for not adopting the EEM was the high expertise required, similarly to A5.

The application of the framework is intended to test its ability to work as an assessment tool. Decision-makers are required to indicate the importance factors have in the adoption process, ranging between 'not important' and 'very important'. Eventually, the relevance in using the framework for the decision-making process and the greater awareness gained from it are asked to the respondents, together with the effort required for its usage and its ease of application.

Table 8. Synoptic of the most recommended EEMs [142] that will be analyzed for the framework application.

ARC Code	Measure	Recommended	% Implementation
2,4236	Eliminate leaks in inert gas and compressed air lines/vales	8138	80.38
2,4221	Install compressor air intakes in coolest locations	5129	46.5
2,4231	Reduce the pressure of compressed air to the minimum required	4446	49.6
2,2434	Recover heat from air compressor	1626	31.86
2,4232	Eliminate or reduce compressed air used for cooling, agitating liquids, moving products, or drying	1450	46
2,4226	Use/purchase optimum sized compressor	692	42.92
2,4224	Upgrade controls on compressors	639	44.6

The independence from other components or EEMs was highly appreciated by the decision-maker of company A5, who was indeed worried about the high involvement of the transportation system in the production processes. Although the same EEM was considered by A9, the decision-maker was at first unaware about the importance of the factor. Rather, he was aware of the high dependency for what concerns the other EEM adopted by the company, i.e., the installation of control systems (two in the specific case), as he recognized how one may influence the proper working of the other. Regarding the repair of leaks in the compressed air lines, the advantage coming from the increased pressure level, which may end up with the reduction of the number of required compressors, was known to A3, A8, and A10. Differently, A6 was sceptic about this potential influence, thus neglected the factor from the analysis and ended up not adopting the EEM. Similarly, the dependency of the considered EEM was not known by A2, which did not take into account the potential risks related to the reduction of the pressure level for other activities to be performed through the same medium. Likewise, the decision-makers within A1 disregarded to resize the air receivers and the possible installation of the central control for the dryers. In both cases, the assessment resulted in the underestimation of the negative sides of the EEMs which could compromise their adoption.

The variation in maintenance effort is considered by almost all the respondents but it was perceived as critical only when the effort would be increased because of the leaks repair activity, i.e., by A3 and A8, which were considering the EEM for the future. Differently, A10, which performs the same EEM regularly, evaluated the effort as manageable.

The accessibility of CAS was widely considered since some companies had issues in the past. A10, e.g., assessed the accessibility as the most critical factor when dealing with the repair of leaks, together with A6 and A8, since parts of their compressed air lines can either be hard to reach or inaccessible (underground). The criticality of the factor was also pointed out by company A9, where the transport system to be replaced is integrated into the process lines, and A4 and A7.

Moving to the compatibility subcategory, technological compatibility was considered a critical factor by many companies. The choice of the controller, for instance, was strictly constrained by the type of compressor installed, as highlighted by A1 and A9. Technological compatibility was also rated as very important by A2, dealing with the reduction of pressure level of the CAS, since the variation in performance depends on the type of compressor. Eventually, A5 and A9 pointed out how the elimination of compressed air from the transportation system is an EEM which cannot be always applied because of technological constraints.

Box 1. Application of the framework to company A5.

Company profile:

- Company A5 is a medium size company, with 105 employees and about €50 million of annual turnover, part of a multinational corporation operating in the food and beverage sector.
- They are specialized in the production and distribution of canned sea food, with six production lines present in the plant. CA is used in the production lines for cleaning activities on the cans, for cutting fish, for the packaging system, and to drive the transportation lines.

Energy profile:

- Energy consumption is around 1% of the total turnover, which makes it a non-energy intensive company [1]. About 15% of the total energy consumption is related to compressed air, with a total power installed of 162 KW, distributed along four compressors located in two separate compressors rooms.
- Company A5 is not certified with ISO 50001.
- The last energy audit was performed in 2016.

Interviewee profile:

- The interviewee is the site manager, who is moreover in charge of the energy management inside the plant.
- The decision-making process is performed by the site manager together with his team, composed of four people. They are also responsible for maintaining the correct conditions, aligned with the indications coming from the installed performance measurement system, during the execution of the production and service processes.

EEM profile:

- Company A5 considered the replacement of CA used for the transportation system for cans and aluminum tubes along the production line with a motor driven vacuum system, aiming at enhancing the performance getting rid of a dated technology.
- The EEM belongs to the past cluster since company A5 eventually did not perform the substitution. The reason lies in the high investment cost and the required shutdown of the entire line which would have meant production disruption, thus losses, since they are continuously operating 24 h per day.

Application of the framework:

Operational factors	**Pressure**	The requirements to be satisfied in terms of pressure were considered by the decision-maker.
	Temperature	Temperature was not perceived as a very influencing factor for the replacement of the CA-based transportation system.
	Flow rate	Together with pressure, the flow rate requirement was considered during the decision-making process, being of paramount importance for the operation of the system.
Economic-energetic factors	**Pay-back time**	The importance of the factor was high, although the decision-maker was more susceptible to costs rather than to the extent of the pay-back period.
	Initial expenditure	The high investment cost required for the EEM, together with the losses due to the stop of production which would have been necessary to perform the substitution of the transportation system, were the main reasons that led to the nonadoption decision.
	Energy savings	Energy savings represent an important factor for the adoption of the EEM, with the decision-makers pointing out the possibility to enhance the energetic performance of the system by replacing a dated technology.

Box 1. *Cont.*

Contextual factors	Complexity	Activity type	The EEM is a new installation.
		Expertise required	The installation of the EEM requires the involvement of experts in the substitution process, negatively affecting the decision according to the decision-maker.
		Independency from other components/EEMs	Considering the pervasive involvement of the transportation system for the proper operation of the production line, the decision-maker pointed out a high dependency for the EEM.
		Change in maintenance effort	No main changes were pointed out by the decision-maker with respect to maintenance efforts.
		Accessibility	For the specific location of the CAS and the transportation system in company A5, the accessibility is not a big issue.
	Compatibility	Technological	The measure cannot be applied on all systems; hence the technological compatibility is a very important factor according to the decision-maker.
		Presence of different pressure loads	Generally, the presence of different pressure loads should usually favor the adoption of the vacuum pumps; however, for the specific situation of company A5, pressure loads differences were almost negligible, reducing the weight of the factor.
		Adaptability to different conditions	The capacity of the EEM to adapt to different operating conditions does not influence the adoption for the specific case of company A5 since a single vacuum pressure level is required.
		Synergy with other activities	Through the exploitation of synergies the installation can be performed when the line is down, taking advantage of a planned production stop; this factor is critical, since for no reason the replacement of the actual transportation system would have been performed in a different time slot, with the risk of influencing and stopping the normal activities.
		Distance from the electric service	For the specific situation of company A5 the factor is not critical due to the installation of the compressors in two rooms, close to the electric service.
		Presence of thermal load	No thermal loads are present for the specific application.
	Observability	Safety	The factor is not highly influential for the adoption of the specific EEM according to the decision-maker.
		Air quality	The variation in the quality of air was not perceived as a very important factor by the decision-maker.
		Wear and tear	The variation in wear and tear of the equipment does not represent a critical factor for the adoption of the specific EEM.
		Noise	The interviewee proved to be almost unaware of the potential improvement in noise level and assigned a low weight to the factor.
		Artificial demand	The factor is not critical for this EEM according to the decision-maker.

Eventually, the framework proved to be able to outline factors not known to the engineer-ing of company A5, although it should be noted that none of the negative ones had been under-estimated. In turn, more aware of the positive consequences of the adoption, the decision-maker could go back to his steps in case of a new stoppage of the line. He admitted that, despite the massive usage of compressed air and its energy consumption, they are not completely aware of the measure which could fit in their context. For this reason, he considered the developed tool as extremely tailored for their case. Moreover, the user-friendliness and the ease of use were positively rated.

Table 9. Assessment of the factors from the application of the model.

Company	EEM	EEM Status	Pressure	Temperature	Flow Rate	Pay-Back Time	Initial Expenditure	Energy Saving	Activity Type	Expertise Required	Independency from Other Components/EEMs	Change in Maintenance Effort	Accessibility	Technological	Presence of Different Pressure Load	Adaptability to Different Conditions	Synergy with Other Activities	Distance to the Electric Service	Presence of Thermal Load	Safety	Air Quality	Wear and Tear	Noise	Artificial Demand
			Operational Factors			Economic-Energetic Factors					Complexity				Compatibility					Observability				
F.1	Install compressor air intakes in coolest location (ARC 2,4221)	past	✓	✓	✓	✓	✓	✓✓	✓	✓						(!!)	✓				(!)			
F.1	Upgrade controls on compressors (ARC 2,4224)	present	✓	(!)	✓	✓	✓	✓✓	✓	✓	(!)			✓		✓✓	✓							✓
F.2	Reduce the pressure of compressed air to the minimum required (ARC 2,4231)	future	✓		✓	✓	✓	✓	✓	✓	(!!)			✓✓								(!)		(!)
F.3	Use/purchase optimum sized compressors (ARC 2,4226)	past	✓		✓	✓	✓✓	✓✓	✓✓	✓	✓				✓									
F.3	Eliminate leaks in inert gas and compressed air lines/valves (ARC 2,4236)	future	✓	✓	✓	✓	✓	✓✓	✓	✓	✓	✓✓	✓	✓✓						✓				✓
F.4	Install compressor air intakes in coolest location (ARC 2,4221)	future	✓	✓	✓	✓✓	✓✓	✓✓	✓✓	✓				✓							(!)			
F.5	Eliminate or reduce the compressed air used for cooling, agitating liquids, moving products, or drying (ARC 2,4232)	past	✓		✓	✓	✓✓	✓✓	✓	✓	✓			✓✓			✓✓							
F.6	Eliminate leaks in inert gas and compressed air lines/valves (ARC 2,4236)	past	✓		✓	✓✓	✓✓	✓✓	✓	✓	(!)		✓✓							✓				(!)
F.7	Recover heat from air compressor (ARC 2,2434)	past		✓	✓	✓✓	✓✓	✓✓	✓	✓✓			✓✓		(!)	✓		✓✓						
F.8	Eliminate leaks in inert gas and compressed air lines/valves (ARC 2,4236)	future	✓		✓	✓✓	✓✓	✓✓	✓	✓	✓	✓✓	✓						✓✓					(!)

Table 9. *Cont.*

| Code | Factors | Categories of Factors | Operational Factors | | | Economic-Energetic Factors | | | Contextual Factors: Complexity | | | | | | Compatibility | | | | | Observability | | | | |
	Factors	Categories of Factors	Pressure	Temperature	Flow Rate	Pay-Back Time	Initial Expenditure	Energy Saving	Activity Type	Expertise Required	Independency from Other Components/EEMs	Change in Maintenance Effort	Accessibility	Technological	Presence of Different Pressure Load	Adaptability to Different Conditions	Synergy with Other Activities	Distance to the Electric Service	Presence of Thermal Load	Safety	Air Quality	Wear and Tear	Noise	Artificial Demand
A9	Upgrade controls on compressors (ARC 2,4224)	present	✓	✓	✓	✓✓	✓✓	✓✓	✓	✓	✓			✓✓		✓✓	✓✓							✓
A9	Eliminate or reduce the compressed air used for cooling, agitating liquids, moving products, or drying (ARC 2,4232)	future	✓		✓	✓✓	✓✓	✓✓	✓	✓✓	(!)		✓	✓			✓✓							
A10	Eliminate leaks in inert gas and compressed air lines/valves (ARC 2,4236)	present	✓		✓			✓✓	✓		✓	✓✓								✓			✓	✓
A10	Install compressor air intakes in coolest location (ARC 2,4221)	future	✓	✓	✓	✓	✓	✓✓	✓	✓							✓✓							
A11	Use/purchase optimum sized compressors (ARC 2,4226)	present	✓		✓	✓✓	✓✓	✓✓	✓✓	✓✓	✓				(!)							✓		

Factors considered as important (✓) and very important (✓✓) by decision-makers and literature; factors important (!) and very important (!!) that should have been considered by decision-makers according to literature, but which were not considered in the decision to adopt EEMs.

The presence of different pressure loads was considered of utmost importance by A3 when dealing with the correct sizing of compressors, since it may influence the decision regarding the number of devices required. However, for the same EEM, A11 did not perceive the criticality of the factor, despite the effective presence of different pressure levels in their lines. The explanation should be researched in the number of pressure reducers installed in the system. Eventually, if the factor had been properly considered, the company would have probably opted for a different and more efficient configuration. Similarly, in A2 the factor was not considered, despite the influence the pressure level has on the heat recovery potential.

The adaptability to different conditions was considered as the most important factor by A1 and A9, both dealing with the adoption of controllers on compressors, which were indeed installed with the specific purpose of changing the operating conditions of the equipment when needed. The factor was, however, underestimated by A1 regarding the assessment of the second EEM, i.e., the displacement of the compressors air intakes in the coolest location, because of a lack of awareness, and this was one of the main reasons hindering the adoption. Moreover, as stated by the decision-maker of company A7, the adaptability to different conditions, related to the variability of requirements in the demand side, is a very important factor when considering the recovery of heat from the compressors.

It should be assessed, however, together with the factor describing the presence of thermal loads, which refers to the availability of the right amount of heat to match the demand side. These are the most important factors to be considered when dealing with that type of EEM according to A7.

The possibility to take advantage of synergies to carry out the installation when the production line is down was considered as a very important point by both A5 and A9 when deciding about the replacement of the old air compressed transportation system with a more efficient technology. Otherwise, this would lead to an additional plant shutdown with related production losses, hence supporting the non-adoption of the EEM. The same factor was rated as critical for the adoption of controllers on compressors carried out by A1 and A9. In particular, the decision-maker of company A1 pointed out that the activity requires a long time to be performed, thus it was done during the summertime when the plant was closed. The synergy is also reported by A1 and A10 considering the displacement of the compressors air intakes in cooler locations.

Regarding the observability factors, all the respondents whose companies performed the repair of leaks in compressed air lines recognized the importance the activity has on the safety.

The air quality was generally not acknowledged as a critical factor, although other authors pointed out its relevance [29]. Companies A1 and A4 considered the displacement of the compressors air inlets from the external environment to the internal one, in a cooler location. Beside a difference in temperature however, the quality of the internal air is usually better: the moisture content is lower, and this may lower the wear of the compressors, extending their lifetime. Differently, for A10 there would be no variation in the air quality but only in air temperature since the EEM would just imply to shift the air inlet indoor.

The variation of CAS wear and tear was considered by A11 in terms of the extended lifetime of the equipment embedded in the installation of the new and correctly sized compressor and, according to the respondents, was a very important factor. Differently, A9 was unaware of the factor when referring to the adoption of a controller, nor A2 when thinking about the reduction of pressure level, although in both cases they agreed on the importance this could have on the decision-making process.

Noise was considered critical by A10 to foster the repair of leaks. A3, A6, and A8, who assessed the same EEM, did not deem the factor important. However, they claimed to perform repair activities as soon as a noise is perceived to limit its effect on the surroundings.

The artificial demand was known and considered very influential only by A3 and A10, both dealing with the repair of leaks. For the same EEM, A6 and A8 did not perceive the criticality. Initially, the decision maker within A2 did not give much importance to the factor. However, he pointed out that the actual compressed air flow was higher than required because of a poorly sized compressor, and the artificial demand phenomenon was further increasing the gap between supply and demand. Therefore, the consideration of this factor could significantly increase the possibilities of a future adoption of the EEM. Moreover, the influence of the artificial demand also affects the adoption of controllers, as pointed out by the decision-makers of companies A1 and A9.

Overall, regardless of the nature of the EEM, i.e., past, present, or future, the framework proved to be able to provide additional information to industrial decision-makers. For instance, the respondent within A1 pointed out that the increased awareness resulting from the framework application would be probably enough to reconsider in the future the displacement of the compressors air intakes in the coolest location. Moreover, using the framework, the decision-maker of company A9 assessed an EEM he was not aware of. The framework resulted effective in A5 to highlight factors unknown to the decision-maker. However, none of the negatives were underestimated, and ultimately the decision not to adopt was due to the high investment costs and the production disruption to carry out the installation. Similarly, A7 acquired more insights from the framework, but the low amount of achievable savings drove the decision not to implement the considered EEM.

Furthermore, all the respondents particularly appreciated the ease of use of the framework and the low efforts required for its application, in particular for being able to completely define the EEMs encompassing only a limited number of factors.

7. Discussion

Comparing the result with the existing models, similarities can be found only regarding energetic and economic factors, since the most widespread and universally accepted indicators are utilized (e.g., pay-back time [26,112,163]) to evaluate the investment from an economic point of view, thus making the tool more user-friendly for the final adopters. On the other hand, differences can be found if considering operative factors, although technical information is widely covered by past literature [39]. The reason lies in the restricted focus of this work, i.e., CAS, being specific enough to enable the analysis of specific characteristics of the technology, which has been rarely investigated to this level of detail concerning characterizing factors. As confirmation of the previous statement, Nehler and Rasmussen [107] indicate that the characteristics of factors may depend on the type of EEMs, as already pointed out by Cagno and Trianni [22] referring to barriers to specific EEMs. Less detailed results come from a variety of studies considering compressed air through a multitechnology analysis [103,106], in many cases not even providing a clustering framework of factors [108,109,113]. Differently, more specific focus is provided by the study conducted by Nehler et al. [27], focused on CAS, which includes among the NEBs an improvement in temperature control, hence indicating the criticality of this factor. Moreover, considerations about pressure and flow rate are listed among the impacts perceived by suppliers concerning specific EEMs, as documented by a wealth of technical manuals and industrial literature extensively covering these aspects, despite neither categorizing the factors into an operative framework nor providing additional insights with respect to the mere technical ones. During the interviews conducted on field, these factors were highly appreciated by industrial decision-makers, given the practicality they confer to the tool; it would be indeed unfeasible to discuss the implementation of EEMs within CAS without taking into account such information. Other differences can be found analyzing those factors which introduce the contextual dimension, making the framework flexible enough to be exploited in all the different situations where the industrial decision-maker is required to operate. The first step toward this path was made by Rogers [99], followed by Tornatzky and Klein [100]; both the studies, however, treat compatibility referring to innovation, thus dealing with society in its entireness rather than a specific technology or field. Although the definition of the category can be adapted to the industrial environment, the details depicted by the single factors are here included for the first time. An exception is represented by the observability factors, i.e., safety, air quality, wear and tear, and noise, which are commonly considered in literature [105,106,164], sometimes clustered in a single element describing the whole working environment [7], given the strict relation with many EEMs, regardless of the technology considered. Industrial respondents were generally aware of such characteristics, despite the fact that they were never considered as the most critical elements leading the adoption of EEMs, with the exception being for A4; however, here compressed air belongs to the production process, which may act as a discriminant for the perceived importance of the role of compressed air. This is aligned with the perspective provided by

Nehler et al. [27], where the importance of NEBs as a driver for the decision-making process is evaluated: enhancements of the working environment and safety conditions are considered. However, they are perceived as of secondary importance with respect to other advantages, e.g., those directly connected to the reliability and lifetime of the equipment. One reason could be the difficulty of their evaluation and monetization, thus the impossibility to include these considerations in the economic assessment of any investment, which represents a critical step of the decision-making process [165]. Nevertheless, according to Nehler and Rasmussen [107] those characteristics that cannot be evaluated from a monetary perspective, may be considered alongside the proposal in the form of comments. Regarding the remaining operational factor, i.e., artificial demand, given the strict dependency with the specific CA technology, it cannot be found in frameworks related to a broader cluster, such as by Trianni et al. [7]. Nevertheless, it should be noted that almost all the interviewees were aware of this phenomenon, despite the technical nature and difficulty of observation make it hard to be recognized by users without deep expertise in CAS.

Apart from observability factors, the complexity ones are partly included in previous literature, despite being categorized differently (e.g., [26,105,126]). Activity type, for instance, is included by Trianni et al. [7], who confined the definition by Rogers [99] and Tornatzky and Klein [100] to a limited field, i.e., industry, to make it practically exploitable. On the other hand, the willingness to focus on more than a single technology prevented them from analyzing all single factors related to compressed air solely. Interestingly, the present framework specifically included for the first time the difficulty in accessing the distribution system (accessibility factor), despite being deemed as important by any decision-maker interviewed. Further, compatibility issues, except for synergies [131], represent a neglected dimension in scientific literature, despite the fact that they are widely recognized in technical manuals or industrial sources (e.g., [29,127,129]). Once more, since the framework is intended for a practical application into companies, these considerations should be encompassed in the decision-making process, as revealed from the investigation where decision-makers acknowledged that some important factors were not always taken into account. This capability was embedded in the design of the framework, thanks to its focus on the single technology of CAS.

The need of a more specific funneled knowledge over relevant factors for EEMs adoption is partially aligned with the specificity of the characteristics but also to the applicability property discussed by Fleiter et al. [26], provided that the efficiency interventions remain confined to CAS. On the other hand, as demonstrated by the different importance attributed to the observability factors during the interviews, the selected factors should not be independent of the context and the adopting company, as stated by [26], but should include the information; the category contextual factors is considered in the present study to fulfil this necessity. In this regard, future research could explore whether such interdependency could be modulated by the different relationships between CA and the core process of the firms. Relationships may also exist among the various factors included in the framework, which are not completely disconnected from each other, confirming the close interactions CAS have with the operations of a company. For instance, the repair of leakages (ARC 2,4236) would lead to a reduction in pressure requirements, which in turn would affect the noise level and the wear and tear of the equipment. Interestingly, preliminary results of the analysis (e.g., Table 4) may suggest that some relationships exist, although more research is needed to shed some light on this. Indeed, an in-depth study of the impacts between factors could make a further contribution to the discussion about impacts on the operations and the other productive resources of a company.

8. Conclusions

The willingness to understand the main factors that rule the adoption of EEMs on CAS represents the driver that pushed toward the definition of the present framework. Aiming at providing a systemic view of the adoption, factors referring to the complexity, compatibility, and observability of the results coming from the adoption of EEMs were included in the model, encompassing, among others, the impacts on the operations and the other productive resources of an industrial firm, together with more traditional considerations regarding the operational and the economic and energetic factors. Results from the empirical application show how these features might prove critical in the path for the adoption, sometimes even capable of reversing the outcome, hence confirming the added knowledge brought by the framework.

In this regard, future longitudinal research could explore the change of awareness in decision-makers when assessing EEMs in CAS and other sustainability practices within industrial operations. Moreover, the focus kept on the specific technology of CAS enabled to point out peculiar factors that might be lost approaching the problem through a more holistic perspective, e.g., difficulties in accessing CAS, which was a recurrent topic in the empirical investigations. Nonetheless, despite its non-negligible importance according to the interviewed decision-makers, the factor has never been approached by previous studies.

Using the framework, industrial decision-makers could tackle the perception of uncertainty they have concerning EEMs, beside finding valuable support to overcome the barriers related to risk, imperfect evaluation criteria, and lack of information, which might represent critical issues preventing a sound decision-making process. These barriers might be particularly present in SMEs, generally characterized by less trained or less skilled decision-makers, who may moreover face difficulties in the use of complex or overly detailed models. However, the structuring resulting from the synthesis process to which the framework was subjected made it possible to obtain a complete framework regarding the factors to be considered in the adoption of CAS EEMs, characterized at the same time by a high ease of use. Indeed, as pointed out by the empirical application, the evaluation of the user-friendliness and the effort required for the usage were overall positive, despite the fact that the greatest share of companies in the sample were SMEs. Policy makers, on the other hand, could take advantage of the framework to design tailored policies for enhancing the efficiency of CAS. Moreover, the assessment of the factors that rule the adoption of EEMs on CAS could lead to a deeper understanding of the specific barriers that affect the technology, which might move away with respect to the issue preventing the adoption of other technologies, assigned to different roles in a plant, e.g., electric motor systems. This deeper knowledge would, in turn, create solid foundations on which to lay the basis for the definition of drivers to overcome these barriers, improving the overall efficiency.

In conclusion, we would like to acknowledge some study limitations, starting from the narrowness of the application sample and its heterogeneity with respect to the industrial sectors. Besides, not all sectors are encompassed in the present study, e.g., textile or metal manufacturing are missing. Moreover, limiting the analysis to the technology of CAS did not enable to consider the entire set of impacts the adoption of an EEM has on the other productive resources or on the operations of a firm. Accordingly, future research could move towards this direction, furtherly extending the analysis to include a broader set of heterogeneous EEMs to better assess the impacts of their adoption. Additionally, further research could effectively develop approaches to measure such impacts more quantitatively, linking the impacts on production and operations performance. Furthermore, research could explore what synergies may be explored by integrating the developed framework into a broader set of tools to improve the sustainability performance of industrial enterprises, also connecting it with assessment tools, maturity models, etc.

Author Contributions: Conceptualization, D.A., A.T. and E.C.; methodology, E.C. and D.A.; validation, A.T., E.C. and D.A.; formal analysis, A.T., D.A. and E.C.; investigation, A.T., E.C. and D.A.; resources, A.T., D.A. and E.C.; data curation, D.A.; writing—original draft preparation, D.A., A.T. and E.C.; writing—review and editing, A.T., E.C. and D.A.; visualization, D.A.; supervision, A.T.; All authors have read and agreed to the published version of the manuscript.

Funding: This research received no external funding.

Conflicts of Interest: The authors declare no conflict of interest.

Appendix A

Table A1. Scores for the theoretical validation of the framework [a].

Analysis	Group	Item	Company V1				Company V2				Company V3				Company V4				Company V5			
			Usefulness	Completeness	Clearness	Absence of Overlapping	Usefulness	Completeness	Clearness	Absence of Overlapping	Usefulness	Completeness	Clearness	Absence of Overlapping	Usefulness	Completeness	Clearness	Absence of Overlapping	Usefulness	Completeness	Clearness	Absence of Overlapping
Top-level analysis	Framework	Structure	-	4	4	-	-	4	4	-	-	4	4	-	-	4	4	-	-	4	4	-
	Framework	Scope	4	-	4	-	4	-	4	-	3	-	4	-	4	-	4	-	4	-	4	-
	Framework	Perspective	4	-	-	-	4	-	-	-	4	-	-	-	3	-	-	-	4	-	-	-
	Categories		-	4	-	-	-	4	-	-	-	4	-	-	-	4	-	-	-	4	-	-
	Subcategories		-	4	-	-	-	4	-	-	-	4	-	-	-	4	-	-	-	4	-	-
	Factors		-	4	-	-	-	4	-	-	-	4	-	-	-	4	-	-	-	3	-	-
Bottom level analysis	Categories	Operational parameters	3	-	4	4	3	-	4	4	4	-	4	4	4	-	4	4	-	4	4	4
	Categories	Economic energetic parameters	4	-	4	4	4	-	4	4	4	-	4	4	4	-	4	4	-	4	4	4
	Categories	Contextual parameters	4	-	4	4	4	-	4	4	4	-	4	3	4	-	4	4	-	4	4	4
	Subcategories	Compatibility	4	-	4	4	4	-	4	4	4	-	4	4	4	-	4	4	-	4	4	4
	Subcategories	Complexity	4	-	4	4	4	-	4	4	4	-	4	4	4	-	4	4	-	4	4	4
	Subcategories	Observability	4	-	4	4	4	-	4	4	4	-	4	4	4	-	4	4	-	4	4	4
	Operational parameters	Pressure	4	-	4	4	4	-	4	4	4	-	4	4	4	-	4	4	-	4	4	4
	Operational parameters	Temperature	4	-	4	4	4	-	4	4	4	-	4	4	4	-	4	4	-	4	4	4
	Operational parameters	Flow rate	4	-	4	4	4	-	4	4	4	-	4	4	4	-	4	4	-	4	4	4
	Economic energetic parameters	Pay back time	4	-	4	4	4	-	4	4	4	-	4	4	4	-	4	4	-	4	4	4
	Economic energetic parameters	Initial expenditure	4	-	4	4	4	-	4	4	4	-	4	4	4	-	4	4	-	4	4	4
	Economic energetic parameters	Energy savings	4	-	4	4	4	-	4	4	4	-	4	4	4	-	4	4	-	4	4	4
	Complexity	Activity type	4	-	4	3	4	-	4	4	4	-	4	4	4	-	4	4	-	4	4	4
	Complexity	Expertise	4	-	4	4	4	-	4	4	4	-	4	4	4	-	4	4	-	4	4	4
	Complexity	Independency from other components/EEMs	4	-	4	4	4	-	4	4	4	-	4	3	4	-	4	4	-	4	4	4

Table A1. *Cont.*

		Company V1				Company V2				Company V3				Company V4				Company V5			
		Usefulness	Completeness	Clearness	Absence of Overlapping	Usefulness	Completeness	Clearness	Absence of Overlapping	Usefulness	Completeness	Clearness	Absence of Overlapping	Usefulness	Completeness	Clearness	Absence of Overlapping	Usefulness	Completeness	Clearness	Absence of Overlapping
Complexity	Change in maintenance effort	4	-	4	4	4	-	4	4	4	-	4	3	4	-	4	4	4	-	4	4
	Accessibility	4	-	4	4	4	-	4	4	4	-	4	4	4	-	4	4	4	-	4	4
Compatibility	Technological	3	-	4	4	4	-	4	4	4	-	4	4	4	-	4	4	4	-	4	4
	Presence of different pressure loads	4	-	4	4	4	-	4	4	4	-	4	4	4	-	4	4	4	-	4	4
	Adaptability to different conditions	4	-	4	4	4	-	4	4	4	-	4	4	4	-	4	4	4	-	4	4
	Synergy with other activities	4	-	3	3	4	-	4	4	4	-	4	3	4	-	4	4	4	-	4	4
	Distance to the electric service	4	-	4	4	3	-	4	4	4	-	4	4	4	-	4	4	4	-	4	4
	Presence of thermal loads	4	-	4	4	4	-	4	4	4	-	4	4	4	-	4	4	4	-	4	4
Observability	Safety	4	-	4	4	4	-	4	4	4	-	4	4	4	-	4	4	4	-	4	4
	Air quality	4	-	4	4	4	-	4	4	4	-	4	4	4	-	4	4	4	-	4	4
	Wear and tear	4	-	4	4	4	-	4	4	4	-	4	4	4	-	4	4	4	-	4	4
	Noise	4	-	4	4	4	-	4	4	4	-	4	4	4	-	4	4	4	-	4	4
	Artificial demand	4	-	3	4	4	-	4	4	4	-	4	4	4	-	3	4	4	-	4	4

(Row group label *Bottom level analysis* spans all rows.)

[a] The green background represents an excellent rating (4 on the Likert scale); the orange background represent a good rating (3 on the Likert scale); no mediocre or poor ratings are present.

References

1. International Energy Agency IEA. *2018 World Energy Outlook—Executive Summary*; OECD/IEA: Paris, France, 2018; p. 11.
2. U.S. Department of Energy. *International Energy Outlook 2019 with projections to 2050*; U.S. Department of Energy: Washington, DC, USA, 2019; p. 85.
3. Berardi, U.; Tronchin, L.; Manfren, M.; Nastasi, B. On the effects of variation of thermal conductivity in buildings in the Italian construction sector. *Energies* **2018**, *11*, 872. [CrossRef]
4. Medved, M.; Ristovic, I.; Roser, J.; Vulic, M. An overview of two years of continuous energy optimization at the Velenje coal mine. *Energies* **2012**, *5*, 2017–2029. [CrossRef]
5. International Energy Agency IEA. *Energy Policies of IEA Countries: Italy 2016 Review*; OECD/IEA: Paris, France, 2016; p. 214. [CrossRef]
6. ENEA. *Rapporto Annuale Efficienza Energetica. Analisi e Risultati delle Policy di Efficienza Energetica del nostro Paese*; ENEA: Rome, Italy, 2018; p. 292.
7. Trianni, A.; Cagno, E.; De Donatis, A. A framework to characterize energy efficiency measures. *Appl. Energy* **2014**, *118*, 207–220. [CrossRef]
8. Cagno, E.; Trianni, A. Analysis of the most effective energy efficiency opportunities in manufacturing primary metals, plastics, and textiles small- and medium-sized enterprises. *J. Energy Resour. Technol. Trans. ASME* **2012**, *134*, 1–9. [CrossRef]
9. Worrell, E.; Martin, N.; Price, L.; Ruth, M.; Elliott, N.; Shipley, A.M.; Thorn, J. Emerging energy-efficient technologies for industry. *Energy Eng.* **2002**, *99*, 36–55. [CrossRef]
10. European Commission. *Reference Document on Best Available Techniques for Energy Efficiency*; European Commission: Bruxelles, Belgium, 2009; p. 430.
11. Nehler, T. Linking energy efficiency measures in industrial compressed air systems with non-energy benefits—A review. *Renew. Sustain. Energy Rev.* **2018**, *89*, 72–87. [CrossRef]
12. Fleiter, T.; Fehrenbach, D.; Worrell, E.; Eichhammer, W. Energy efficiency in the German pulp and paper industry—A model-based assessment of saving potentials. *Energy* **2012**, *40*, 84–99. [CrossRef]
13. Russell, C.H. Energy management pathfinding: Understanding manufacturers' ability and desire to implement energy efficiency. *Strateg. Plan. Energy Environ* **2009**, *25*, 20–54. [CrossRef]
14. Wheeler, G.M.; Bessey, E.G.; McGill, R.D.; Vischer, K. Compressed air system audit software. In Proceedings of the ACEEE Summer Study on Energy Efficiency in Industry, Saratoga Springs, NY, USA, 8–11 July 1997; pp. 343–353.
15. International Energy Agency IEA. *Energy Efficiency 2018—Analysis and Outlooks to 2040*; Market Report Series; OECD/IEA: Paris, France, 2018; p. 174.
16. Weber, L. Some reflections on barriers to the efficient use of energy. *Energy Policy* **1997**, *25*, 833–835. [CrossRef]
17. Backlund, S.; Thollander, P.; Palm, J.; Ottosson, M. Extending the energy efficiency gap. *Energy Policy* **2012**, *51*, 392–396. [CrossRef]
18. Trianni, A.; Cagno, E.; Farnè, S. An empirical investigation of barriers, drivers and practices for energy efficiency in primary metals manufacturing SMEs. In Proceedings of the 6th International Conference on Applied Energy, Taipei, Taiwan, 30 May–2 June 2014; Volume 61, pp. 1252 1255. [CrossRef]
19. Cagno, E.; Trianni, A.; Spallina, G.; Marchesani, F. Drivers for energy efficiency and their effect on barriers: Empirical evidence from Italian manufacturing enterprises. *Energy Effic.* **2017**, *10*, 855–869. [CrossRef]
20. Trianni, A.; Cagno, E.; Worrell, E.; Pugliese, G. Empirical investigation of energy efficiency barriers in Italian manufacturing SMEs. *Energy* **2013**, *49*, 444–458. [CrossRef]
21. Johansson, I.; Mardan, N.; Cornelis, E.; Kimura, O.; Thollander, P. Designing policies and programmes for improved energy efficiency in industrial SMEs. *Energies* **2019**, *12*, 1338. [CrossRef]
22. Cagno, E.; Trianni, A. Evaluating the barriers to specific industrial energy efficiency measures: An exploratory study in small and medium-sized enterprises. *J. Clean. Prod.* **2014**, *82*, 70–83. [CrossRef]
23. Trianni, A.; Cagno, E.; Farné, S. Barriers, drivers and decision-making process for industrial energy efficiency: A broad study among manufacturing small and medium-sized enterprises. *Appl. Energy* **2016**, *162*, 1537–1551. [CrossRef]
24. Sorrell, S.; Mallett, A.; Nye, S. *Barriers to Industrial Energy Efficiency: A Literature Review*; United Nations Industrial Development Organization: Vienna, Austria, 2011; p. 98.

25. DeCanio, S.J. Barriers within firms to energy-efficient investments. *Energy Policy* **1993**, *21*, 906–914. [CrossRef]
26. Fleiter, T.; Hirzel, S.; Worrell, E. The characteristics of energy-efficiency measures—A neglected dimension. *Energy Policy* **2012**, *51*, 502–513. [CrossRef]
27. Nehler, T.; Parra, R.; Thollander, P. Implementation of energy efficiency measures in compressed air systems: Barriers, drivers and non-energy benefits. *Energy Effic.* **2018**, *11*, 1281–1302. [CrossRef]
28. Unger, M.; Radger, P. Energy Efficiency in Compressed Air Systems—A review of energy efficiency potentials, technological development, energy policy actions and future importance. In Proceedings of the 10th International Conference on Energy Efficiency in Motor Driven Systems, Rome, Italy, 6–8 September 2017; pp. 207–233.
29. U.S. Department of Energy. *Improving Compressed Air System Performance. A Sourcebook for Industry*; U.S. Department of Energy: Washington, DC, USA, 2003; p. 128.
30. Barbieri, D.; Jacobson, D. Operations and maintenance in the glass container industry. In Proceedings of the 3rd ACEEE Summer Study on Energy Efficiency in Industry: Industry and Innovation in the 21st Century, New York, NY, USA, 15–18 June 1999; pp. 655–665.
31. Muller, M.R.; Pasi, S.; Baber, K.; Landis, S. *Industrial Assessment Center—Assessment Recommendation Code*; U.S. Department of Energy: Washington, DC, USA, 2019; p. 44.
32. Cagno, E.; Trucco, P.; Trianni, A.; Sala, G. Quick-E-scan: A methodology for the energy scan of SMEs. *Energy* **2010**, *35*, 1916–1926. [CrossRef]
33. Harris, P.; Nolan, S.; O'Donnell, G.E. Energy optimisation of pneumatic actuator systems in manufacturing. *J. Clean. Prod.* **2014**, *72*, 35–45. [CrossRef]
34. Beyene, A. Energy efficiency and industrial classification. *Energy Eng.* **2005**, *102*, 59–80. [CrossRef]
35. Kaya, D.; Phelan, P.; Chau, D.; Sarac, H.I. Energy conservation in compressed-air systems. *Int. J. Energy Res.* **2002**, *26*, 837–849. [CrossRef]
36. Heat Recovery and Compressed Air Systems. Available online: https://www.airbestpractices.com/technology/air-compressors/heat-recovery-and-compressed-air-systems (accessed on 31 July 2020).
37. Inlet Air Temperature Impacts on Air Compressor Performance. Available online: https://www.airbestpractices.com/system-assessments/compressor-controls/inlet-air-temperature-impacts-air-compressor-performance (accessed on 31 July 2020).
38. Ming, H.C. Experimental Investigation of Inlet Air Temperature on Input Power in an Oil-Flooded Rotary Screw Air Compressor. Master's Thesis, The University of Alabama, Tuscaloosa, AL, USA, 2015.
39. Saidur, R.; Rahim, N.A.; Hasanuzzaman, M. A review on compressed-air energy use and energy savings. *Renew. Sustain. Energy Rev.* **2010**, *14*, 1135–1153. [CrossRef]
40. Saidur, R.; Mekhilef, S. Energy use, energy savings and emission analysis in the Malaysian rubber producing industries. *Appl. Energy* **2010**, *87*, 2746–2758. [CrossRef]
41. Zhang, Y.; Cai, M.; Kong, D. Overall energy efficiency of lubricant-injected rotary screw compressors and aftercoolers. In Proceedings of the Asia Pacific Power and Energy Engineering Conference, Wuhan, China, 28–31 March 2009; pp. 1–5. [CrossRef]
42. Dindorf, R. Estimating potential energy savings in compressed air systems. In Proceedings of the XIIIth International Scientific and Engineering Conference on Hermetic Sealing, Vibration Reliability and Ecological Safety of Pump and Compressor Machinery, Sumy, Ukraine, 6–9 September 2011; Volume 39, pp. 204–211. [CrossRef]
43. Alavi, S.; Alavi, S.; Gholami, S. Investigation of compression power in a three-stage centrifugal compressor by thermodynamic modeling and neural network modeling algorithm. *Indian J. Sci. Res.* **2014**, *1*, 866–879.
44. La, T. Don't Let Compressed Air Blow Away Your Profits. *Energy Eng.* **2013**, *111*, 7–15. [CrossRef]
45. Terrell, R.E. Improving compressed air system efficiency—Know what you really need. *Energy Eng.* **1999**, *96*, 7–15. [CrossRef]
46. Energy Savings Result from Compressed Air Dryer Selection. Available online: https://www.airbestpractices.com/technology/air-treatment/n2/energy-savings-result-compressed-air-dryer-selection (accessed on 31 July 2020).
47. Marshall, R.C. Optimization of Single-unit Compressed Air Systems. *Energy Eng.* **2012**, *109*, 10–35. [CrossRef]
48. Neale, J.R.; Kamp, P.J.J. Compressed air system best practice programmes: What needs to change to secure long-term energy savings for New Zealand? *Energy Policy* **2009**, *37*, 3400–3408. [CrossRef]

49. Gordon, F.; Peters, J.; Harris, J.; Scales, B. Why is the treasure still buried? Breaching the barriers to compressed air system efficiency. In Proceedings of the 3rd ACEEE Summer Study on Energy Efficiency in Industry: Industry and Innovation in the 21st Century, New York, NY, USA, 15–18 June 1999; pp. 709–718.
50. Scot Foss, R. Optimizing the compressed air system. *Energy Eng.* **2005**, *102*, 49–60. [CrossRef]
51. Yang, M. Air compressor efficiency in a Vietnamese enterprise. *Energy Policy* **2009**, *37*, 2327–2337. [CrossRef]
52. Anderson, K.J.; Collins, B.; Cortese, A.; Ekman, A. Summary of Results for Six Industrial Market Transformation Projects. Funded by the Northwest Energy Efficiency Alliance. In Proceedings of the ACEEE Summer Study on Energy Efficiency in Industry, New York, NY, USA, 24–27 July 2001; Volume 1, pp. 255–266.
53. Carbon Trust. *Variable Speed Drives. Introducing Energy Saving Opportunities for Business*; The Carbon Trust: London, UK, 2011; p. 27. Available online: https://www.academia.edu/15350420/Variable_speed_drives_Introducing_energy_saving_opportunities_for_business_Technology_guide (accessed on 23 September 2020).
54. De Almeida, A.T.; Ferreira, F.J.T.E.; Fonseca, P.; Chretien, B.; Falkner, H.; Reichert, J.C.C.; West, M.; Nielsen, S.B.; Both, D. *VSDs for Electric Motor Systems*; European Commission: Bruxelles, Belgium, 2000; p. 109. Available online: https://silo.tips/download/vsds-for-electric-motor-systems (accessed on 23 September 2020).
55. Barringer, F.L.; Woodbury, K.A.; Middleton, B. *Efficiency Improvements for Compressed Air Systems*; SAE International: Warrendale, PA, USA, 2012; p. 9. [CrossRef]
56. AlQdah, K.S. Prospects of energy savings in the national meat processing factory. *Int. J. Sustain. Energy* **2013**, *32*, 670–681. [CrossRef]
57. Galvão, J.; Moreira, L.; Leitão, S.; Silva, É.; Neto, M. Sustainable energy for plastic industry plant. In Proceedings of the 4th International Youth Conference on Energy, Siófok, Hungary, 6–8 June 2013. [CrossRef]
58. Widayati, E.; Nuzahar, H. Compressed Air System Optimization: Case Study Food Industry in Indonesia. In Proceedings of the International Conference on Engineering and Technology for Sustainable Development, Yogyakarta, Indonesia, 11–12 November 2015; Volume 105. [CrossRef]
59. Abdelaziz, E.A.; Saidur, R.; Mekhilef, S. A review on energy saving strategies in industrial sector. *Renew. Sustain. Energy Rev.* **2011**, *15*, 150–168. [CrossRef]
60. Šešlija, D.; Ignjatovi, I.; Dudi, S.; Lagod, B. Potential energy savings in compressed air systems in Serbia. *Afr. J. Bus. Manag.* **2011**, *5*, 5637–5645. [CrossRef]
61. Energy efficiency best practice program. In *Compressing Air Costs*, 1th ed.; Good Practice Guide N126; ETSU: Oxon, UK; March Consulting Group: Manchester, UK, 1997; p. 51. Available online: http://air-safe.org.uk/pdf/carbontrust/GPG126.pdf (accessed on 31 July 2020).
62. Capehart, L.B.; Turner, C.W.; Kennedy, J.W. *Guide to Energy Management*, 7th ed.; The Fairmont Press: Lilburn, GA, USA, 2012; p. 659. ISBN 0881736724.
63. Akbaba, M. Energy conservation by using energy efficient electric motors. *Appl. Energy* **1999**, *64*, 149–158. [CrossRef]
64. How to Size and Select an Air Compressor. Available online: https://www.quincycompressor.com/images/pdf/how%20to%20size%20and%20select%20an%20air%20compressor.pdf (accessed on 31 July 2020).
65. Rotary Screw Air Compressor Basics. Available online: https://www.plantengineering.com/articles/rotary-screw-air-compressor-basics/ (accessed on 31 July 2020).
66. Petracca, G. *Energy conversion and management. Principles and Applications*; Springer: Berlin/Heidelberg, Germany, 2016; p. 356.
67. Air Compressor Guide. Available online: https://www.air-compressor-guide.com/articles/the-air-compressor-inlet-filter-more-important-than-you-think/ (accessed on 23 September 2020).
68. U.S. Department of Energy. *DOE Assessment of the Market for Compressed Air Efficiency Services*; U.S. Department of Energy: Washington, DC, USA, 2001; p. 63.
69. CEA Technologies Inc. *(CEATI) Compressed Air. Energy Efficiency Reference Guide*; CEA Technologies Inc.: Fyshwick, ACT, Australia, 2007; p. 118.
70. D'Antonio, M.; Epstein, G.; Bergeron, P. Compressed air load reduction approaches and innovations. In Proceedings of the ACEEE Summer Study on Energy Efficiency in Industry, New York, NY, USA, 24–27 July 2001; Volume 2, pp. 317–327.
71. Technical Brief on Pressure Loss. Available online: https://cagi.org/news/assets/PressureLossTechnicalBrief.pdf (accessed on 31 July 2020).
72. ATLAS COPCO. *Compressed Air Manual*; ATLAS COPCO: Wilrijk, Belgium, 2015; p. 146. ISBN 9789081535809.

73. OEM Machine Design. Pneumatics: Sizing Demand Users. Available online: https://www.airbestpractices.com/system-assessments/end-uses/oem-machine-design (accessed on 31 July 2020).

74. Systems Approach Cuts Fabric Mill Energy Costs. Available online: https://www.airbestpractices.com/system-assessments/end-uses/systems-approach-cuts-fabric-mill-energy-costs (accessed on 31 July 2020).

75. Kroger Company Bakery Saves Energy. Available online: https://www.airbestpractices.com/technology/blowers/kroger-company-bakery-saves-energy (accessed on 31 July 2020).

76. Compressed Air versus Blowers—The Real Truth. Available online: http://2015pdfs.s3.amazonaws.com/Nexflow-Air/compressed_air_vs_blowers.pdf (accessed on 31 July 2020).

77. Food Industry Factory Saves $154,000 in Annual Energy Costs. Available online: https://www.airbestpractices.com/system-assessments/vacuum/blowers/food-industry-factory-saves-154000-annual-energy-costs (accessed on 31 July 2020).

78. Laouar, S.; Molodtsof, Y. Experimental characterization of the pressure drop in dense phase pneumatic transport at very low velocity. *Powder Technol.* **1998**, *95*, 165–173. [CrossRef]

79. Pump or Venturi? Available online: https://fluidpowerjournal.com/pump-or-venturi/ (accessed on 31 July 2020).

80. Seslija, D.; Milenkovic, I.M.; Dudic, S.P.; Sulc, J.I. Improving energy efficiency in compressed air systems—Practical experiences. *Therm. Sci.* **2016**, *20*, 355–370. [CrossRef]

81. Bottler Best Practices in California. Available online: https://www.airbestpractices.com/industries/plastics/bottler-best-practices-california (accessed on 31 July 2020).

82. U.S. Department of Energy. *Industrial Technologies Program. Eliminate Inappropriate Uses of Compressed Air*; DOE/GO-102004-1963; U.S. Department of Energy: Washington, DC, USA, 2004; p. 2. Available online: https://iac.tamu.edu/files/doe/2_-_Inappropriate_Uses.pdf (accessed on 23 September 2020).

83. Reduce Energy Costs in Compressed Air Systems by up to 60%. Available online: https://www.festo.com/net/SupportPortal/Files/300860/WhitePaper_EnergySavingServices_EN.pdf (accessed on 23 September 2020).

84. Are Compressed Air Leaks Worth Fixing? Available online: https://www.airbestpractices.com/system-assessments/leaks/are-compressed-air-leaks-worth-fixing (accessed on 23 September 2020).

85. U.S. Department of Energy. *Industrial Technologies Program Minimize Compressed Air Leaks*; DOE/GO-102004-1964; U.S. Department of Energy: Washington, DC, USA, 2004; p. 2. Available online: https://www.energy.gov/sites/prod/files/2014/05/f16/compressed_air3.pdf (accessed on 23 September 2020).

86. Specifying an Air Compressor Cooling System. Available online: https://www.airbestpractices.com/technology/cooling-systems/specifying-air-compressor-cooling-system (accessed on 31 July 2020).

87. Control Panel Cooling Change Saves Compressed Air Electrical Costs. Available online: https://www.airbestpractices.com/technology/cooling-systems/control-panel-cooling-change-saves-compressed-air-electrical-costs (accessed on 31 July 2020).

88. Calculating the Water Costs of Water-Cooled Air Compressors. Available online: https://www.airbestpractices.com/technology/air-compressors/calculating-water-costs-water-cooled-air-compressors-part-1 (accessed on 31 July 2020).

89. Turning Air Compressors into an Energy Source. Available online: https://www.kaeserknowhow.co.nz/turn-your-compressors-into-an-energy-source-with-heat-recovery/ (accessed on 24 September 2020).

90. Carbon Trust. *How to Recover Heat from a Compressed Air System*; The Carbon Trust: London, UK, 2018; p. 4. Available online: https://www.carbontrust.com/?id=GPG238 (accessed on 23 September 2020).

91. Recupero del Calore dall'Aria Compressa: Quanto Conviene. Available online: http://www.progettoenergiaefficiente.it/recupero-del-calore-dallaria-compressa-quanto-conviene/ (accessed on 31 July 2020).

92. Hepke, J.; Weber, J. Energy saving measures on pneumatic drive systems. In Proceedings of the 13th Scandinavian International Conference on Fluid Power, Linköping, Sweden, 3–5 June 2013; pp. 475–483. [CrossRef]

93. Gryboś, D.; Leszczyski, J.S. Implementation of Energy Harvesting System of Wastes of Compressed Air Wastes for Electrical Steel Cutting Line. In Proceedings of the E3S Web of Conference—Energy and fuel, Krakow, Poland, 19–21 September 2018; Volume 108. [CrossRef]

94. Cummins, J.J.; Nash, C.J.; Thomas, S.; Justice, A.; Mahadevan, S.; Adams, D.E.; Barth, E.J. Energy conservation in industrial pneumatics: A state model for predicting energetic savings using a novel pneumatic strain energy accumulator. *Appl. Energy* **2017**, *198*, 239–249. [CrossRef]

95. Ferraresi, C.; Franco, W.; Quaglia, G.; Scopesi, M. Energy Saving in Pneumatic Drives: An Experimental Analysis. In Proceedings of the 9th International Workshop on Research and Education in Mechatronics, Bergamo, Italy, 18–19 September 2008.

96. Trujillo, A. Energy Efficiency of High Pressure Pneumatic Systems. Ph.D. Thesis, Polytechnic University of Catalonia, Barcelona, Spain, 2015.

97. Leszczynski, J.S.; Kastelik, K.; Kaminski, R.; Tomasiak, B.; Grybos, D.; Olszewski, M.; Plewa, A.; Mermer, P.; Rygal, R.; Soinski, M. Some harvesting system transforming energy wastes of compressed air to electricity. In Proceedings of the E3S Web of Conference—Energy and Fuel, Krakow, Poland, 21–23 September 2016; Volume 14. [CrossRef]

98. Doll, M.; Neumann, R.; Sawodny, O. Energy efficient use of compressed air in pneumatic drive systems for motion tasks. In Proceedings of the 2011 International Conference on Fluid Power and Mechatronics, Beijing, China, 17–20 August 2011; pp. 340–345. [CrossRef]

99. Rogers, E.M. *Diffusion of Innovation*, 3rd ed.; Collier Macmillan Publishers: New York, NY, USA, 1983; p. 512.

100. Tornatzky, L.G.; Klein, K.J. Innovation Characteristics and Innovation Adoption- Implementation: A Meta-Analysis of Findings. *IEEE Trans. Eng. Manag.* **1982**, *29*, 28–45. [CrossRef]

101. Roberts, S.J.F.; Ball, P.D. Developing a library of sustainable manufacturing practices. In Proceedings of the 5th CIRP Conference on Assembly Technologies and Systems, Dresden, Germany, 13–14 November 2014; Volume 15, pp. 159–164. [CrossRef]

102. Mills, E.; Rosenfelds, A.R.T. Consumer non-energy benefits as a motivation for making energy-efficiency improvements. *Energy* **1996**, *21*, 707–720. [CrossRef]

103. Skumatz, L.A.; Gardner, J. Methods and results for measuring non-energy benefits in the commercial and industrial sectors. In Proceedings of the ACEEE Summer Study on Energy Efficiency in Industry: Cutting the High Cost of Energy, West Point, NY, USA, 19–22 July 2005; Volume 6, pp. 163–176.

104. Piette, M.A.; Nordman, B. Costs and Benefits from Utility-Funded Commissioning of Energy-Efficiency Measures in 16 Buildings. In Proceedings of the ASHRAE Winter Meeting, Atlanta, GA, USA, 17–21 February 1996.

105. Worrell, E.; Laitner, J.A.; Ruth, M.; Finman, H. Productivity benefits of industrial energy efficiency measures. *Energy* **2003**, *28*, 1081–1098. [CrossRef]

106. Lung, R.B.; Mckane, A.; Leach, R.; Marsh, D. Ancillary Savings and Production Benefits in the Evaluation of Industrial Energy Efficiency Measures. In Proceedings of the ACEEE Summer Study on Energy Efficiency in Industry: Cutting the High Cost of Energy, West Point, NY, USA, 19–22 July 2005; Volume 6, pp. 103–114.

107. Nehler, T.; Rasmussen, J. How do firms consider non-energy benefits? Empirical findings on energy efficiency investments in Swedish industry. *J. Clean. Prod.* **2016**, *113*, 472–482. [CrossRef]

108. Hall, N.; Roth, J. Non-Energy Benefits from Commercial & Industrial Programs: What Are the Benefits and Why Are They Important to Participants? In Proceedings of the ACEEE Summer Study on Energy Efficiency in Buildings: Breaking out of the box, Pacific Grove, CA, USA, 22–27 August 2004; Volume 4, pp. 123–135.

109. Lilly, P.; Pearson, D. Determining the full value of industrial efficiency programs. In Proceedings of the 3rd ACEEE Summer Study on Energy Efficiency in Industry: Industry and Innovation in the 21st Century, New York, NY, USA, 15–18 June 1999; pp. 349–362.

110. Doyle, F.; Cosgrove, J. An approach to optimising compressed air systems in production operations. *Int. J. Ambient Energy* **2018**, *39*, 194–201. [CrossRef]

111. Sustainability Victoria. *Energy Efficiency Best Practice Guide. Compressed Air Systems*; Sustainability Victoria: Melbourne, VIC, Australia, 2009; p. 27. Available online: https://www.sustainability.vic.gov.au/Business/Energy-efficiency-for-business/Business-energy-efficiency/Energy-efficiency-best-practice-guidelines (accessed on 23 September 2020).

112. Finman, H.; Laitner, J.A. Industry, Energy Efficiency and Productivity Improvement. In Proceedings of the ACEEE Summer Study on Energy Efficiency in Industry, New York, NY, USA, 24–27 July 2001; Volume 1, pp. 561–570.

113. Pye, M.; McKane, A. Making a stronger case for industrial energy efficiency by quantifying non-energy benefits. *Resour. Conserv. Recycl.* **2000**, *28*, 171–183. [CrossRef]

114. Skumatz, L.A.; Dickerson, C.A.; Coates, B. Non-energy benefits in the residential and non-residential sectors: Innovative measurements and results for participant benefits. In Proceedings of the ACEEE Summer Study on Energy Efficiency in Buildings: Energy Efficiency & Sustainability, Pacific Grove, CA, USA, 17–21 August 2000; Volume 8, pp. 353–364.

115. Parra, R.; Thollander, P.; Nehler, T. Barriers to, drivers for and non-energy benefits for industrial energy efficiency improvement measures in compressed air systems. In Proceedings of the Eceee Industrial Summer Study: Going beyond energy efficiency to deliver savings, competitiveness and a circular economy, Berlin, Germany, 12–14 September 2016; Volume 2, pp. 293–304.

116. Rasmussen, J. The additional benefits of energy efficiency investments—A systematic literature review and a framework for categorisation. *Energy Effic.* **2017**, *10*, 1401–1418. [CrossRef]

117. Cagno, E.; Moschetta, D.; Trianni, A. Only non-energy benefits when adopting an EEM? Cases from industry. In Proceedings of the Eceee Industrial Summer Study: Going beyond energy efficiency to deliver savings, competitiveness and a circular economy, Berlin, Germany, 12–14 September 2016; Volume 2, pp. 281–292.

118. Thollander, P.; Ottosson, M. Energy management practices in Swedish energy-intensive industries. *J. Clean. Prod.* **2010**, *18*, 1125–1133. [CrossRef]

119. Kemp, R.; Volpi, M. The diffusion of clean technologies: A review with suggestions for future diffusion analysis. *J. Clean. Prod.* **2008**, *16* (Suppl. 1), S14–S21. [CrossRef]

120. Cagno, E.; Worrell, E.; Trianni, A.; Pugliese, G. A novel approach for barriers to industrial energy efficiency. *Renew. Sustain. Energy Rev.* **2013**, *19*, 290–308. [CrossRef]

121. Millar, H.H.; Russell, S.N. The adoption of sustainable manufacturing practices in the Caribbean. *Bus. Strateg. Environ.* **2011**, *20*, 512–526. [CrossRef]

122. Damanpour, F. Innovation type, radicalness, and the adoption process. *Communic. Res.* **1988**, *15*, 545–567. [CrossRef]

123. Dahlin, K.B.; Behrens, D.M. When is an invention really radical? Defining and measuring technological radicalness. *Res. Policy* **2005**, *34*, 717–737. [CrossRef]

124. Sandberg, P.; Soderstrom, M. Industrial energy efficiency: The need for investment decision support from a manager perspective. *Energy Policy* **2003**, *31*, 1623–1634. [CrossRef]

125. Thollander, P.; Palm, J. *Improving Energy Efficiency in Industrial Energy Systems. An Interdisciplinary Perspective on Barriers, Energy Audits, Energy Management, Policies and Programs*; Springer: Berlin/Heidelberg, Germany, 2013; p. 151. ISBN 144714161X.

126. Hellström, T. Dimensions of Environmentally Sustainable Innovation: The Structure of Eco-Innovation Concepts. *Sustain. Dev.* **2007**, *15*, 148–159. [CrossRef]

127. Request for Proposals: Resource Assessment, Emerging Technology, Measure Characterization. Available online: https://www.energytrust.org/wp-content/uploads/2017/08/Resource-Assessment-Emerging-Technologies-RFP-FINAL.pdf (accessed on 31 July 2020).

128. Arensberg, C.M.; Niehoff, A.H. Introducing Social Change: A Manual for American Overseas. *Science* **1968**, *147*, 500. [CrossRef]

129. Designing Your Compressed Air System. How to Determine the System You Need. Available online: https://pdf.directindustry.com/pdf/kaeser-compressors/designing-your-compressed-air-system-guide-3/68092-177031.html (accessed on 31 July 2020).

130. Pascual, R.; Meruane, V.; Rey, P.A. On the effect of downtime costs and budget constraint on preventive and replacement policies. *Reliab. Eng. Syst. Saf.* **2008**, *93*, 144–151. [CrossRef]

131. Cagno, E.; Moschetta, D.; Trianni, A. Only non-energy benefits from the adoption of energy efficiency measures? A novel framework. *J. Clean. Prod.* **2019**, *212*, 1319–1333. [CrossRef]

132. The Ins and Outs of Vacuum Generators. Available online: https://www.airbestpractices.com/technology/vacuum/ins-and-outs-vacuum-generators (accessed on 31 July 2020).

133. Health and Safety Executive, HSE. *Compressed Air Safety (HSG39)*, 2nd ed.; HSE: Bootle, UK, 1998; p. 50. Available online: https://www.hse.gov.uk/pubns/books/hsg39.htm (accessed on 23 September 2020).

134. Technical Brief—Sizing Compressed Air Equipment. Available online: https://www.cagi.org/pdfs/Sizing-Technical-Brief-Final.pdf (accessed on 31 July 2020).

135. Sorrell, S.; Schleich, J.; Scott, S.; O'Malley, E.; Trace, F.; Bode, U.; Kowener, D.; Mannsbart, W.; Ostertag, K.; Radgen, P. *Reducing Barriers to Energy Efficiency in Private and Public Organisations*; Science Policy Research Unit SPRU, University of Sussex: Falmer, UK, 2000; p. 197.

136. Worrell, E.; Biermans, G. Move over! Stock turnover, retrofit and industrial energy efficiency. *Energy Policy* **2005**, *33*, 949–962. [CrossRef]

137. Matasniemi, T. *Operational Decision Making in the Process Industry. Multidisciplinary Approach*; VTT Technical Research Centre of Finland: Espoo, Finland, 2008; p. 143.

138. Yin, R. *Case Study Research: Design and Methods*, 5th ed.; Sage: London, UK, 2014; p. 219.

139. Voss, C.; Tsikriktsis, N.; Frohlich, M. Case research in operations management. *Int. J. Oper. Prod. Manag.* **2002**, *22*, 195–219. [CrossRef]

140. Eisenhardt, K.M.; Graebner, M.E. Theory building from cases: Opportunities and challenges. *Acad. Manag. J.* **2007**, *50*, 25–32. [CrossRef]

141. Trianni, A.; Cagno, E.; Bertolotti, M.; Thollander, P.; Andersson, E. Energy management: A practice-based assessment model. *Appl. Energy* **2019**, *235*, 1614–1636. [CrossRef]

142. Industrial Assessment Center (IAC) Database. Available online: https://iac.university/download (accessed on 23 August 2020).

143. U.S. Department of Energy. *Effect of Intake on Compressor Performance*; DOE/GO-102004-1933; U.S. Department of Energy: Washington, DC, USA, 2004; p. 2.

144. Testing and Troubleshooting Drain Traps on Compressed Air Systems. 1996. Available online: https://www.armstronginternational.com/sites/default/files/resources/documents/testingandtroubleshootingdraintrapsoncompressedairsystems11-96.pdf (accessed on 23 September 2020).

145. Compressed Air Dryer Efficiency. 2015. Available online: https://fluidpowerjournal.com/compressed-air-dryer-efficiency/ (accessed on 23 September 2020).

146. Cycling Refrigerated Air Dryers—Are Savings Significant? Available online: https://www.compressedairchallenge.org/data/sites/1/media/library/articles/2011-11-CABP.pdf (accessed on 31 July 2020).

147. Seven Rules for Improving the Quality of Compressed Air. Available online: https://www.hydraulicspneumatics.com/technologies/air-filters-and-frls/article/21883468/seven-rules-for-improving-the-quality-of-compressed-air (accessed on 31 July 2020).

148. Reliable Operation of High Purity Desiccant Compressed Air Dryers. Available online: https://www.airbestpractices.com/technology/air-treatment/n2/reliable-operation-high-purity-desiccant-compressed-air-dryers (accessed on 31 July 2020).

149. Controls Increase Compressor Efficiency. Available online: https://www.plantengineering.com/articles/controls-increase-compressor-efficiency/ (accessed on 31 July 2020).

150. Qin, H.; McKane, A. *Improving Energy Efficiency of Compressed Air System Based on System Audit*; Lawrence Berkeley National Laboratory: Berkeley, CA, USA, 2008; p. 10. Available online: https://escholarship.org/uc/item/13w7f2fc (accessed on 23 September 2020).

151. Neelis, N.; Worrell, E.; Masanet, E.R. *Energy Efficiency Improvement and Cost Saving Opportunities for the Petrochemical Industry*; Lawrence Berkeley National Laboratory: Berkeley, CA, USA, 2008; p. 151. Available online: https://ies.lbl.gov/publications/energy-efficiency-improvement-and-8 (accessed on 23 September 2020).

152. Compressed Air System Design. 2016. Available online: https://compressedair.net.au/wp-content/uploads/2016/02/003_CAAA_Fact-Sheet-Copy_Compressed-Air-System-Design_290416.pdf (accessed on 23 September 2020).

153. U.S. Department of Energy. *Evaluation of the Compressed Air Challenge Training Program*; DOE/GO-102004-1919; U.S. Department of Energy: Washington, DC, USA, 2004; p. 98. Available online: https://www.energy.gov/sites/prod/files/2014/05/f16/eval_cactp_cd.pdf (accessed on 23 September 2020).

154. Compressed Air Versus Blower Systems. Available online: https://www.stanmech.com/articles/compressed-air-versus-blower-systems (accessed on 31 July 2020).

155. Optimize a Compressed Air System with Highly Fluctuating Air Demand. Available online: https://www.airbestpractices.com/system-assessments/end-uses/optimize-compressed-air-system-highly-fluctuating-air-demand (accessed on 31 July 2020).

156. Bernard, H.R. *Research Methods in Anthropology: Qualitative and Quantitative Approaches*, 4th ed.; Rowman Altamira: Lanham, MD, USA, 2006; p. 680. ISBN 0759108692.

157. Evaluating Air Compressor Cooling System. Available online: https://www.airbestpractices.com/sustainability-projects/cooling-systems/evaluating-air-compressor-cooling-systems (accessed on 31 July 2020).

158. U.S. Department of Energy. *Maintenance of Compressed Air Systems for Peak Performance*; U.S. Department of Energy: Washington, DC, USA, 1998; p. 3.

159. Air Compressors—Care & Maintenance Tips. 2009. Available online: https://www.mobil.com.de/de-de/industrial/lubricant-expertise/resources/air-compressor-maintenance-tips (accessed on 23 September 2020).

160. Vortex A/C. Available online: https://www.vortextube.co.uk/pdf/Vortex-AC-VT.pdf (accessed on 31 July 2020).

161. European Commission. *Observatory of European SMEs Analytical Report*; European Commission: Bruxelles, Belgium, 2007; p. 267.

162. Trianni, A.; Cagno, E. Dealing with barriers to energy efficiency and SMEs: Some empirical evidences. *Energy* **2012**, *37*, 494–504. [CrossRef]

163. Cooremans, C. Investment in energy efficiency: Do the characteristics of investments matter? *Energy Effic.* **2012**, *5*, 497–518. [CrossRef]

164. Nehler, T.; Thollander, P.; Ottosson, M.; Dahlgren, M. Including non-energy benefits in investment calculations in industry—Empirical findings from Sweden. In Proceedings of the Eceee Industrial Summer Study: Retool for a competitive and sustainable industry, Arnhem, The Netherlands, 2–5 June 2014; Volume 2, pp. 711–719.

165. Skumatz, L.A. *Non-Energy Benefits/Non-Energy Impacts (NEBs/NEIs) and their Role & values in Cost-Effectiveness Tests: State of Maryland*; Skumatz Economic Research Associates, Inc.: Superior, CO, USA, 2014; p. 71. Available online: https://www.energyefficiencyforall.org/resources/non-energy-benefits-non-energy-impacts-nebs-neis-and-their-role-and-values/ (accessed on 23 September 2020).

Article

Establishing Energy Efficiency—Drivers for Energy Efficiency in German Manufacturing Small- and Medium-Sized Enterprises

Werner König [1,*], Sabine Löbbe [1], Stefan Büttner [2] and Christian Schneider [2]

[1] REZ—Reutlingen Energy Center for Distributed Energy Systems and Energy Efficiency, Reutlingen University, 72762 Reutlingen, Germany; Sabine.Loebbe@Reutlingen-University.de

[2] EEP—Institute for Energy Efficiency in Production, University of Stuttgart, 70569 Stuttgart, Germany; Stefan.Buettner@eep.uni-stuttgart.de (S.B.); Christian.Schneider@eep.uni-stuttgart.de (C.S.)

* Correspondence: Werner.Koenig@Reutlingen-University.de; Tel.: +49-7121-271-7136

Received: 9 June 2020; Accepted: 29 September 2020; Published: 2 October 2020

Abstract: Despite strong political efforts in Europe, industrial small- and medium-sized enterprises (SMEs) seem to neglect adopting practices for energy efficiency. By taking a cultural perspective, this study investigated what drives the establishment of energy efficiency and corresponding practices in SMEs. Based on 10 ethnographic case studies and a quantitative survey among 500 manufacturing SMEs, the results indicate the importance of everyday employee behavior in achieving energy savings. The studied enterprises value behavior-related measures as similarly important as technical measures. Raising awareness for energy issues within the organization, therefore, constitutes an essential leadership task that is oftentimes perceived as challenging and frustrating. It was concluded that the embedding of energy efficiency in corporate strategy, the use of a broad spectrum of different practices, and the empowerment and involvement of employees serve as major drivers in establishing energy efficiency within SMEs. Moreover, the findings reveal institutional influences on shaping the meanings of energy efficiency for the SMEs by raising attention for energy efficiency in the enterprises and making energy efficiency decisions more likely. The main contribution of the paper is to offer an alternative perspective on energy efficiency in SMEs beyond the mere adoption of energy-efficient technology.

Keywords: industrial energy efficiency; energy efficiency culture; energy efficiency practices; energy management

---

1. Introduction

Increased industrial energy efficiency has been a highlighted objective in political agendas in Europe, aiming at productivity gains and ecological sustainability. Small- and medium-sized enterprises (SMEs) hold a special position in this context and they are often considered the backbone of the European industrial structure [1]. In Germany in 2017, 184,667 SMEs represented about 96.9% of industrial enterprises [2]. Despite strong political efforts in Europe, SMEs seem to be neglecting to adopt effective measures for energy saving and efficiency. Thollander et al. [3] estimate the energy efficiency potential of manufacturing SMEs in the European Union at more than 25%. Why this potential remains untapped has kept policy makers and scientists occupied since the notion of the "energy efficiency gap" [4,5] emerged; academia struggles with another empirical phenomenon often referred to as the energy "efficiency-paradox" [6]. Despite high profitability, energy efficiency measures are often not implemented.

The question of what constrains and drives decisions for energy efficiency in industrial organizations represents a vast research field in energy literature [7–10]. Barriers and drivers can be defined as all factors that hamper or foster the adoption of cost-effective, energy-efficient technologies and their diffusion [11,12]. Accounting for the fact that measures representing high rates of return, or requiring no capital investment, are often not undertaken by SMEs [13,14]. A perspective solely focusing on economically rational decisions appears insufficient for a thorough understanding of the situation of SMEs.

Studies on the adoption and implementation of energy-efficient practices represent an overlapping key area of research on industrial energy efficiency [15]. Recent analyses emphasize the benefits and characteristics of measures [16], the potentials of particular technical processes [17], or beneficial intersections to other management aspects such as supply chain management [18]. Despite the theoretical importance and practical value of these approaches, a one-sided view of technical measures has been increasingly criticized in recent publications [19–23] on the barriers and adoption of energy efficiency measures. To date, practices other than technical measures have received inadequate attention in empirical studies [24], neglecting the material, social, cultural, and institutional aspects framing the decision-making processes [25,26].

Rejecting an atomistic perspective on decision making and technology, scholars of sociology, ethnology, and anthropology have drawn increasing attention to the cultural aspects of energy-related behavior in enterprises. By looking at the values, norms, laws, and everyday practices, these approaches emphasize the embeddedness of organizational decisions on energy efficiency in cultural, social, and material contexts [27–31]. Ethnographic case studies have since shown the significance of SME owners' and managers' personal values in terms of environmental decision making [32] and how energy management practices are influenced by organizational cultures, team dynamics, and individual's aspirations [33]. Despite these efforts, Andrews and Johnson [34] call for an increase in studies addressing the rules, norms, beliefs, and logics embedded in the organization's context. Fawcett and Hampton similarly argue that a "more complex understanding of SMEs, as organizations operating in a socio-technical landscape, and with varied capabilities, objectives and values" could provide a more effective policy design [35] (p. 3).

By adopting a cultural perspective [36], this empirical study on German manufacturing SMEs examined the energy efficiency climate, the energy efficiency practices, and the intersections between the enterprises and their members and their institutional environment. The purpose of this study was to investigate the establishment of energy efficiency within the SMEs and identify general drivers in promoting energy efficiency decisions, energy-saving behavior, and the general establishment of energy efficiency in SMEs. The study followed a mixed-methods approach and utilized qualitative (single case studies) and quantitative (survey) data of SMEs situated in the federal state of Baden-Wuerttemberg, southern Germany.

The study indicates the importance of everyday employee behavior in achieving energy savings. The studied enterprises value behavior-related measures as similarly important as technical measures. Raising awareness for energy issues within the organization, therefore, constitutes an essential leadership task, which is oftentimes perceived as challenging and frustrating. The results suggest the embedding of energy efficiency in corporate strategy, the use of a broad spectrum of different practices, and the empowerment and involvement of employees as major drivers in establishing energy efficiency within SMEs. Furthermore, the findings reveal external influences on shaping the meanings of energy efficiency for the SMEs by raising attention for energy efficiency in the enterprises and making energy efficiency decisions more likely.

The remainder of this study is organized as follows: The next section sets out the theoretical background and research focus. Section 3 provides a brief overview of the research strategy and the methods used. Section 4 is devoted to the main results of the study. The following section presents the discussion of the results and Section 6 provides the conclusions and implications of the study.

2. Theoretical Background and Focus

2.1. The Concept of Culture in the Context of Energy Efficiency Research

Culture is frequently explained as a constraining soft factor from an organizational perspective [37,38]. Culture can be roughly defined as the mix of knowledge, ideology, norms, values, laws, and everyday rituals that characterize a social system [39]. Sorrell et al. [40] presume that environmental values embedded in organizational cultures and practices have an essential effect on organizational decisions and behavior. However, they do not view culture to be a barrier, "but an important explanatory variable" [40] (p. 15). Although a cultural perspective on industrial energy efficiency promises useful findings for science and politics, only a few researchers have transferred this complex to a strategy for empirical inquiry. Criticizing that "within the energy literature, the concept of culture has generally been more implied than overt", Stephenson et al. [41] (p. 6123) developed a conceptual framework of energy cultures. The so-called Energy Cultures Framework (EFC) is based on sociological theory [42,43] and represents a heuristic approach to investigate influences on energy-related behaviors in social systems to identify the "levers for change towards more energy-efficient behaviors" [41] (p. 6123). According to the ECF, the energy cultures are constituted via the interactions between material culture, practices, and norms, all of which are affected by external influences and embedded in wider cultural spheres [44]. Since its development, the ECF has been applied to different contexts, ranging from household energy-related behavior [45] to industrial sectors such as the timber processing industry [46].

König [36] introduced a similar framework addressing specifically industrial organizations. Taking a sociological perspective, he views organizations as cultural systems embedded in wider social contexts and he developed a theoretical framework addressing individual organizational and institutional dimensions, shaping decisions on energy efficiency. The framework combines multidisciplinary concepts and theoretical approaches of organizational theory. It integrates concepts of sociological neo-institutional theory [47,48], the translation perspective on diffusion [49], the attention-based view of the firm [50], and organizational [51] and energy culture research [41,44]. The energy efficiency culture of an industrial organization is defined as unconscious, shared understandings, which are mutually dependent from the organizational structures, practices, environment, and individual members. Decisions on energy efficiency in industrial enterprises are, therefore, based on a multilevel process shaped by individuals, organizations, and the environment. Referring to the attention-based view of the firm [50], the organization structures the situational context of, distributes the attention to, and shapes the focus of attention on energy efficiency issues. This framework supports the identification of drivers and served as an anchor in conceptualizing the research design of the study.

2.2. Theoretical Perspective of the Study

The data collection and analysis were structured by the theoretical concept developed by König (Figure 1), which assumes that decisions and actions on energy efficiency emerge at the intersection between three levels.

1. The Macro level encompasses the institutional issue field of which organizations and actors have emerged around the issue of energy efficiency. This field and its actors exert regulative (e.g., through policies, laws, and discourse), economic–financial (e.g., through prizes, grants, subsidies), normative (e.g., work roles, habits, professional, social, and scientific norms), and cognitive–cultural (e.g., constitutive schemes, values, beliefs, and assumptions) influences on the organization's decisions.

2. The Meso level encompasses the industrial organization with its material environment, energy efficiency climate, energy efficiency practices, and basic energy assumptions and beliefs. The material environment of industrial enterprises has been a focal point of empirical studies on barriers, such as energy intensity [52,53] or firm size [54–56], and must be considered as a

crucial factor of decision making. Following Denison's concept of organizational climate [57], the energy efficiency climate represents the interpretation of the situations related to energy efficiency within the organization. The energy efficiency practices are understood as the totality of all practices toward energy efficiency and energy conservation by an enterprise and represent outcomes as well as inputs to decisions on energy efficiency measures. Referring to Fiedler and Mircea [58], who view energy management as "the sum of all measures and activities which are planned or executed in order to minimize the energy consumption of a company", the energy efficiency practices synonymously represent the energy management of an enterprise. Following Schein's concept of organizational culture [51], the basic energy assumptions and beliefs within an industrial organization are mutually dependent from the organizational structures, practices, environment, and individual members.

3. The Micro level incorporates the decision makers and members of the organization with their individual characteristics (e.g., attitudes, interests, competencies). These characteristics are mutually dependent of the positioning and socialization of individuals within the organization.

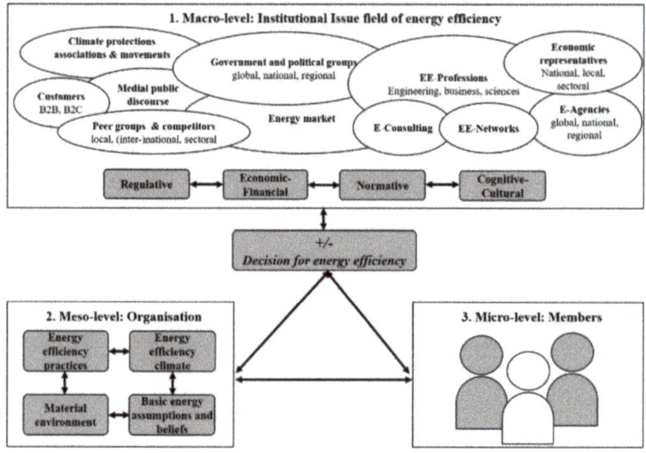

Figure 1. Energy efficiency culture framework following König [36].

Decisions on energy efficiency represent processes of theorization and problematization, linking together the issue-field (1. Macro level), the organization (2. Meso level), and the members (3. Micro level). In this sense, decision makers are not considered as atomistic units; they are members of professional groups, work groups, milieus, or families in the case of family businesses [34] (p. 198).

2.3. Research Focus and Research Questions

The research focused on manufacturing SMEs in Baden-Württemberg, Germany. This federal state constitutes the most industrial area in Germany, with about 1.3 million employees in the industry sector and around 7500 manufacturing SMEs with around 505,833 employees [59] representing the industrial backbone. Baden-Württemberg is ahead of all G-10 states (a group of the 11 leading industrial countries) with a share of manufacturing industry in the gross value added and exceeds the benchmark for industry defined by the EU for 2020 (20% share of industry in gross value added) by around 60% with 32.5% [60,61].

Referring to the framework described above, the research concentrated on four areas that were derived as crucial in answering the overarching research question: What drives the establishment of energy efficiency in SMEs in everyday work life? The study focused on examining the energy efficiency climate, the energy efficiency practices, the intersection between the enterprises and their members, and the intersection to its relevant environment. These four areas were assigned the guiding research questions, which structured the data collection and analysis.

1. Energy efficiency climate:

According to the theoretical perspective taken, the energy efficiency climate reflects the interpretation of the situations related to energy efficiency within the organization, which structure the attention of its members. Correspondingly, the investigation concentrated on the following research questions:

- What importance does energy efficiency have for the SMEs?
- What meanings and importance does energy efficiency have for and within the SMEs?
- How is energy efficiency perceived as being established within the enterprises and what aspects drive the establishment of energy efficiency within the SMEs?

2. Energy efficiency practices:

As outlined in the introduction, the study should not only take technical measures into account. Following the classification of energy efficiency practices by König [36] (p.6), six different forms of energy efficiency practices were investigated: Technology investment-related practices (e.g., the purchase and implementation of energy-efficient technical equipment), technology organization-related practices (e.g., the enhancement and optimization of existing support or process technology), organization-related practices (e.g., corporate energy strategy, the implementation of an energy management system), information-related practices (e.g., energy monitoring, internal technical meetings), competence-related practices (e.g., workshops, trainings), and behavior-related practices (e.g., raising awareness for energy saving by personal encouragement and explicit behavior guidelines). The following research questions were focused on:

- What importance does energy efficiency have for corporate strategy of the SMEs?
- What importance do different energy efficiency practices have for the enterprises?
- What importance does energy management have for the SMEs?

3. Interface between the enterprise and its members:

Organizational procedures and structures potentially regulate the energy-related behavior of their members and subunits [34]. Aiming at everyday work life in the enterprises, the following research questions were targeted:

- What importance does the everyday behavior of the employees have for energy conservation and energy efficiency?
- Who is perceived as responsible for energy efficiency and energy conservation within the enterprises?
- How do the SMEs and their leaders attempt to raise awareness among their workforce?

4. Interface between the enterprise and its environment:

As described above, it was assumed that external actors and organizations potentially exert regulative, economic–financial, normative, and cognitive–cultural influences on the SMEs regarding energy efficiency practices and decisions. Accordingly, the study focused on these questions:

- Regulative: How do the SMEs perceive external imperatives for energy efficiency?
- Economic–financial: To what extent is the financing of measures considered as an obstacle by the SMEs?
- Normative: What information sources do the SMEs use and how actively do they search for information?
- Cultural–cognitive: What importance does energy efficiency have for the environment of the SMEs and to what extent does it influence the decisions of the SMEs?

3. Materials and Methods

The study followed a sequential exploratory mixed-methods approach [62] and combined ethnographic case studies with a subsequent quantitative survey (Figure 2). To gain an understanding of how decisions on energy efficiency are constituted and how the enterprises deal with energy efficiency issues in everyday work life, firstly, ethnographic case studies [63] were carried out on 10 industrial SMEs using qualitative interviews and observations as the main methods. Secondly, and based on their key results and orienting on the discussed framework, the questionnaire was conceptualized, which was addressed to 500 SMEs.(An SME is here intended as an enterprise according to the 2003 recommendation of the European Council [64]).

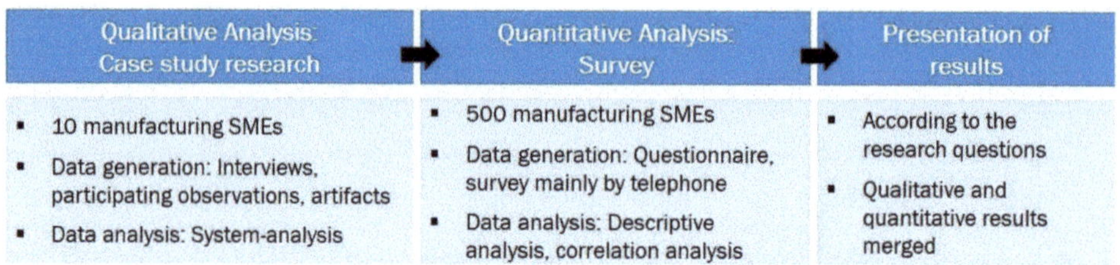

Figure 2. Research design.

3.1. Qualitative Analysis: Case Studies

The ethnographic organizational analyses formed the starting point of the research work, which primarily aimed at the observation and reconstruction of situations in everyday work life in order to find out "how work is organized and how that organizing organizes people". [63] (p. 1). This qualitative approach is particularly suitable for studies of organizational cultures [65] (p. 20). Exploratively designed case studies focused on several basic questions: How do decisions for energy efficiency come about in the SMEs? Which driving (or constraining) processes and aspects can be identified? How are energy efficiency issues treated, organized, and communicated in everyday work life? The sample (Table 1) comprised 10 manufacturing enterprises from different industrial sectors (chemicals, minerals, engineering, and machinery). The cases were selected by theoretical sampling [66] according to the premise of "minimum/maximum contrast", especially with regard to energy intensity, sector, and number of employees of the enterprises. All participating enterprises are family businesses (ownership and/or control). (Ninety-one percent of all enterprises in Baden-Württemberg are family controlled and 88% of all enterprises in the manufacturing sector are family controlled [67].). The data generation was mainly based on qualitative interviews [68] with members from different divisions within the enterprises. Around seven to 10 interviews per SME (one-on-one and multi-person) were carried out in each enterprise. Interviews were conducted with managing directors, owners, and energy managers as well as production workers and controlling, marketing, and human resources staff (Table 2). In addition to the interviews, participating observations [69] and artefacts (e.g., company presentations, homepages, work instructions) were included in the analysis. Depending on what was appropriate from the perspectives of the enterprises and their members, the observations were either performed as fly-on-the-wall (e.g., at meetings of formal or informal energy teams or meetings with external energy efficiency consultants) or following the daily routines throughout the work day. The data (primarily interview text and observation protocols) were analyzed by system analysis [68,69]. This hermeneutic approach focuses on the interpretation of the data in two steps. In the first step, hypotheses are formed, from which subjective and organizational meanings and conditions could lead to a statement (e.g., "Only the boss is responsible for energy.") or observation. In the second step, hypotheses are formed, from which these meanings and conditions could have effects for the organization (e.g., centralization of competences and responsibility for measures). The field research was carried out by one person of the University of Reutlingen, taking about one year all together and spending around one work week in each SME.

Table 1. List of enterprises participating in the case studies.

Case	Number of Employees	Sector	Energy Intensity *
Enterprise A	110	Surface engineering	Average
Enterprise B	90	Mechanical engineering	Low
Enterprise C	70	Foundry industry	Very high
Enterprise D	135	Manufacture of products of wood, synthetics, and metal	Average
Enterprise E	115	Mineral industry	Very high
Enterprise F	240	Pulp and paper industry	High
Enterprise G	85	Mechanical engineering and service	Average
Enterprise H	45	Surface engineering	High
Enterprise I	20	Mechanical engineering	Low
Enterprise J	85	Manufacture of chemical products	High

* Self-evaluation of the enterprises.

Table 2. Shares on the roles of the interviewed persons of the SMEs.

Director/ Owner	Energy Management	Controlling/ Accounting	Production Workers	Engineering/ Maintenance	Human Resource	Marketing	Trainees
10	4	7	28	14	6	3	3

3.2. Quantitative Analysis: Survey

Based on the case study research, a quantitative survey was conceptualized through a questionnaire comprising questions on topics such as the importance of energy efficiency, measures, support measures, the influence of the business environment, the relevance of employee behavior, financing, and others. The questionnaire consisted of 20 different types of questions including multiple choice questions, Likert scale questions, and matrix questions as well as single-choice questions (see Appendix A). The survey took place from May to June 2018 and around 500 SMEs from the federal state of Baden-Württemberg, Germany, were surveyed. A market research institute was commissioned with the survey itself while the analysis was conducted by the Institute for Energy Efficiency in Production, Universität Stuttgart. On the basis of available data bases and selected by company size (micro-, small-, middle-sized) and sectors, the SMEs were reached by telephone. The energy demand of the surveyed SMEs is shown in Figure 3. The distribution of the responding SMEs with respect to the number of employees is shown in Table 3. Analogous to the sample procedure for the survey of the Energy Efficiency Index of German Industry [10], the distribution of the sample size did not correspond to the real distribution of SMEs by enterprise size in Germany [70]. In order to allow valid results for small and medium sized enterprises, the share of micro enterprises was reduced. Furthermore, a thorough coverage of the manufacturing sector including subsectors of particular importance for the state of Baden-Wuerttemberg (such as the manufacturing of metal products and processing, mechanical engineering, and the automotive sector) was targeted. The respondents were either owners, managing directors, technical managers, production managers, energy or environmental managers, controlling, or other persons of the SMEs. Naturally, not all enterprises were open to be interviewed. Therefore, a self-selection bias can be assumed. In addition to descriptive data analysis, a correlation analysis (using SPSS) investigating the factors driving the internal establishment of energy efficiency was conducted. The correlation analysis was performed using ordinally scaled variables, with the Spearman rho rank correlation coefficient as an indicator of correlation.

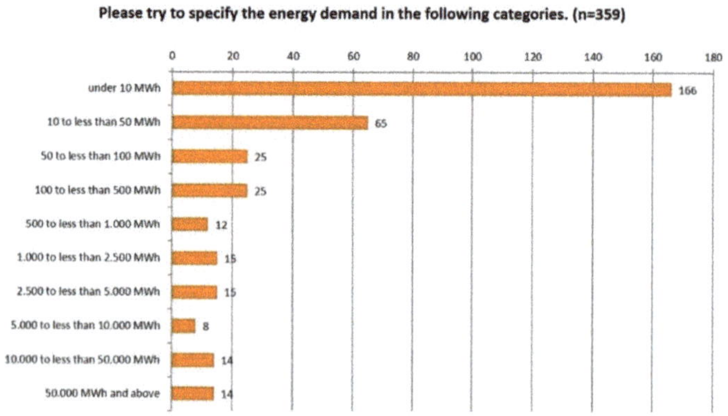

Figure 3. Energy demand of the surveyed SMEs.

Table 3. Sample composition, according to size.

Company Size	Number of Employees	Respondents	Percentage	Distribution of Manufacturing SMEs in Germany
micro	0–9	282	56.7%	62.8%
small	10–49	135	27.2%	27.8%
medium	50–249	80	16.1%	9.3%

4. Results

The presentation of the results concentrates on those four areas that were identified as crucial in constituting decisions on energy efficiency and establishing an effective energy efficiency culture within the SMEs.

4.1. Energy Efficiency Climate

4.1.1. What Importance Does Energy Efficiency Have for the SMEs?

As part of the questionnaire, the SMEs were queried as to how they assess the current meaning of energy efficiency for the enterprise. The results show that energy efficiency is perceived as an important issue for the SMEs, although equally important with other factors. While there were only minor differences between the enterprises when looking at their energy demand, the answers differed considerably according to the size of the enterprise (Figure 4). The analysis suggests that the size of enterprises influences the perceived importance of energy efficiency—the smaller the SMEs are, the less pronounced the importance of energy efficiency seems to be. In this respect, the energy efficiency climate in micro-enterprises appears less positive than in medium-sized enterprises. Almost one-third of micro-enterprises considers the current importance of energy efficiency to be relatively low.

The SMEs were additionally asked how they assess the importance of energy efficiency and energy saving in day-to-day work for the company's work force. This showed that almost half of the enterprises (46%) rate the meaning of energy saving in everyday working life as high or very high. On the other hand, only 19% rate energy efficiency as low or very low. With regard to the energy demand, hardly any differences were noticeable. However, differences could be observed according to the size of the enterprises. As seen above, a similar pattern was observed, although less pronounced—the larger the enterprises are, the higher the importance of energy efficiency in everyday work life seems to be.

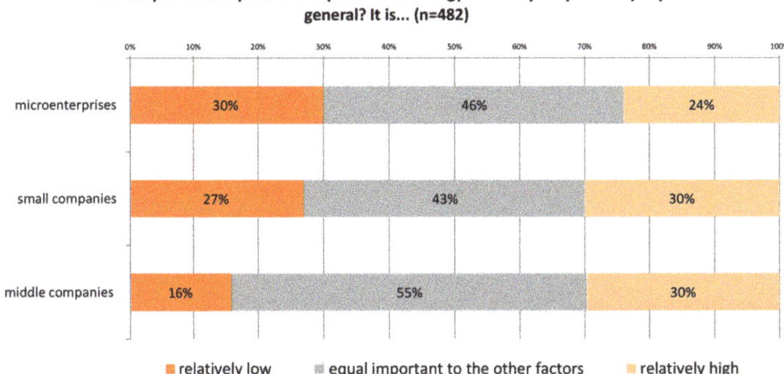

Figure 4. Current importance of energy efficiency by enterprise size.

4.1.2. What Meanings Does Energy Efficiency Have for the SMEs?

In the course of the interviews during the ethnographic fieldwork, it became apparent that the associations of the organizational members with energy efficiency issues are manifold. Subjective and manifest assessments on energy efficiency issues and individual experiences may range from "necessity" to "annoyance" within an enterprise or even within a division. As every enterprise has its own "energy efficiency history", the interview partners provided individual experiences as well as shared collective stories, revealing the meanings associated with energy efficiency issues. For example, the interviewed persons reported on the introduction of a management system, reflected on their role in initiating measures, or critically questioned the management's intentions when introducing new work rules. Therefore, personal experiences, corporate values, and motivations as well as realized measures constitute the meaning of energy efficiency. To untangle the bubble of meanings [71], the articulated meanings by the interviewed persons (Table 4) were differentiated by whether they refer to the organizational discourse (meaning for the enterprise) or individual experiences, ideas, and beliefs of individual employees and decision makers (meanings within the enterprise).

Table 4. Meanings of energy efficiency for and within the enterprise (energy efficiency climate).

Meanings of Energy Efficiency for the Enterprise	Meanings of Energy Efficiency within the Enterprise
<ul><li>Cost reduction and competitiveness</li><li>Differentiation from others (peer organizations or competitors)</li><li>Securing the long-time future of the enterprise</li><li>Aspiration of a positive public image</li><li>Conformity to a perceived industrial state of the art (e.g., progressiveness, modernity)</li><li>Conformity to local expectations (e.g., engagement in programs or initiatives)</li><li>Risk mitigation (e.g., by combined heat and power generation)</li><li>Source and symbol of corporate identity</li><li>Fulfillment of enforced external expectations</li></ul>	<ul><li>Accordance with individual interests (e.g., in specific issues or in general)</li><li>Increasing complexity of production and auxiliary processes</li><li>Addition of troublesome tasks or problems</li><li>Normality and everyday routines</li><li>Career opportunity within the enterprise</li><li>Limiting autonomous behavior</li><li>Increasing bureaucracy</li><li>Corporate profit seeking (depreciative)</li><li>Fulfillment of enforced external expectations</li></ul>

The fieldwork research indicates that the meaning of energy efficiency for the enterprise is not only shaped by internal criteria such as cost reduction, profitability, risk mitigation, and future safety, but also by external criteria. Meanings, such as social or ecological responsibility, modernity, progress, or the desire for a positive external image, are unthinkable without recourse to the expectations of the corporate environment and broader society; the meaning of energy efficiency is a social product.

For example, in several cases the general societal discourse on sustainability together with regional energy efficiency programs or initiatives of contractors can have a considerable influence in shaping the attention on energy efficiency issues within an enterprise. On the one hand, this indicates the potential influence of the organizational environment on the enterprises (which will be further elucidated in Section 4.4). Furthermore, those external criteria and expectations represent aspirations for the future and the development of the enterprises.

The meanings for and within the enterprise do not necessarily intersect and have both positive and negative (or rather destructive) connotations. Energy efficiency can, thus, be treated ambivalently by employees despite its positive meanings for an enterprise. Nevertheless, members should not necessarily be accused of lack of understanding, resentment, or bad intentions. On the contrary, the interviews revealed that the employees oftentimes do have much more personal thoughts about issues of increasing energy efficiency than the management staff sometimes expects. Nevertheless, the individual frames of references do differ. The case studies strongly indicate that the positioning within the enterprise, individual attitudes, interests, and experiences shape how energy efficiency issues are perceived and evaluated by the individual members. For instance, in the case of a managing director of one studied enterprise, individual experiences with energy efficiency technology (e.g., in private households) and a strong interest in sustainability proved as a strong driver for the implementation and consideration of any measures. In another case, the energy manager similarly expressed a strong individual interest in sustainability as well as in energy-efficient technology and measures in general. However, feeling under-supported by the top management and decision makers on the shop floor level, he has become more and more frustrated with the role as an energy manager he initially was eager to fill out in the enterprise. In the context of the enterprise, energy efficiency became a burden loaded with negative associations.

4.1.3. How Is Energy Efficiency Perceived as Being Established within the Enterprises and What Aspects Drive the Establishment of Energy Efficiency within the SMEs?

To what extent energy efficiency is established in the SMEs was one of the key questions of the entire study. With regard to the participating enterprises in the case study research, the establishment appears fairly strong. Although in almost all SMEs potentials for improvement were indicated by the management staff, the topic was, nevertheless, considered sufficiently established by most of them. From a critical perspective, this finding could be attributed to a sampling bias (due to the fact that participating in an extensive case study research requires rather unusual high interest for energy efficiency, in general, and a rather strong establishment, after all). This risk is much lower for the survey in which the companies are asked the same question. However, energy efficiency appears to be fairly well established in the surveyed SMEs as well. Rather surprisingly, the energy demand of the enterprises did not seem to have a particular influence on how energy efficiency is established. Conversely, the size of enterprises appeared to have a more significant influence (Figure 5). In the case of micro-enterprises, around 30% perceive energy efficiency to be strongly to very strongly established, compared with around 50% for medium-sized enterprises.

To further validate the interpretation of the descriptive analysis and to investigate potential positive factors, a correlation analysis was carried out (Table 5). Although energy efficiency seems less established in smaller enterprises, the analysis showed only a minor correlation with regard to firm size. The size of the enterprise as well as the energy demand hardly seemed to determine the extent to which energy efficiency was established in the enterprises. In contrast, the embeddedness of energy efficiency in the corporate strategy and the variety of past measures appeared to have a considerably stronger correlation.

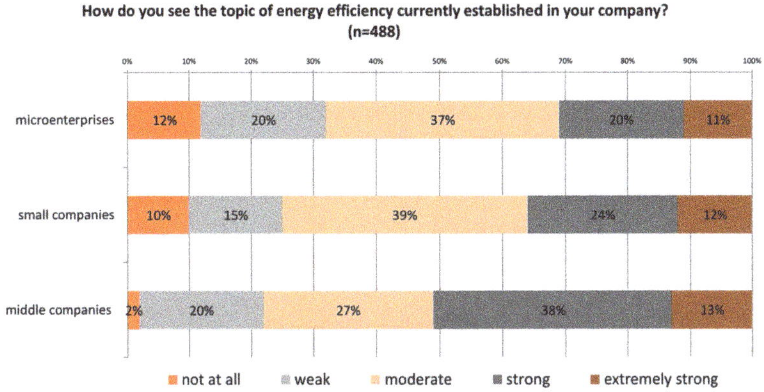

Figure 5. Establishment of energy efficiency in the SMEs.

Table 5. Factors correlating with the establishment of energy efficiency within the SMEs.

Variables	1	2	3	4	5	6	7
1. Establishment of energy efficiency	1.000						
2. Importance for corporate strategy	0.475 **	1.000					
3. Variety past energy efficiency practices	0.328 **	0.295 **	1.000				
4. Importance of employee behavior for energy savings	0.226 **	0.338 **	0.124 **	1.000			
5. Importance of energy efficiency for the environment	0.204 **	0.223 **	0.290 **	0.216 **	1.000		
6. Energy demand	0.116 *	0.119 **	0.218 **	0.071	0.161 **	1.000	
7. Firm size	0.140 **	0.094 *	0.137 **	0.232	0.116	−0.140 **	1.000

Note: n = 488; Spearman correlation. * Correlation significant at $p < 0.05$ (two sided). ** Correlation significant at $p < 0.01$ (two sided).

4.2. Energy Efficiency Practices

4.2.1. What Importance Does Energy Efficiency Have for Corporate Strategy of the SMEs?

As indicated above, the environment of an enterprise represents important frame of reference, which appears to be particularly crucial in terms of the strategic approach on energy efficiency. The case study research showed a clear link between the meanings of energy efficiency for an enterprise and the strategic approaches an enterprise makes in this respect. In one studied SME, for example, it became apparent that a positive external image, in particular, was interpreted as the most important function of all undertaken and planned efforts. Accordingly, the enterprise concentrates particularly on measures that are salient and can be distinctively presented to the outside world (e.g., photovoltaic system, e-mobility).

Nevertheless, the case study research indicated that embedding energy efficiency into corporate strategy has a positive effect on implementing technology and practices regardless of the dominant orientation. In the course of the survey, the SMEs were, therefore, asked about the importance of energy efficiency for their corporate strategy. For almost half of the surveyed SMEs, energy efficiency occupies a high (33%) or very high priority (16%) for the corporate strategy. On the other hand, only 13% of the enterprises surveyed consider energy efficiency having a low priority for their general corporate strategy, whereas 6% assume a very low priority. Considerable differences were observed regarding the enterprise size and energy demand. Both appear to have a positive effect on the embedding of energy efficiency in the corporate strategy. This result can certainly be explained by the prevalence of energy management systems according to the international standard ISO 50001, which are quite common, especially in larger or more energy-intensive enterprises in Germany [72], and explicitly require the definition of explicit strategic energy goals and policy.

4.2.2. What Importance Do Different Energy Efficiency Practices Have for the Enterprises?

The case study research indicates that the implementation of a broad range of practices—ranging from technical investments to raising awareness measures—proved to be particularly effective in establishing energy efficiency within the enterprises and tapping the energy efficiency potentials adequately. The fieldwork revealed that the enterprises undertake a variety of measures in different contexts: Simultaneously, sequentially, and sometimes even unintentionally. In the course of the specific "energy efficiency history" of the enterprises, the focus on particular practices necessarily changes over time. Thus, an issue can be treated in different contexts with different measures. The interplay of different practices that may emerge over time can be well illustrated by an example of the case study research.

A medium-sized engineering company draws its attention to its compressed air supply and starts problematizing the technical equipment. The enterprise first turns to compressed air generation, invests in new compressors, and starts monitoring energy consumption. After attending a regional information event, a maintenance employee suggests that the piping system should be checked for leakages and optimized. Top management decides to redesign the compressed air system and commissions a service provider. Although the enterprise can report a significant reduction in energy consumption, the management is not sufficiently satisfied. At a production meeting, the records of savings and energy consumption of the compressed air supply are discussed. The practical use of compressed air becomes a focal point, and the enterprise begins to inform production employees about the sensitive use of compressed air. Half a year later, the results of energy consumption show hardly any differences, and top management wonders why the measures for raising awareness have little effect and what further measures are appropriate. Under the impression that the employees are ignoring the previous measures, the company changes its approach. The quality manager is instructed to formulate working rules for the use of compressed air. At the same time, the technical team is instructed to look for ways to automate the use of air-operated machines.

As the example above indicates, the exploitation of energy efficiency potentials requires a broad spectrum of energy efficiency measures over time. Within the scope of the survey, the SMEs were asked what type of energy efficiency practices they implemented in the last three years, what measures they were currently focusing on, what measures they plan for the future (in the following three years), and which ones they do not plan to carry out at all. In the past, the SMEs mostly focused on technical-investment measures, and in the future the focus will also be placed on technical measures. Measures for raising awareness have had a high priority for SMEs and will also be held as important in the near future. Furthermore, the current focus was mostly drawn to such measures, and all other types seemed to have considerably less importance for the enterprise (Figure 6).

The relatively low importance of organizational, information-, and competence-related practices becomes even more distinct when considering the size of the enterprises. The smaller the enterprises are, the less they seem to value these measures. Additionally, the percentage of measures being not planned is noticeably higher the smaller the enterprises are. Fewer measures were implemented and will not likely be carried out in the future, particularly in micro-enterprises A similar picture appears when looking at the energy demand of the SMEs. The more energy-intensive the enterprises are, the more important the measures, other than technical investments and awareness measures, are. Organizational measures have particularly been a focal point of more energy-intensive enterprises in the past. Financial incentives by the government for the implementation of an energy audit or an energy management system (such as ISO 50001) might be a plausible explanation for this peculiarity.

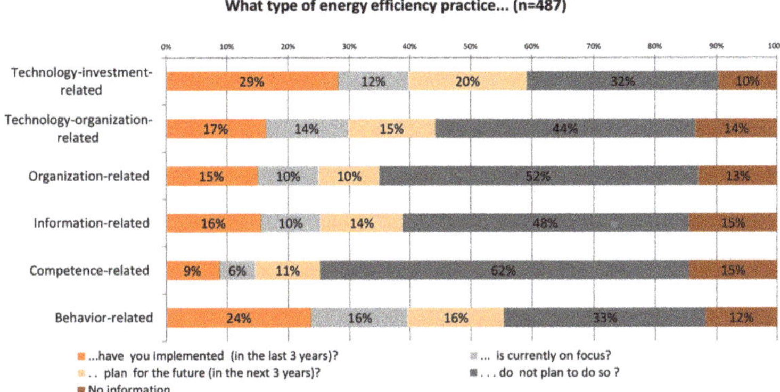

Figure 6. Importance of different types of practices.

4.2.3. What Importance Does Energy Management Have for the SMEs?

Energy management is often regarded as synonymous with the norm ISO 50001. The case studies showed that such a classification is not necessarily tenable. Although five of the 10 enterprises investigated within the case study research operated an energy management system according to the standard ISO 50001, this does not mean that the remaining enterprises did not practice energy management. On the contrary, the case studies show that those enterprises successfully conduct energy management without committing themselves to a standardized system. They embed energy efficiency issues in their corporate strategy, set up energy efficiency goals, appoint energy managers, digitize and monitor their energy consumption, plan and implement measures, train their employees, and research possible technical measures and their financing. The difference only lies in the formal structure. For example, in one case, a company did not appoint a formal energy team in the enterprise, yet an informal network of people regularly meets to discuss energy efficiency issues. In another case, employees are aware of general premises regarding energy efficiency decisions or energy saving behavior, and yet no energy policy has ever been documented. It is also noteworthy that those SMEs do not aspire to implement a standard energy management system in the future at all. Due to a lack of personnel resources and administrative and certification costs, an implementation is not a goal or viable option, especially for small SMEs.

The analysis of the individual cases indicates that the implementation of a formal management system does not necessarily guarantee effectiveness. For example, in one case, the enterprise has established formal responsibilities and an explicit energy policy, although a lack of authority to take action and employees who are unfamiliar with energy issues constrain the implementation of measures. In addition, the implementation of an energy management system can cause unintended effects. In one case, energy efficiency was mostly perceived by members of the enterprise as a forced external expectation due to the implementation process of ISO 50001. During the interviews the respondents either directly ("our management/competitors'/the customers' expectations forced us to implement … ") or rather vaguely ("we had to do it") referred to strong expectations instead of providing hardly any other motivation. This finding allows the interpretation that complying with the paragraphs of the norm and pleasing the auditors became the dominant frame of reference for interpreting energy efficiency issues, despite diametrical intentions of the top management. Additionally, and despite the rational intent of top management to institutionalize energy efficiency within the enterprise, another unintended issue became apparent in the same case. When asked about energy efficiency issues or measures, almost all interviewed persons referred to the designated energy manager, while the interviewed energy manager complained about the lacking support, especially of the production personnel, despite the establishment of an energy team consisting of such members. Roughly speaking, energy management became reduced to the face of the energy manager, who, in turn, got overwhelmed by the responsibility of managing everything on his own. The observations and interviews within the scope of the case study research indicate that those enterprises without formal energy management sometimes take much more effective measures and establish energy management effectively within the organization.

4.3. Interface between the Enterprise and Its Members

4.3.1. What Importance Does the Everyday Behavior of the Employees Have for Energy Conservation and Energy Efficiency?

The case studies showed that the everyday behavior of the employees is perceived as an important influencing factor for improving energy efficiency. Throughout every interview with management staff and personnel charged with energy efficiency tasks, the impact of everyday behavior was valued as a vital factor in achieving energy savings. In one extreme case, the energy savings were even almost exclusively attributed to changes in employee behavior. Accordingly, within the scope of the survey, the SMEs were asked how they consider the behavior of the employees in the enterprise to contribute to the success of energy savings. Almost two-thirds of SMEs (63%) consider the importance of energy-saving behavior to be important. On the other hand, only 13% of the SMEs surveyed rate the importance of employee behavior as rather or completely unimportant. No considerable differences regarding enterprise size and energy demand could be observed.

4.3.2. Who Is Perceived as Responsible for Energy Efficiency and Energy Conservation within the SMEs?

As part of the questionnaire survey, the enterprises were queried as to which organizational actors in the enterprise are responsible for energy saving and energy efficiency. Throughout the entire sample, owners and management are seen to be most responsible (Figure 7). Differences could be observed according to enterprise size and energy demand. Both factors seem to have a positive effect on the perceived distribution of responsibility. This circumstance might be attributed to the fact that larger and more energy-intensive SMEs more likely employ dedicated personnel (e.g., energy manage or environmental manager). However, the data give evidence that the lower the energy demand and the size of the SMEs are, the greater the centralization of responsibility is perceived to be.

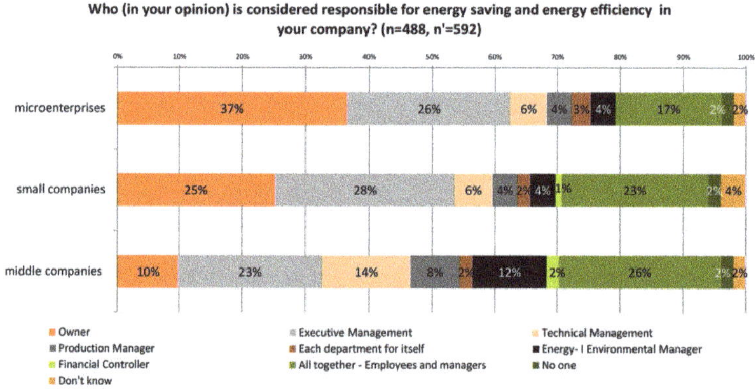

Figure 7. Responsibility for energy saving and energy efficiency in the SMEs.

4.3.3. How Do the SMEs and Their Leaders Attempt to Raise Awareness among Their Workforce?

As the case study research shows, the top management of the enterprises oftentimes puts a lot of effort into defining energy efficiency goals, structures, and processes as well as narratives as to why investments, changes, and new practices are necessary. Energy efficiency, therefore, proves to be a demanding management task from the perspective of top management. Hence, top management usually either tries to outsource this task (e.g., to a designated energy manager) or to involve key persons in support. These key persons act as energy efficiency agents within the organization and oftentimes play an important role to the organizational institution of energy efficiency. These agents do not necessarily have to be explicitly appointed energy managers. The case studies showed that the top management of the SMEs usually searches for individual personnel aware and interested in mediating, communicating, and spreading energy efficiency issues within the enterprise. Oftentimes, these persons act only informally—as informal energy managers or as an energy team.

However, both formally and informally appointed energy efficiency personnel usually require a wide set of skills and knowledge, ranging from technical, economic, and social skills as well as to knowledge about legal requirements and external actors. Considering the complexity of energy efficiency in the context of industrial organizations, this finding seems rather obvious. Less obvious seems another aspect, which was frequently expressed by the interviewed members of the investigated SMEs. From their points of view, the integrity of those in charge with managing and spreading energy efficiency within the enterprise is valued even more highly than their competences. In other words: Whoever is in place has to walk the talk.

In the context of the case study research, it was investigated which strategies the SMEs pursue in order to promote and enforce energy-saving behavior within the enterprise. Four different strategic approaches were identified: (1) Raising awareness (e.g., creation of consciousness by trainings, empowerment, or speech); (2) motivation (e.g., promotion of self-interest by sanctions, incentives, or job roles); (3) regulation (e.g., establishment of conformity by formal or informal work rules); and (4) automation (e.g., avoidance of human risks by technical measures). These approaches represent "ideal types" [73], which do not occur in pure form in the enterprises. Rather, the enterprises mix and complement, for example, raising awareness practices with formal rules or automation measures. In addition, it was not necessarily possible to determine which typical approach would be the most effective—the individual competencies, qualifications, and corporate cultures (meaning general norms, beliefs, ideas, and routines) considerably shape what is feasible.

However, the case studies showed that raising awareness among the employees is the most important strategic approach to foster energy efficiency decisions and energy-saving behavior. Actions for raising awareness usually mark the starting point for actions in the enterprises to address energy efficiency issues. The focus of how the enterprises attempt to shape the individual behavior and decision making of its members might shift over time. For example, an enterprise perceives the undertaken raising of awareness measures as ineffective and decides to set up rigorous work rules instead. In another case, formal work rules seem not to work sufficiently and the company theorizes that monetary incentives might generate more self-interest and motivation among its employees to save energy. In a rather extreme cases, the focus shifts solely to the automation of processes, as no other strategy has proven effective in the past. Mainly relying on automation, therefore, represents an avoidance strategy.

Top management and key personnel often spend a lot of time and effort situating attention on energy issues among the workforce. Occasionally, they feel themselves becoming "energy educators" within the enterprise. At the same time, top management often experiences encouraging energy saving among the employees to be a daunting task. Actions for raising awareness are sometimes perceived as "Sisyphus work", as one managing director described it graphically. Similarly, many of the interviewed top management personnel or energy managers (formal and informal) complained of the challenging nature of raising awareness for energy-saving behavior. From their points of view, those tasks are frequently associated with high affectivity (e.g., incomprehension, frustration, annoyance). Through formal speech, discussion, and storytelling, they facilitate knowledge, values, and beliefs about energy efficiency issues. Drawing attention to energy issues in everyday interactions proves to be particularly important to establish an alert energy efficiency climate. However, not every enterprise or manager is willing or able (e.g., due to lack of time, competence, or patience) to perform these educational tasks. In their defense, the enterprises often claim the lack of competent personnel as an obstacle to raising awareness ambitions. Additionally, the extent to which raising awareness measures might succeed depends strongly on the individual characteristics attributed to the "energy educators" in charge. As mentioned above, not only do they have to demonstrate sufficient knowledge (e.g., technical, practical, social knowledge), but the employees' perception of integrity seems to be equally important.

4.4. Interface between the Enterprise and the Environment

4.4.1. Regulative: How Do the SMEs Perceive External Imperatives for Energy Efficiency?

Undoubtedly, increasing energy efficiency represents a rising political and social expectation, which is an expectation that can be perceived by enterprises as a manifest regulative demand (such as large enterprises by statutory energy audits) or rather latent imperative (such as political or medial discourse by spreading values, ideas, and beliefs). While practical or legal imperatives were hardly mentioned, more diffuse imperatives became particularly evident during the interviews in the course of the field research. The need for increased energy efficiency was frequently expressed as a rather vague expectation an industrial organization has to live up to nowadays. More concrete, the interview partners referred to the expectations of customers, national policies, or the local communities as reasons for an increased attention on energy efficiency issues.

As part of the questionnaire survey, the enterprises were asked to which external actors they attribute the demand for energy efficiency. As Figure 8 shows, the SMEs largely attribute the demand for energy efficiency to national and global political actors, followed by the society as a whole and industry associations at a distance. Hardly any noticeable differences in answering could be identified regarding the size of the enterprises or their energy demand.

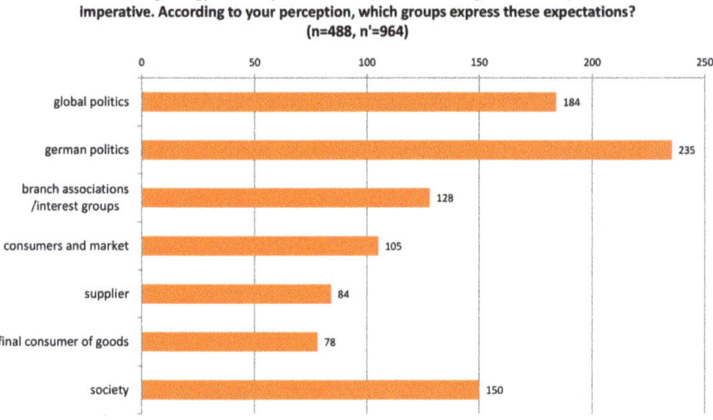

Figure 8. The attribution of the external imperative for energy efficiency.

Even more important than finding out to which actors the expectations are attributed is the question of their acceptance. As far as the participating SMEs of the case study research is concerned, the acceptance was usually positively rated. Nevertheless, in such a face-to-face setting the risk of a socially desirable response is certainly significantly higher than in anonymized questionnaire survey. As part of the questionnaire survey, the SMEs were further asked how the expectation for energy efficiency is perceived by the enterprises. Most enterprises perceive the general imperative for energy efficiency positively (39.2%) or rather positively (30.6%). Only about 4% of the SMEs perceive this imperative as negative. In terms of enterprise size or energy demand, no significant differences between the enterprises could be observed.

4.4.2. Economic–Financial: To What Extent Is the Financing of Measures Considered as an Obstacle by the SMEs?

As discussed above, technical measures have a high priority for the SMEs in general. At the same time, especially investment in capital¬–intensive technical measures (e.g., combined heating and power station) can mean a considerable financial outlay for small enterprises. In the course of the ethnographic field research, the financing of energy-efficient technology was a broadly discussed topic on many occasions. Either in interviews with decision makers, controlling staff, or even in meetings with consultants, which could be observed, the financing of measures was surprisingly never articulated as an obstacle. Although the decision makers usually pointed out that investments must pay off, mostly between a range of two to five years, the financing was presented as an uncritical endeavor.

In the survey, the enterprises were accordingly asked if the financing of energy efficiency measures is an obstacle. From the points of view of the SMEs surveyed, the financing of energy efficiency measures is not a clear obstacle. While about 28% agree that it is, about 35% of the interviewees do not perceive the financing as an obstacle and 37% neither agree nor disagree. No considerable differences could be observed regarding the size of the enterprises or the energy demand.

4.4.3. Normative: What Information Sources Do the SMEs Use and How Actively Do They Search for Information?

In the context of the field research, the question of how the participating SMEs obtain information for measures and to which actors and normative guidelines they orient themselves was investigated. From their perspectives, at least basic information on energy efficiency measures are rather easily available from various sources. On the contrary, the perception of an inflation of information and consulting services prevails among the enterprises. In every SME at least one person reported of being literally bombarded, usually several times a week, with inquiries or advertising, especially from consultancy firms. This abundance oftentimes leads to the fact that most of it is ignored (and must be ignored) and a skeptical view of the trustworthiness of the entire field develops overall. This aspect is closely related to previous experiences with consultants. A wide spread of experiences can be stated from very good to especially bad experiences. From the points of view of the enterprises, "black sheep" affect the general trustworthiness in the consulting and service industry. Hence, in addition to general qualifications and skills, it is above all the trustworthiness of the actors that the enterprises question and expect. As part of the questionnaire, SMEs were surveyed as to where they gain information on energy efficiency measures (Figure 9). The professional journals are clearly the most important sources of information for SMEs. Information by consultancy firms is more valued by medium-sized enterprises than by smaller enterprises. Subsequently, the enterprises were queried about their research behavior on measures. About one-third search actively, one-third search moderately, and one-third search only rarely. Differences with regard to energy demand were hardly noticeable, but clear differences regarding the size of the enterprises were. The larger the enterprises are, the more actively information on measures seems to be sought. Around 15% of micro-enterprises do not even look for information on energy efficiency measures. This result can most likely be explained by lack of human resources in smaller enterprises.

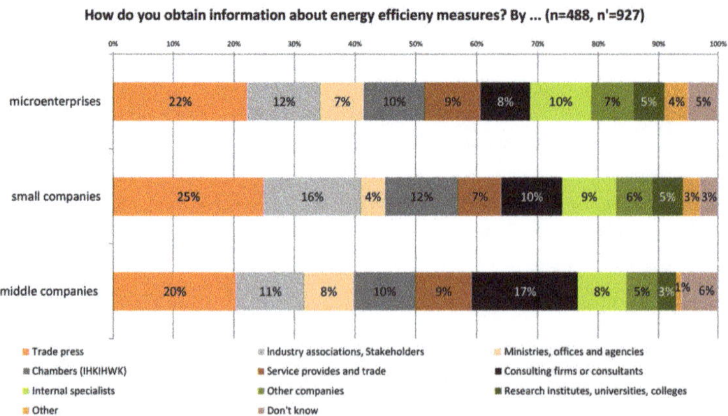

Figure 9. Information sources about energy efficiency measures.

4.4.4. Cultural–Cognitive: What Importance Does Energy Efficiency Have for the Environment of the SMEs and to What Extent Does it Influence the Decisions of the SMEs?

In order to capture the influence of external expectations on energy efficiency decisions, the SMEs were confronted with a two-step question. First, the companies were queried as to how they perceive the importance of energy efficiency for their business environment. As Figure 10 shows, customers are most likely attributed to valuing energy efficiency as very important by the enterprises. The importance for the local environment, competitors, owners, and professional groups is perceived considerably lower, but at a similar level.

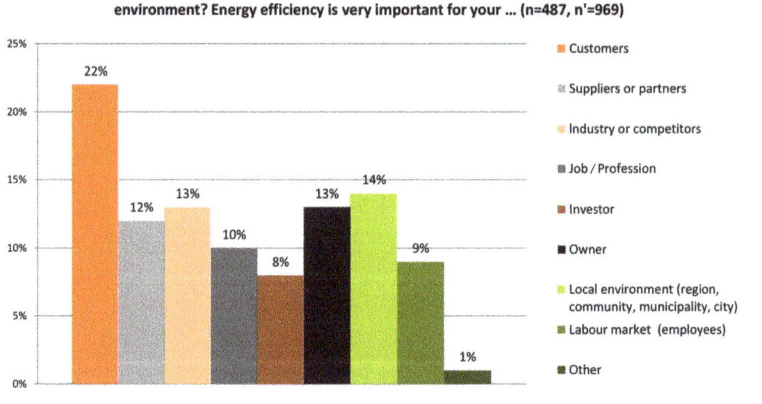

Figure 10. The perception of the importance of energy efficiency for the environment of the SMEs.

Subsequently, the SMEs were asked what influence these external groups and actors would have on the decisions about energy efficiency measures (Figure 11). Customers appear to have the greatest influence on energy efficiency decisions, followed at a clear distance by the local environment, competitors, and the owners.

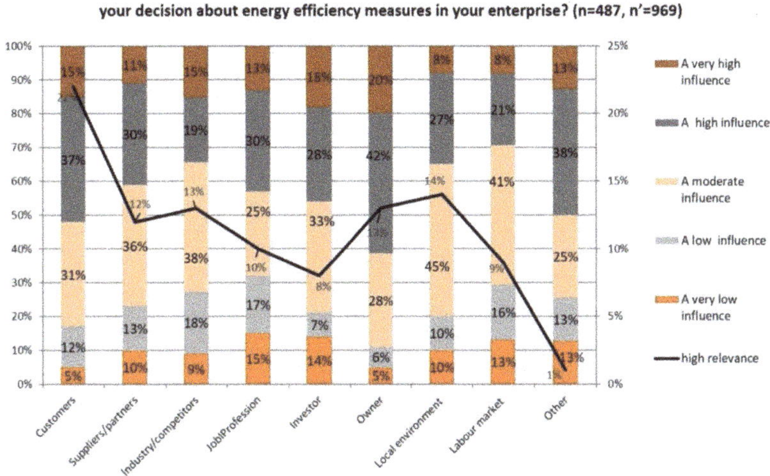

Figure 11. The perception of the importance of energy efficiency for the environment of the SMEs.

5. Discussion

Overall, this study shows that energy efficiency seems fairly well established among the surveyed SMEs. The descriptive survey data and correlation analysis indicate no considerable driving influence of the energy demand of the SMEs. Regarding firm size, the descriptive analysis signifies a lesser importance of energy efficiency, less establishment within the enterprises, a higher centralization of responsibility, and less usage of organizational energy efficiency practices among smaller SMEs. However, the correlation analysis indicates only a minor influence on the establishment of energy efficiency within the enterprises and rather the embedding of energy efficiency in corporate strategy, the usage of a broad spectrum of practices, and the strong importance of employee behavior for energy conservation appear as crucial aspects. One possible interpretation of this result is a higher energy demand or a larger firm size do not guarantee these aspects. It seems more likely that the social and cultural contexts shape how the SMEs and their decision makers approach energy efficiency issues. In this sense, the individual results can be read as a description of this contexts.

Starting by analyzing the energy efficiency climate, the meanings of energy efficiency for and within the SMEs were elucidated. Energy efficiency is first and foremost associated with cost reduction—undoubtedly this serves as a major motivator for enterprises to draw attention to energy efficiency. However, energy efficiency can have much broader meanings for the enterprises and their members. Meanings such as differentiation from competitors, conformity to a perceived industrial state of the art, or the aspiration of a positive public image are crucial aspects on how energy efficiency issues are treated by the enterprises, especially with regards to internal competition against other issues, and project a dense web of meanings of energy efficiency that seems vital to generate attention and decisions for measures and new practices. Those additional meanings of energy efficiency refer to four aspects.

First, it indicates that meanings of energy efficiency do not solely depend on internal criteria such as cost reduction or profitability, but also on external criteria. Meanings, such as social or ecological responsibility, modernity, and progressiveness, or the desire for a positive external image are implausible without recourse to expectations of society and the corporate environment. Similarly, an increasing number of studies have already shown that issues such as climate change, energy conservation, and environmental pollution more and more influence practices in SMEs. [14,33,74,75]. Second, it indicates that energy efficiency is perceived as an increasing social obligation – an obligation endorsed by the majority of the surveyed SMEs. Third, the meanings of energy efficiency can sometimes have destructive connotations for individual members of the enterprises due to negative experiences. Fourth, the meanings reflect the benefits of energy efficiency practices, oftentimes distinguished between energy and non-energy benefits [76,77]. As these benefits constitute the source of strategic decisions on energy efficiency [78], their consideration is particularly important for practitioners as well as future research. Overall, the results from fieldwork and survey data indicate that the denser and broader the (positive) meanings for energy efficiency are, the more likely an enterprise will embed energy efficiency in its strategy and the more likely decision makers will opt for energy efficiency measures. In other words: Energy efficiency must accommodate multifunctional meanings for the enterprises [79] to generate attention.

By focusing on the practices, it was shown that the SMEs consider, plan, and carry out a variety of energy efficiency practices in everyday work life. Although the majority of the surveyed enterprises concentrate primarily on technical measures, behavioral measures are rated as equally important. In comparison, organizational measures are perceived to be substantially less relevant according to the survey results, even though the case-study research indicated the driving aspects of organizational practices in establishing energy efficiency within the enterprises. The underestimation of organizational practices by the SMEs might be explained by the informal way the enterprises oftentimes manage their staff and energy use. This might also explain the remarkably high importance of behaviorally related practices.

Energy management is often considered a vital means for enterprises to overcome barriers and to improve energy efficiency [80,81]. According to Thollander and Palm, industrial enterprises that adopt energy management practices may reduce their energy use by up to 40 % [82] (p. 102). Similarly, the results indicate that the embedding of energy efficiency in corporate strategy combined with a broad spectrum of different practices and the distribution of attention by organizational measures considerably drive the establishment of energy efficiency within enterprises and foster the improvement of energy savings. From this point of view, energy management is the key factor in the institutionalization of energy efficiency within enterprises. Yet, the implementation of an energy management system (ISO 50001) is oftentimes not a viable option or aspiration, especially for small SMEs. However, as the case studies have shown, SMEs can operate effective energy management without being bound to a norm, as long as they follow basic principles.

According to the analysis, the existence of an energy efficiency strategy appears more effective than typical structural characteristics, which are identified in other studies as major influencing variables in raising attention for energy efficiency issues. This includes firm size [38,56] or the energy demand [52]. The finding of the importance of an energy efficiency strategy essentially coincides with the investigations of industrial enterprises (SMEs and large enterprises) in Sweden [83,84]. According to the authors, the existence of a "long-term strategy" in the enterprises is a key driver for energy efficiency measures. Similarly, the analysis underlines the importance of developing and embedding an energy efficiency strategy in SMEs. Additionally, the results indicate the importance of embedding energy efficiency in corporate strategy for establishing energy efficiency within enterprises—not only by structuring goals and measures, but also by providing symbolic and cognitive frames of references.

The interviews with top management personnel and the informal or formal members of the energy teams in the context of the case study research showed one thing very clearly: Establishing energy efficiency within the enterprise is by no means a trivial task and usually means initiating a permanent change process, which is a process particularly challenging to the top management and responsible personnel. Due to their decisions, actions, and interactions, they inevitably convey the meanings of energy efficiency for the enterprise, thus providing a frame of reference for the organizational members. If, like in one studied case, energy efficiency is framed by the top management only as the fulfilment of an external and unpleasant requirement, it is highly probable that the employees will also interpret corresponding tasks as an annoying duty. This aspect represents the symbolic aspect of leadership, which should not be underestimated.

Looking at the intersection between the enterprise and its members, the study first and foremost shows how important most of the investigated SMEs value the everyday efforts of the employees to increase the energy savings in the SMEs. The results show that raising awareness among the employees is an important issue for the surveyed SMEs. Raising awareness is the first and most relevant strategy of the SMEs to create energy savings. On the other hand, it is also oftentimes perceived as a frustrating task by the management personnel. Nevertheless, it is considered as a necessary task to put energy issues on broad shoulders. In virtually all cases the top management of the SMEs at least tried to establish a broad attention for energy efficiency among their staff. Either via the installation of formal energy teams or the involvement and empowerment of informal key personnel, the enterprises distribute attention for energy issues. It seems essentially irrelevant whether these networks of responsibilities exist formally or informally. For instance, and with regard to the case study research, informal energy teams can be equally effective as formal energy teams in driving energy efficiency measures or energy-saving behavior. Similarly, the empowerment of the production personnel by granting authorities (e.g., for internal trainings) and responsibilities (e.g., for the implementation of measures or monitoring tasks) can sometimes be far more effective than leaving all issues to a single explicit energy manager. From the perspectives of the studied enterprises and their top management, practices to stem energy efficiency issues on broad shoulders are a necessity to make the increasing complexity of energy efficiency manageable.

Energy efficiency as leadership process must, therefore, not be characterized by the centralization of decisions. On the contrary, the general complexity of industrial energy efficiency requires the decentralization of attention, responsibility, and authority. The involvement of key persons ("energy efficiency agents") is, therefore, of particular importance. Thollander and Palm have already pointed out that a "strong leadership in combination with delegated authority is crucial" [82] (p. 102) for effective energy management. The distribution of attention and responsibility is also significant for another reason: Exploiting energy efficiency potentials will not necessarily become less complex in the long term, for example, due to new technologies, legal frameworks, or energy market dynamics. The general complexity of industrial energy efficiency requires a decentralization of attention, responsibility, and authority in SMEs. The involvement of key personnel ("energy efficiency agents") and organizational measures, therefore, holds particular importance and will likely become an increasing necessity for SMEs in the long run.

Financial aspects are often cited as key barriers to the adoption of energy efficient technology, particularly the access to capital [54]. An extensive study from Anderson and Newell among manufacturing SMEs participating on a volunteer assessment program in the US frequently mentioned an insufficient cash flow as a barrier [85]. Similarly, a survey among 50 Greek industrial firms shows that two-thirds of the respondents stated no access to capital and the high cost of implementation as a barrier [86]. However, the results indicate that the financing of energy efficiency measures is not a considerable obstacle by a majority of the SMEs studied. This finding raises the question of whether financial incentives are an effective policy instrument for the promotion of energy-efficient technologies in the case of SMEs in Germany.

Although SMEs in Germany are not (yet) politically forced to implement energy efficiency practices, they nevertheless perceive a discursive imperative to do so. The analysis shows that SMEs attribute this imperative to national and global political discourse and societal expectations in general. Furthermore, the results show that the majority of the SMEs accept this political and societal imperative. Looking at recent movements, such as Fridays for Future, the expectations on industrial enterprises will likely increase in the future. Fawcett and Hampton assume that the increase in public concern and discourse related to environmental issues becomes more salient for SMEs [35] (p.4). Drawing on interviews with 20 SME owners in Liverpool, UK, North and Nurse identified morality—meaning social and pro-environmental values in line with regional cultural mores—as a key driver for the engagement in efforts to reduce carbon emissions [74].

The lack of information on technology, costs, and benefits as well as the trustworthiness of the information sources are frequently identified as barriers to energy efficiency measures [56,87]. According to the presented results, two aspects seem particularly relevant with regard to informational issues. First, information sources close to the core operations of the enterprises (such as trade press and industrial associations) are more likely to be listened to than professional groups such as service providers. Second, the analysis indicates that SMEs perceive an inflation of consulting services, which seems to compromise the trustworthiness of external service providers.

6. Conclusions and Future Research

The main contribution of this research article is to offer a perspective on energy efficiency in industrial SMEs beyond the mere adoption of energy-efficient technology. With this study, light was shed on how energy efficiency is established within SMEs in the context of their organizational context and institutional environment. Furthermore, it was presented which practices the enterprises undertake, how energy efficiency is managed, how the SMEs are challenged to raise awareness for energy savings among their members, and how institutional expectations are perceived.

This study points out the meanings of energy efficiency as being socially produced by the industrial organizations, its members, and environment. According to sociological neo-institutionalism theory, organizations adopt practices and structures that are perceived as desirable or appropriate within some socially constructed systems of norms, values, beliefs, and definitions [47,88]. Hence, organizational decisions are considered legitimate if they appear desirable and appropriate when measured against the social values, norms, and beliefs of their environment. Apparently, as energy efficiency is perceived as holding strong importance for the environment of the enterprises, decisions on energy efficiency practices are more likely to be constituted. This result can be interpreted as an opportunity for policy making toward developing more value-driven narratives of energy efficiency, emphasizing the moral obligation equivalent to the economic benefits. Particularly in the context of the recent rise in public concern about climate change, a stronger political debate on energy efficiency issues would increase the moral imperatives the SMEs are open to oblige, as the results indicate.

In addition, this article shows how important the everyday actions of employees in the SMEs are considered for increasing energy efficiency. On the other hand, the sometimes frustratingly perceived efforts to raise awareness within the enterprises were also highlighted. Nevertheless, these practices point to the importance of SMEs and their decision makers as change agents. Their everyday efforts in establishing energy efficiency constitute important "institutional work" [89]. A strengthening of the social and political discourse would, therefore, also strengthen the legitimacy of the efforts of change agents within the enterprises.

Considering the increasing complexity of energy efficiency issues and the conclusion about the driving effects of using a broad repertoire of practices, the institutional facilitation of knowledge seems crucial to us in the long run. The situation of SMEs, in particular, requires attention because, as Cagno et al. state, around two-thirds of the SMEs in Europe "do not implement even simple rules to manage the energy use" [90] (p. 1256). The establishment of basic knowledge and awareness for energy efficiency issues should become a mandatory part of professional education (e.g., industrial job profiles, trainings). Considering that the financing of measures does not present a difficult obstacle for the surveyed SMEs, this approach would possibly be more effective than financial incentives in the long run. The support of chambers and industry associations that work directly with SMEs could prove beneficial in spreading information and knowledge, as Fresner et al. [91] showed in terms of engaging SMEs in energy efficiency audits.

It is in the nature of scientific studies that their results reveal limitations or open up new questions. The present study is no exception in this respect. A larger sample size could be the subject of further research to validate and enrich the results. For instance, we were unable to make sufficient comparisons of different sectors. The focus on individual sectors could bring interesting questions to light. Additionally, a comparison between large enterprises and SMEs could reveal considerable differences in practices, motivations, or needs. Furthermore, in-depth comparisons between federal or national states could provide insights into how different institutional contexts shape the focus on different energy practices and the establishment of energy efficiency within industrial enterprises. However, a methodological prerequisite for this would be a parallel analysis of the specific institutional conditions and the specific discourse around energy efficiency.

Author Contributions: Conceptualization, W.K., S.L., S.B., and C.S.; Methodology, W.K. and S.B.; Investigation: Ethnographic research, W.K.; Investigation: Survey, C.S.; Validation, S.L.; Visualization, C.S.; Resources, S.L. and S.B.; Software, S.B.; Data curation, W.K. and S.B.; Writing—Original draft, W.K.; Writing—Review and editing, S.L., S.B., and C.S.; Project administration, W.K. and S.L.; Supervision, funding acquisition, S.L. All conclusions, errors, or oversights are solely the responsibility of the authors. All authors have read and agreed to the published version of the manuscript.

Funding: This research was funded by Ministry of Science, Research and Culture Baden-Württemberg (Research program: Innovative projects, 2016) within the scope of the "Decisions for Energy Efficiency" project (project no. 2015.01-Löbbe). The article processing charge was funded by the Baden-Württemberg Ministry of Science, Research and Culture Baden-Württemberg in the funding program Open Access Publishing.

Conflicts of Interest: The authors declare no conflict of interest.

Appendix A

Questionnaire of the Study

1. What position do you have in the company? (if you have several positions please list the highest ranking, (Single selection)

   - ☐ Managing Director
   - ☐ Owner
   - ☐ Energy Manager or Energy Officer
   - ☐ Technical Manager
   - ☐ Production Manager
   - ☐ Controlling
   - ☐ Other

2. How many employees does your company have? (Single selection)

   - ☐ 1–9
   - ☐ 10–49
   - ☐ 50–249

3. What was the turnover in your company in the last financial year (million EUR)? If you are not sure, please estimate it. If you cannot estimate it, please try to indicate the turnover by selecting below. (Single selection)

   - ☐ under 250.000 Euros
   - ☐ 250.000 to less than 500.000 Euros
   - ☐ 500.000 to less than 1 million Euros
   - ☐ 1 million to less than 2 millions Euros
   - ☐ 2 million to less than 5 millions Euros
   - ☐ 5 millions to less than 10 millions Euros
   - ☐ 10 millions to less than 25 millions Euros
   - ☐ 25 millions to less than 50 millions Euros
   - ☐ 50 millions to less than 100 millions Euros

    ☐   100 millions to less than 500 millions Euros

    ☐   500 millions Euros and above

4. What was the energy demand (all energy sources such as electricity, gas, oil, etc.) in your company in the last 12 months in megawatt hours (MWh)? If you are not sure, you can estimate it or give the composition of your energy needs. _______________________MWh If you cannot estimate the energy demand, please try to specify the energy demand in the following categories. (Single selection)

    ☐   under 10 MWh

    ☐   10 to less than 50 MWh

    ☐   50 to less than 100 MWh

    ☐   100 to less than 500 MWh

    (a)   500 to less than 1.000 MWh

    ☐   Less than 2.500 MWh

    ☐   2.500 to less than 5.000 MWh

    ☐   5.000 to less than 10.000 MWh

    ☐   10.000 to less than 50.000 MWh

    ☐   50.000 MWh and above

or give the composition of your energy needs.

    ☐   Electricity approx.___________________(in MWh)

    ☐   Coal approx.___________________(stating an unit_______)

    ☐   Oil approx.___________________(stating an unit_______)

    ☐   Gas approx.___________________(stating an unit_______)

    ☐   District heating approx.___________________(stating an unit_______)

    ☐   Biomass approx.___________________(stating an unit_______)

    ☐   Other approx.___________________(stating an unit)_______)

5. In which year was your company founded?

    ______ (Year)

6. What type of energy efficiency measures have you "implemented" (in the last 3 years), "is currently on focus", "plan for the future"(in the next 3 years) and "do not plan to do so" (in the next 3 years? (please mark one answer per question)

7. How do you see the subject of energy efficiency currently being established in your company? (Single selection)
(1 = Very strong, 3 = moderate, 5 = not all)

        ☐1 ☐2 ☐3 ☐4 ☐5 ☐ Don't know

8. How do you estimate the importance of energy efficiency and energy saving in everyday work for the workforce in the company? (Single selection)
(1 = Very high, 3 = moderate, 5 = very low)

        ☐1 ☐2 ☐3 ☐4 ☐5 ☐ Don't know

	... "Have You Implemented ?" (*Multiple Selection*)	... "Is Currently on Focus ?" (*Multiple Selection*)	... "Plan for the Future ?" (*Multiple Selection*)	... "Do not Plan to Do so ?" (*Multiple Selection*)
Technical investment (*e.g. procurement of energy-efficient technology*)	□	□	□	□
Technical and organizational (*e.g. energetic optimized process control*)	□	□	□	□
Organizational (*e.g. energy audit, energy team*)	□	□	□	□
Information-related (*e.g. energy monitoring, energy consulting*)	□	□	□	□
Competence-related (*e.g. workshops, training courses*)	□	□	□	□
Awareness and behavior related (*e.g. employee sensibilities, rules of conduct*)	□	□	□	□

9. How important is energy efficiency currently for the general corporate strategy? (Single selection)
   (1 = Very high priority, 3=a moderate priority, 5=a very low priority)

   □1   □2   □3   □4   □5   □ Don't know

10. How important do you rate the behavior of employees in the company as a contribution to the achievement of energy savings? (Single selection)
    (1 = Very important, 3 = moderate, 5 = unimportant)

    □1   □2   □3   □4   □5   □ Don't know

11. Who (in your opinion) is responsible for energy saving and energy efficiency in your company? (Multiple selection)

    □   Managing director or owner
    □   Energy manager or energy officer
    □   Technical management or single department
    □   Everyone—Employees and directors
    □   No one
    □   Don't know

12. Increasing energy efficiency in industry represents an increasing social and political demand. Which groups or actors do you attribute to these demands? (Multiple selection)

    □   Global politics/world politics
    □   National/German politics
    □   Industry associations and stakeholders
    □   Customers and sales market
    □   Suppliers
    □   End customers
    □   Society
    □   Don't know

13. How is this demand for energy efficiency perceived in your company? (Single selection)
    (1 = Positive, 3 = indifferent, 5 = negative)

    □1 □2 □3 □4 □5 □ Don't know

14. Energy efficiency measures often require investment. Which economic possibilities of the realization of measures are suitable for your enterprise? (Multiple selection)

    □   Equity capital
    □   Bank credit
    (a)   Sponsored loan and credit
    □   Funding and subsidies
    □   Contracting
    □   Other
    □   Don't know

15. The financing of energy efficiency measures is an obstacle for my company. (Single selection)
    (1 = Strongly agree, 3 = partly agree, 5 = strongly disagree)

    □1 □2 □3 □4 □5 □ Don't know

16. How do you obtain information about energy efficiency measures? (Multiple selection)

    □   Trade press
    □   Industry associations, stakeholders
    □   Ministries, offices and agencies
    □   Commerce Chambers (IHK/HWK)
    □   Service providers and trade
    □   Consulting firms or consultants
    □   Internal specialists
    □   Other companies
    □   Research institutes, universities, colleges
    □   Other
    □   Don't know

17. How active is your company looking for information about energy efficiency measures? *(Single selection)*
    (1 = Very active, 3 = moderate, 5 = inactive)

    □1 □2 □3 □4 □5 □ Don't know

18. Importance of your environment: (1) "How do you rate the importance of energy efficiency for your x(see below)-environment?" (2) "Which influence does this setting have on your decision on energy efficiency measures in your company?" (Matrix question)

	(1) "How Do You Rate the Importance of Energy Efficiency for Your Environment? " Energy Efficiency Is Very Important for Your (*Multiple Selection*)	(2) "Which Influence Does This Environment Have on Your Decision on Energy Efficiency Measures in Your Company?" (1 = A Very High Influence, 3 = a Moderate Influence, 5 = a Very Low Influence) (*Selection Only for Those Mentioned in (1)*)					
Customers	□	□1	□2	□3	□4	□5	□Don't know
Suppliers or partners	□	□1	□2	□3	□4	□5	□Don't know
Industry or competitors	□	□1	□2	□3	□4	□5	□Don't know
Job/Profession	□	□1	□2	□3	□4	□5	□Don't know
Investor	□	□1	□2	□3	□4	□5	□Don't know
Owner	□	□1	□2	□3	□4	□5	□Don't know
Local environment (region, community, municipality, city)	□	□1	□2	□3	□4	□5	□Don't know
Labor market (employees)	□	□1	□2	□3	□4	□5	□Don't know
Other	□	□1	□2	□3	□4	□5	□Don't know
Don't know	□	□1	□2	□3	□4	□5	□Don't know

19. How important is energy saving to you personally? (Single selection)
    (1 = Very important, 3 = moderate, 5 = unimportant)

    □1　□2　□3　□4　□5　□ Don't know

20. To which extension does the topic of energy efficiency affect you in your everyday life in the company? (Multiple selection)

    □　decide on measures and actions
    □　am involved in decisions
    □　am looking for information about energy efficiency
    □　measure and monitor energy flows or consumption
    □　try to save energy personally
    □　I'm hardly or not affected by the topic
    □　am affected in a different way
    □　Don't know

References

1. Trianni, A.; Cagno, E.; Farne, S. Barriers, drivers and decision-making process for industrial energy efficiency: A broad study among manufacturing small and medium-sized enterprises. *Appl. Energy* **2016**, *162*, 1537–1551. [CrossRef]
2. Destatis. Statista Dossier on Small and Medium-Sized Enterprises (SMEs) in Germany. 2019. Available online: https://de.statista.com/statistik/studie/id/46952/dokument/kleine-und-mittlere-unternehmen-kmu-in-deutschland (accessed on 28 February 2020).
3. Thollander, P.; Backlund, S.; Trianni, A.; Cagno, E. Beyond barriers—A case study on driving forces for improved energy efficiency in the foundry industries in Finland, France, Germany, Italy, Poland, Spain, and Sweden. *Appl. Energy* **2013**, *111*, 636–643. [CrossRef]
4. Jaffe, A.B.; Stavins, R.N. The energy-efficiency gab. What does it mean? *Energy Policy* **1994**, *22*, 804 810. [CrossRef]
5. Gerarden, T.G.; Newell, R.G.; Stavins, R. *Addressing the Energy-Efficiency Gap*; Harvard Environmental Economics Program: Cambridge, MA, USA, 2015.

6. DeCanio, S.; Dibble, C.; Amir-Atefi, K. The importance of organizational structures for the adoption of innovations. *Manag. Sci.* **1998**, *46*, 1285–1299. [CrossRef]

7. Weber, L. Some reflections on barriers to the efficient use of energy. *Energy Policy* **1997**, *25*, 833–835. [CrossRef]

8. Cagno, E.; Worrell, E.; Trianni, A.; Pugliese, G. A novel approach for barriers to industrial energy efficiency. *Renew. Sustain. Energy Rev.* **2013**, *19*, 290–308. [CrossRef]

9. Sudhakara Reddy, B. Barriers and drivers to energy efficiency—A new taxonomical approach. *Energy Convers. Manag.* **2013**, *74*, 403–416. [CrossRef]

10. Buettner, S.M.; Bottner, F.; Sauer, A.; Koenig, W.; Loebbe, S. Barriers to and decisions for energy efficiency: What do we know so far? A theoretical and empirical overview. In Proceedings of the 2018 Eceee Industrial Summer Study, Berlin, Germany, 10–13 June 2018.

11. Sorrell, S.; O'Malley, E.; Schleich, J.; Scott, S. *The Economics of Energy Efficiency*; Edward Elgar Publishing: Cheltenham, UK, 2004.

12. Fleiter, T.; Worrell, E.; Eichhammer, W. Barriers to energy efficiency in industrial bottom-up energy demand models—A review. *Renew. Sustain. Energy Rev.* **2011**, *15*, 3099–3111. [CrossRef]

13. Parker, C.M.; Redmond, J.; Simpson, M. A review of interventions to encourage SMEs to make environmental improvements. *Environ. Plan. C Gov. Policy* **2009**, *27*, 279–301. [CrossRef]

14. Revell, A.; Stokes, D.; Chen, H. Small businesses and the environment: Turning over a new leaf? *Bus. Strategy Environ.* **2010**, *19*, 273–288. [CrossRef]

15. Wilhite, H. Energy Consumptions as Cultural Practice: Implications for the Theory and Policy of Sustainable Energy Use. In *Cultures of Energy: Power, Practices, Technologies*; Strauss, S., Rupp, S., Love, T., Eds.; Routledge, Taylor & Francis Group: London, UK; New York, NY, USA, 2013; pp. 60–72.

16. Trianni, A.; Cagno, E.; De Donatis, A. A framework to characterize energy efficiency measures. *Appl. Energy* **2014**, *118*, 207–220. [CrossRef]

17. Marchi, B.; Zanoni, S.; Zavanella, Z.E. Energy efficiency measures for refrigeration systems in the cold chain. In Proceedings of the XXIV Summer School "Francesco Turco"—Industrial Systems Engineering, Brescia, Italy, 11–13 September 2019. [CrossRef]

18. Marchi, B.; Zanoni, S. Supply Chain Management for Improved Energy Efficiency: Review and Opportunities. *Energies* **2017**, *10*, 1618. [CrossRef]

19. Perroni, M.G.; Gouvea da Costa, S.E.; Pinheiro de lima, E.; Vieira da Silva, W. The relationship between enterprise efficiency in resource use and energy efficiency practices adoption. *Int. J. Prod. Econ.* **2017**, *190*, 108–119. [CrossRef]

20. McCardell, S. *Energy Effectiveness: Strategic Objectives, Energy and Water at the Heart of the Enterprise*; Springer International Publishing AG: Cham, Switzerland, 2018.

21. Palm, J.; Thollander, P. An interdisciplinary perspective on industrial energy efficiency. *Appl. Energy* **2010**, *87*, 3255–3261. [CrossRef]

22. Backlund, S.; Thollander, P.; Palm, J.; Ottosson, M. Extending the energy efficiency gap. *Energy Policy* **2012**, *51*, 392–396. [CrossRef]

23. Brunke, J.-C.; Johannsson, M.; Thollander, P. Empirical investigation of barriers and drivers to the adoption of energy conservation measures, energy management practices and energy services in the Swedish iron and steel industry. *J. Clean. Prod.* **2014**, *84*, 509–525. [CrossRef]

24. Sa, A.; Paramonova, S.; Thollander, P.; Cagno, E. Classification of Industrial Energy Management Practices—A case study of a Swedish foundry. *Energy Procedia* **2015**, *75*, 2581–2588. [CrossRef]

25. Banks, N.; Fawcett, T.; Redgrove, Z. *What Are the Factors Influencing Energy Behaviours and Decision-Making in the Non-Domestic Sector? A Rapid Evidence Assessment*; CSE: Bristol, UK, 2012.

26. Cooremans, C. Investment in energy efficiency: Do the characteristics of investments matter? *Energy Effic.* **2012**, *5*, 497–518. [CrossRef]

27. Perrow, C. Organizing for environmental destruction. *Organ. Environ.* **1997**, *10*, 66–72. [CrossRef]

28. Palm, J. Placing barriers to industrial energy efficiency in a social context: A discussion of lifestyle categorization. *Energy Effic.* **2009**, *2*, 263–270. [CrossRef]

29. Shwom, R. Strengthening Sociological Perspectives on Organizations and the Environment. *Organ. Environ.* **2009**, *22*, 271–292. [CrossRef]

30. Shwom, R. A middle range theorization of energy politics: The struggle for energy efficient appliances. *Environ. Politics* **2011**, *22*, 705–726. [CrossRef]

31. Zierler, R.; Wehrmayer, W.; Murphy, R. The energy efficiency behaviour of individuals in large organisations: A case study of a major UK infrastructure operator. *Energy Policy* **2017**, *104*, 38–49. [CrossRef]

32. Williams, S.; Schaefer, A. Small and medium-sized enterprises and sustainability: Managers' values and engagement with environmental and climate change issues. *Bus. Strategy Environ.* **2013**, *22*, 173–186. [CrossRef]

33. Hampton, S. Making Sense of Energy Management Practice: Reflections on Providing Low Carbon Support to Three SMEs in the UK. *Energy Effic.* **2018**. [CrossRef]

34. Andrews, R.N.L.; Johnson, E. Energy use, behavioral change, and business organizations: Reviewing recent findings and proposing a future research agenda. *Energy Res. Soc. Sci.* **2016**, *11*, 195–208. [CrossRef]

35. Fawcett, T.; Hampton, S. Why & how energy efficiency policy should address SMEs. *Energy Policy* **2020**, *140*, 111337. [CrossRef]

36. König, W. Energy efficiency in industrial organizations—A cultural-institutional framework of decision-making. *Energy Res. Soc. Sci.* **2020**, *60*, 101314. [CrossRef]

37. O'Malley, E.; Scott, S.; Sorrell, S. *Barriers to Energy Efficiency: Evidence from Selected Sectors*; Policy Research Series 47; The Economic and Social Research Institute: Dublin, Ireland, 2003.

38. Trianni, A.; Cagno, E. Dealing with barriers to energy efficiency and SMEs: Some empirical evidences. *Energy* **2012**, *37*, 494–504. [CrossRef]

39. Hatch, M.J. *Organization Theory: Modern, Symbolic, and Postmodern Perspectives*; Oxford University Press: Oxford, UK, 1997.

40. Sorrell, S.; Schleich, J.; Scott, S.; O'Malley, E.; Trace, F.; Boede, U.; Ostertag, K.; Radgen, P. *Reducing Barriers to Energy Efficiency in Public and Private Organizations*; Final Report to the European Commission; University of Brighton: Brighton, UK, 2000.

41. Stephenson, J.; Barton, B.; Carrington, G.; Gnoth, D.; Lawson, R.; Thorsnes, P. Energy cultures: A framework for understanding energy behaviours. *Energy Policy* **2010**, *38*, 6120–6129. [CrossRef]

42. Bourdieu, P. *Le Sens Pratique*; Polity Press: Cambridge, UK, 1992.

43. Giddens, A. *Central Problems in Social Theory: Action, Structure and Contradiction in Social Analysis*; University of California Press: Berkeley, CA, USA; Los Angeles, CA, USA, 1979.

44. Stephenson, J.; Barton, B.; Carrington, G.; Doering, A.; Ford, R.; Hopkins, D.; Lawson, R.; MacCarthy, A.; Rees, D.; Scott, M.; et al. The energy cultures framework: Exploring the role of norms, practices and material culture in shaping energy behaviour in New Zealand. *Energy Res. Soc. Sci.* **2015**, *7*, 117–123. [CrossRef]

45. Mirosa, M.; Lawson, R.; Gnoth, D. Linking personal values to energy-efficient behaviors in the home. *Environ. Behav.* **2011**, *27*, 1–21. [CrossRef]

46. Bell, M.; Carrington, G.; Lawson, R.; Stephenson, J. Socio-technical barriers to the use of low-emission timber drying in New Zealand. *Energy Policy* **2014**, *67*, 747–755. [CrossRef]

47. Scott, R.W. *Institutions and Organizations: Ideas and Interests*, 3rd ed.; SAGE Publ.: Thousand Oaks, CA, USA, 2008.

48. Hoffman, A. Institutional evolution and change: Environmentalism and the US chemical industry. *Acad. Manag. J.* **1999**, *42*, 351–371. [CrossRef]

49. Wæraas, A.; Nielsen, J.A. Translation Theory "Translated": Three Perspectives on Translation in Organizational Research. *Int. J. Manag. Rev.* **2016**, *18*, 236–270. [CrossRef]

50. Ocasio, W. Towards an attention-based view of the firm. *Strateg. Manag. J.* **1997**, *18*, 187–206. [CrossRef]

51. Schein, E.H. *Organizational Culture and Leadership*, 3rd ed.; Jossey-Bass: San Francisco, CA, USA, 2004.

52. Phylipsen, G.J.M.; Blok, K.; Worrell, E. International comparisons of energy efficiency—Methodologies for the manufacturing industry. *Energy Policy* **1997**, *25*, 715–725. [CrossRef]

53. Trianni, A.; Cagno, E.; Farné, S. An empirical investigation of barriers, drivers and practices for energy efficiency in primary metals manufacturing SMEs. *Energy Procedia* **2014**, *61*, 1252–1255. [CrossRef]

54. Fleiter, T.; Hirzel, S.; Worrell, E. The characteristics of energy-efficiency measures–a neglected dimension. *Energy Policy* **2012**, *51*, 502–513. [CrossRef]

55. Costa-Campi, M.T.; García-Quevedo, J.; Segerra, A. Energy efficiency determinants: An empirical analysis of Spanish innovative firms. *Energy Policy* **2015**, *83*, 229–239. [CrossRef]

56. Cagno, E.; Trianni, A. Exploring drivers for energy efficiency within small- and medium-sized enterprises: First evidences from Italian manufacturing enterprises. *Appl. Energy* **2013**, *104*, 276–285. [CrossRef]

57. Denison, D.R. What is the difference between organizational culture and organizational climate? A native's point of view on a decade of paradigm wars. *Acad. Manag. Rev.* **1996**, *21*, 619–654. [CrossRef]

58. Fiedler, T.; Mircea, P.M. Energy management systems according to the ISO 50001 standard—Challenges and benefits. In Proceedings of the International Conference on Applied and Theoretical Electricity (ICATE), Craiova, Romania, 25–27 October 2012. [CrossRef]

59. Federal Statistical Office Baden-Württemberg. Structural Data Industry by Employee Size Classes. 2019. Available online: https://www.statistik-bw.de/Industrie/Struktur/VG-GK-BBEU.jsp (accessed on 28 February 2020).

60. European Commission. *Industrial Revolution Brings Industry Back to Europe*; Press Release Database; European Commission: Brussels, Belgium, 2012. Available online: http://europa.eu/rapid/press-release_IP-12-1085_en.htm?locale=en (accessed on 1 August 2020).

61. Ministry of Economics and Finance Baden-Württemberg. Industry Perspective Baden-Württemberg 2025. Available online: http://www.kmu-digital.eu/de/publikationen/tags/fuehrung-management/15-gemeinsam-in-die-zukunft-industrieland-baden-wuerttemberg/file (accessed on 1 August 2020).

62. Creswell, J.W. *Research Design—Qualitative, Quantitative, and Mixed Methods Approaches*, 3rd ed.; SAGE Publ.: Thousand Oaks, CA, USA, 2009.

63. Ybema, S.; Yanow, D.; Wels, H.; Kamsteeg, F. *Organizational Ethnography: Studying the Complexities of Everyday Life*; SAGE Publications Ltd.: Thousand Oaks, CA, USA, 2009.

64. European Council. Commission recommendation of 6 May 2003 concerning the definition of micro, small and medium-sized enterprises. *Off. J. Eur. Union* **2003**, *46*, 36–41.

65. Kühl, S.; Strodtholz, P.; Teffertshofer, A. *Handbuch Methoden der Organisationsforschung*; VS Verlag für Sozialwissenschaften: Wiesbaden, Germany, 2009.

66. Glaser, B.G.; Strauss, A.L. *Grounded Theory: Strategien Qualitativer Forschung*; H. Huber: Göttingen, Germany, 1998; pp. 51–83.

67. Mannheimer Company Panel. The Economical Importance of Family Businesses in Germany. 2019. Available online: https://www.familienunternehmen.de/media/public/pdf/publikationen-studien/studien/Die-volkswirtschaftliche-Bedeutung-der-Familienunternehmen-2019_Stiftung_Familienunternehmen.pdf (accessed on 20 February 2018).

68. Froschauer, U.; Lueger, M. *Das Qualitative Interview: Zur Praxis Interpretativer Analyse Sozialer Systeme*; Facultas Verlags- und Buchhandels AG: Wien, Austria, 2003.

69. Froschauer, U.; Lueger, M. *Interpretative Sozialforschung: Der Prozess*; Facultas Verlags- und Buchhandelshandels AG: Wien, Austria, 2009.

70. Statista. Number of Companies in Germany by Company Size and Economic Sector in 2017. 2018. Available online: https://de.statista.com/statistik/daten/studie/731993/umfrage/unternehmen-in-deutschland-nach-unternehmensgroesse-und-wirtschaftszweigen/ (accessed on 10 August 2018).

71. Czarniawska, B. Culture is the Medium of Life. In *Reframing Organizational Culture*; Frost, P.J., Ed.; SAGE Publ.: Thousand Oaks, CA, USA, 1992; pp. 285–297.

72. FEA—Federal Environment Agency. Data on Environmental and Energy Management Systems in Germany. 2019. Available online: https://www.umweltbundesamt.de/daten/umwelt-wirtschaft/umwelt-energiemanagementsysteme (accessed on 22 March 2019).

73. Weber, M. *Gesammelte Aufsätze zur Wissenschaftslehre*, 6th ed.; Winckelmann, J. (Hrsg.); Mohr Verlag: Tübingen, Germany, 1985.

74. North, P. The business of the Anthropocene? Substantivist and diverse economies perspectives on SME engagement in local low carbon transitions. *Prog. Hum. Geogr.* **2016**, *40*, 437–454. [CrossRef]

75. North, P.; Nurse, A. 'War Stories': Morality, curiosity, enthusiasm and commitment as facilitators of SME owners' engagement in low carbon transitions. *Geoforum* **2014**, *52*, 32–41. [CrossRef]

76. Pye, M.; McKane, A. Making a stronger case for industrial energy efficiency by quantifying non-energy benefits. *Resour. Conserv. Recycl.* **2000**, *28*, 171–183. [CrossRef]

77. Cagno, E.; Moschetta, D.; Trianni, A. Only non-energy benefits from the adoption of energy efficiency measures? A novel framework. *J. Clean. Prod.* **2019**, *212*, 1319–1333. [CrossRef]

78. Finster, M.P.; Hernke, M.T. Benefits organizations pursue when seeking competitive advantage by improving environmental performance. *J. Ind. Ecol.* **2014**, *18*, 652–662. [CrossRef]

79. Trianni, A.; Cagno, E.; Marchesani, F.; Spallina, G. Classification of drivers for industrial energy efficiency and their effect on the barriers affecting the investment decision –making process. *Energy Effic.* **2017**, *10*, 199–215. [CrossRef]

80. Caffal, C. *Learning from Experiences with Energy Management in Industry*; Analysis Series 17; Centre for the Analysis and Dissemination of Demonstrated Energy Technologies: Sittard, The Netherlands, 1995.

81. Thollander, P.; Palm, J.; Rohdin, P. Categorizing Barriers to Energy Efficiency—An Interdisciplinary Perspective. In *Energy Efficiency*; Palm, J., Ed.; InTechOpen: London, UK, 2010. Available online: https://www.intechopen.com/books/energy-efficiency/categorizing-barriers-to-energy-efficiency-an-interdisciplinary-perspective (accessed on 28 February 2020).

82. Thollander, P.; Palm, J. *Improving Energy Efficiency in Industrial Energy Systems: An Interdisciplinary Perspective on Barriers, Energy Audits, Energy Management, Policies, and Programs*; Springer: London, UK, 2013.

83. Thollander, P.; Ottosson, M. An energy efficient Swedish pulp and paper industry: Exploring barriers to and driving forces for cost-effective energy efficiency investments. *Energy Effic.* **2008**, *1*, 21–34. [CrossRef]

84. Thollander, P.; Solding, P.; Söderström, M. Energy management in industrial SMEs. In Proceedings of the 5th European Conference on Economics and Management of Energy in Industry (ECEMEI-5), Vilamoura, Portugal, 14–17 April 2009; pp. 1–9.

85. Anderson, S.T.; Newell, R.G. Information programs for technology adoption: The case of energy efficiency audits. *Resour. Energy Econ.* **2004**, *26*, 27–50. [CrossRef]

86. Sardianou, E. Barriers to industrial energy efficiency investments in Greece. *J. Clean. Prod.* **2008**, *16*, 1416–1423. [CrossRef]

87. Sanstad, A.; Howarth, R. 'Normal' markets, market imperfections and energy efficiency. *Energy Policy* **1994**, *10*, 811–818. [CrossRef]

88. Suchmann, M.C. Managing legitimacy: Strategic and institutional approaches. *Acad. Manag. Rev.* **1995**, *20*, 571–610. [CrossRef]

89. Hwang, H.; Colyvas, J.A. Problematizing actors and institutions in Institutional Work. *J. Manag. Inq.* **2011**, *20*, 62–66. [CrossRef]

90. Cagno, E.; Trianni, A.; Worrell, E.; Miggiano, F. Barriers and drivers for energy efficiency: Different perspectives from an exploratory study in the Netherlands. *Energy Procedia* **2014**, *61*, 1256–1260. [CrossRef]

91. Fresner, J.; Morea, F.; Krenn, C.; Uson, J.A.; Tomasi, F. Energy efficiency in small and medium enterprises: Lessons learned from 280 energy audits across Europe. *J. Clean. Prod.* **2017**, *142*, 1650–1660. [CrossRef]

Article

Potential and Impacts of Cogeneration in Tropical Climate Countries: Ecuador as a Case Study

Manuel Raul Pelaez-Samaniego [1], Juan L. Espinoza [2,*], José Jara-Alvear [3], Pablo Arias-Reyes [4], Fernando Maldonado-Arias [5], Patricia Recalde-Galindo [6], Pablo Rosero [6] and Tsai Garcia-Perez [1]

[1] Department of Applied Chemistry and Systems of Production, Faculty of Chemical Sciences, Universidad de Cuenca, Cuenca 010107, Ecuador; manuel.pelaez@ucuenca.edu.ec (M.R.P.-S.); tsai.garcia@ucuenca.edu.ec (T.G.-P.)

[2] Faculty of Engineering, DEET, Universidad de Cuenca, Cuenca 010107, Ecuador

[3] Corporación Eléctrica del Ecuador CELEC E.P., Cuenca 010109, Ecuador; jose.jara@celec.gob.ec

[4] Faculty of Electrical Engineering, Smart Grid Energy Lab., Universidad Católica de Cuenca, Cuenca 010107, Ecuador; pariasr@ucacue.edu.ec

[5] Faculty of Economic Sciences, Universidad de Cuenca, Cuenca 010107, Ecuador; fernando.maldonado@ucuenca.edu.ec

[6] Ministry of Energy and Natural Non-Renewable Resources, Quito 170135, Ecuador; patricia.recalde@recursosyenergia.gob.ec (P.R.-G.); pablo.rosero@recursosyenergia.gob.ec (P.R.)

* Correspondence: juan.espinoza@ucuenca.edu.ec

Received: 29 August 2020; Accepted: 24 September 2020; Published: 10 October 2020

Abstract: High dependency on fossil fuels, low energy efficiency, poor diversification of energy sources, and a low rate of access to electricity are challenges that need to be solved in many developing countries to make their energy systems more sustainable. Cogeneration has been identified as a key strategy for increasing energy generation capacity, reducing greenhouse gas (GHG) emissions, and improving energy efficiency in industry, one of the most energy-demanding sectors worldwide. However, more studies are necessary to define approaches for implementing cogeneration, particularly in countries with tropical climates (such as Ecuador). In Ecuador, the National Plan of Energy Efficiency includes cogeneration as one of the four routes for making energy use more sustainable in the industrial sector. The objective of this paper is two-fold: (1) to identify the potential of cogeneration in the Ecuadorian industry, and (2) to show the positive impacts of cogeneration on power generation capacity, GHG emissions reduction, energy efficiency, and the economy of the country. The study uses methodologies from works in specific types of industrial processes and puts them together to evaluate the potential and analyze the impacts of cogeneration at national level. The potential of cogeneration in Ecuador is ~600 MW$_{el}$, which is 12% of Ecuador's electricity generation capacity. This potential could save ~18.6 × 10^6 L/month of oil-derived fuels, avoiding up to 576,800 tCO$_2$/year, and creating around 2600 direct jobs. Cogeneration could increase energy efficiency in the Ecuadorian industry by up to 40%.

Keywords: cogeneration; trigeneration; sustainability; industrial energy efficiency; tropical climate country; biomass

---

1. Introduction and Literature Review

Energy is key for people's well-being and for a countries' development. Still, current global energy use and production heavily relies on fossil derived fuels and electricity produced using this type of fuel. For instance, in 2018, 85% of the worldwide fuel consumption had its origin in fossil fuels. The total petroleum, coal, and natural gas consumption reached 4714 MTOE/year (Million tons of oil equivalent per year), 3744 MTOE/year, and 3328 MTOE/year, respectively [1]. One of the negative consequences of the large consumption of fossil fuels is the raising of greenhouse gas (GHG) emissions that are responsible for global warming. In addition, for several developing countries (especially tropical climate countries), there are pending tasks to fully meet energy needs and make energy generation and use more sustainable. Low energy efficiency, poor diversification of energy sources, low rate of access to electricity service, and necessity to make the energy systems less dependent on fossil fuels are among those pending tasks. The necessity of reducing the use of fossil fuels is critical as these countries may suffer the impact of climate change more intensively (in part due to energy-related activities). The associated costs to mitigate such impacts are very high [2–4]. Although tropical climate countries possess a benign weather and a diversity of energy resources, balancing electricity generation with weather conditions and the reduction of energy sources (e.g., hydropower) are forcing those countries to look for new options for electricity generation and management. This is the case for Ecuador.

The Ecuadorian energy matrix highly depends on oil and oil-derived fuels, which are used in the transportation and industrial sectors, as well as in households (mainly as fuel for cooking) and for electricity generation (in smaller amounts) [5–7]. The transportation and the industrial sectors are responsible for 42% and 18% of the total fuel consumption, respectively [8,9]. Lack of natural gas (NG) sources and insufficient oil refining capacity force the country to import part of the fuels used. The high expenses to import these fuels and the resulting negative environmental consequences are driving Ecuador to look for alternatives to imported fuels and to make the energy sector more sustainable. The Ecuador's GHG inventory shows that the energy sector in the country is responsible for 46.6% of the total of CO_{2eq} emissions [10]. Heat for running industrial processes is produced mostly by burning subsidized oil-derived fuels, especially diesel and fuel oil [5,7,9] and only a few companies use renewable energies (particularly biomass) to produce heat and power. Recent attempts made by the Ecuadorian government to reduce or eliminate subsidies to fuels have failed due to political and social pressure.

The electricity generation in Ecuador, on the other hand, is almost entirely based on hydropower. The current hydropower installed capacity in the country is ~5000 MW, from which 88% corresponds to power plants located in rivers that discharge into the Amazonian river basin, while the rest corresponds to plants located in rivers that discharge into the Pacific Ocean. Hydropower generation, however, has problems to adjust to the country's seasonal rains, which negatively impacts electricity production. Locating hydropower plants on both sides of the Andes Mountains has been a strategy for partially balancing the seasonality of rains. Figure 1 shows the variation of water inflow in hydropower plants located in the Amazonian River and the Pacific Ocean basins in Ecuador. The power generation is proportional to water inflow in the plants. It is seen that from October to January, the water inflow is reduced as a consequence of lower rainfalls [11,12]. Since the seasonality of hydropower generation could jeopardize the electricity supply and its sustainability in the mid-term, Ecuador is currently looking for options to ensure electricity generation in coming years, especially during the dry season. The adoption of the National Plan of Energy Efficiency 2016–2035 (known as PLANEE 2016–2035) is expected to have a positive impact on the energy demand and use [7,8]. In addition, the Ecuadorian State aims to increase the incipient participation of other renewable energy sources (i.e., wind, solar, and biomass) in the electricity sector [7]. In 2017, hydroelectricity contributed with more than 80% of the total electricity generated in the country, but the share of other renewable energy sources was only 0.5% (16.5 MW wind, 24 MW photovoltaic) [13], whereas in 2019, the hydropower share was 85% [14]. In the following years, wind farms (160 MW total) and solar photovoltaic (200 MW) projects will start operating. Nevertheless, although the electricity generation capacity in Ecuador has shown improvements, the negative effects of rains seasonality are unavoidable in coming years, and new electricity generation methods are sought. The PLANEE 2016–2035 foresees that the industry can play an important role by becoming more energy efficient and by generating its own electricity (at least partially) through cogeneration [8]. Besides, the substitution and/or better use of fossil fuels to produce heat in the industrial sector is a pending task.

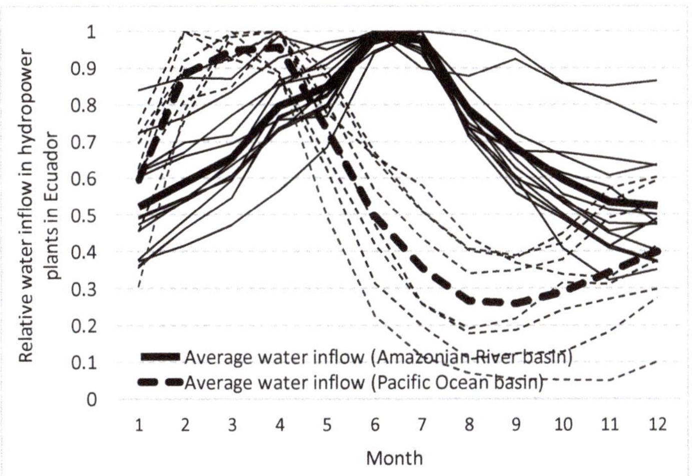

Figure 1. Monthly variation of water inflow in hydropower plants located in the Amazonian River and the Pacific Ocean basins in Ecuador. Thick lines show mean values from 1964 to 2016 [14,15].

Cogeneration has been recognized as a key element for the diversification of the electricity generation matrix (to help balancing the seasonal hydropower generation), for the reduction of the costs of subsidies to energy in the Ecuadorian industry (by making a better use of fuels for heat production), for the increase in energy efficiency, and for reducing GHG emissions [8]. However, further work is required to determine how much the potential of cogeneration in the Ecuadorian industry is and to define strategies for implementing cogeneration in this sector. Year-round tropical climate, subsidies of the state to fossil fuels and electricity, and insufficient energy policies to promote investments in the energy sector are factors that have hindered the penetration of cogeneration in the country. Because of the relatively constant year-round temperature conditions, indoor heating is not required, even in the Andean highlands (where temperature normally varies between 7 and 23 °C). Thus, cogeneration has been adopted only marginally in the industrial sector. Our field work (see Section 2.1.2 for details) and [8,9] have identified that Ecuador's current installed cogeneration capacity is 172 MW_{el}, which represents only 2% of the total (nominal) electricity generation capacity (i.e., 7361 MW_{el}) [7]. Lignocellulosic biomass is the main fuel employed for cogeneration due to the utilization of bagasse in the sugarcane industry (Table 1). Although there are abundant lignocellulosic biomass resources in the country (e.g., oil palm, rice, banana, and wood residues), the use of these energy sources for cogeneration in the country is very low [7]. For example, in Ecuador, there are currently 35 companies that process oil palm fruit and 4 companies that produce oil from oil palm kernel, of which only 2 currently use cogeneration. Because of the positive impacts of biomass for cogeneration [16], the use of this fuel deserves more attention in the country. In addition to the existing installed cogeneration capacity in the country, there is a thermal power plant (Termogas Machala, 132 MW_{el} of installed capacity) [15] that is currently being retrofitted for operating as a combined cycle (CC) plant by adding heat recovery steam generators (HRSG) and steam turbines. This plant runs with natural gas—NG (obtained from the Gulf of Guayaquil) and gas turbines.

Despite the positive reputation and the extended use of cogeneration worldwide (especially in temperate climate countries), there are not enough studies showing the potential of cogeneration of whole industrial sectors or how cogeneration, in the conditions of tropical climate countries, could contribute to meet energy requirements, help to increase energy efficiency, reduce national GHG emissions, and, thus, contribute to sustainable development. For some tropical climate countries, there exists some studies focused on cogeneration in specific industrial sectors, such as the sugarcane industry [17–25], the oil palm industry [26–28], and the wood processing industry [16,29–33]. The methodologies and learnings from those works can be used to conduct a wider analysis on the impacts of cogeneration in a whole country or geographic region, although more research overall is necessary. Thus, the objective of this paper is two-fold: first, to compute the potential of cogeneration in the Ecuadorian industry, and, second, to show the positive impacts of cogeneration on power generation capacity, GHG emission reduction, industrial energy efficiency, and the economy of the country. The presence of subsidies from the state to both electricity and fuels in Ecuador, the seasonality of rains to run hydropower plants, and its year-round tropical weather are particular challenges considered in the study.

Table 1. Ecuador's current cogeneration installed capacity.

Type of Industry	Technology /Process	Type of Fuel	Year Operation Started	Installed Capacity (MW_{el})	Electricity Generation (GWh/year)
Sugarcane industry	Rankine cycle (3 plants)	Sugarcane bagasse	2004	136.6	408.3
Food industry	Diesel engine (1 plant)	Diesel	2007	1.0	N/A
Oil palm industry	Rankine cycle (2 plants)	Oil palm solid residues	1983	2.2	N/A
Wood industry	Rankine cycle (1 plant)	Wood residues	2003	1.0	N/A
Oil refining	Rankine cycle (1 plant)	Fuel oil	N/A	30.75	N/A
Ethanol production	Rankine cycle (1 plant)	Fuel oil	N/A	0.3	N/A
	TOTAL:			172 MW_{el}	

N/A—information is not available.

2. Materials and Methods

Our literature review suggests that there are not standardized methods for computing the potential of cogeneration/trigeneration in a specific geographical region or country, which is understandable since each country and its industrial sector have specific conditions that need to be taken into account. There are different aspects that need to be analyzed to determine the most suitable methodology to compute cogeneration potential at a country level (e.g., weather, types of energy sources available, altitude above the sea level, energy policies and incentives). In tropical climate countries such as Ecuador, the weather is an important factor that determines specific types of cogeneration schemes because, as previously mentioned, there is no need for indoor heating (an important energy requirement in tempered climate countries), but air conditioning is required instead [34–36]. Consequently, cogeneration projects are more suitable in the industrial sector and in other places where hot and cold fluids are used (e.g., hospitals, hotels, airports, shopping malls). These are the target places for cogeneration projects in tropical climate countries.

Another factor to consider for computing the potential of cogeneration is the pattern of energy consumption in the industrial sector, which in Ecuador is relatively constant throughout the year, reflecting a common feature of energy consumption in the industry of tropical countries. For Ecuador, and to illustrate this important point, Figure 2 shows two examples of energy consumption curves (both electricity and fuel) corresponding to two large Ecuadorian industrial companies (herein referred to as companies M and N) devoted to the production of tires (M) and pulp and paper (N). This energy consumption pattern of the industrial sector in Ecuador suggests that cogeneration plants in tropical climate countries could operate at approximately constant capacity year-round, which makes the sizing process of the cogeneration plants easier. The methodology adopted herein considers these elements.

2.1. Methodology

The potential of cogeneration in the whole industrial sector of a country can be obtained if the potential of cogeneration of each industrial plant in which cogeneration can be adopted is determined. The methods for sizing cogeneration plants for specific types of industries are based on their annual energy requirements (normally, heat for the industrial process and/or plant operation, since producing surplus heat will otherwise be wasted). Furthermore, producing electricity is not a priority in the industrial plants in the country due to its relatively low cost (i.e., due to subsidies). Table 2 presents a list of works devoted to determining the cogeneration capacity in specific types of industrial plants. These works served as the basis to compute the potential cogeneration capacity in industrial plants in

Ecuador. In addition, a report on the potential of cogeneration in Spain [37] and a report by the Office of Environment and Heritage New South Wales [38] were used. Moreover, for sizing cogeneration plants, it is necessary to define the cogeneration schemes suitable to specific types of industries and the respective fuels available. In this study, such schemes are shown in Appendix A, while the main equations used are provided in Appendix B. Then, the potential of each industrial plant was added to obtain the potential of cogeneration by cluster of industries and the whole country's potential. The methodology adopted consisted of five stages (summarized in Figure 3) that are detailed in the following subsections.

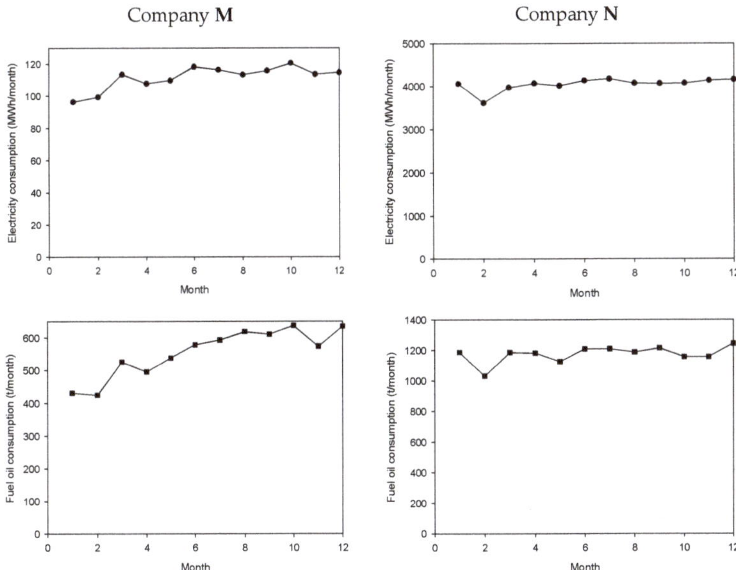

Figure 2. Typical curves of electricity (above) and fuel (below) consumption of two industrial companies (**M** and **N**), taken as examples of yearly (approximately constant) energy demand in most Ecuadorian industrial plants.

Table 2. Some works on cogeneration computing methods for different types of industries.

Type of Industry/Plant	Reference(s)
Hospitals	[39–41]
Small- and medium-sized industries and services	[42–44]
Large-sized industry and commercial sector	[45–47]
Sugarcane/ethanol	[17–19,21–25,48–50]
Oil palm	[27,28,31,51–54]
Wood and wood-derived products	[16,29–33]
Pulp and paper	[32]
Cement industry	[55]
Hotels	[56]
Chemical industry	[57]
Breweries	[58]
Food industry	[59]
Greenhouse gas emissions from cogeneration	[60]
Biogas/renewable energy	[61,62]
Others	[63–71]

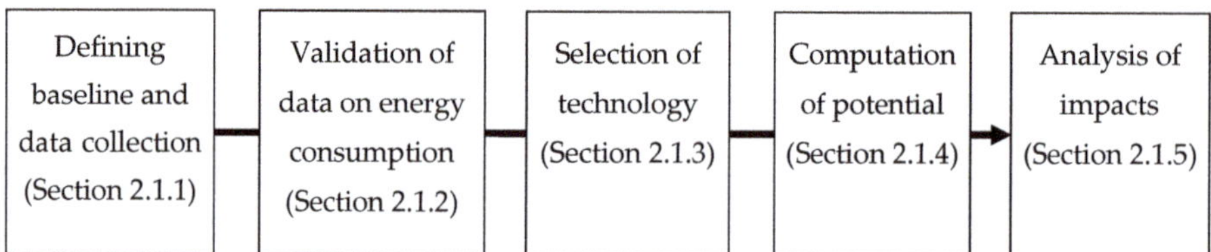

Figure 3. Methodology framework to compute the potential of cogeneration and the resulting impacts in Ecuador.

2.1.1. Data Collection and Energy Consumption Baseline

The tasks described in Sections 2.1.1.1 and 2.1.1.2 aimed to determine which industrial plants could adopt cogeneration (or trigeneration) in the country. For this, information on electricity and fuel consumption was used to define a baseline that allows selecting prospective industrial companies. This information was obtained from two official sources, the Agency of Regulation and Control of Electricity—ARCONEL (in Spanish Agencia de Regulación y Control de Electricidad) and the Agency of Regulation and Control of Hydrocarbon Fuels—ARCH (in Spanish Agencia de Regulación y Control de Hidrocarburos), which are the institutions in charge of regulating and controlling the distribution and use of electricity and fossil-derived fuels, respectively. The data used corresponded to 2015 and were the information available at the time that this study was conducted (2017 and 2018).

2.1.1.1. Electricity Consumption Baseline

The initial list on electricity consumption from the ARCONEL contained clients/consumers reporting electricity consumption above 20,000 kWh/month. This electricity consumption baseline was established after analyzing the energy demand of a small food processing company with installed capacity of approximately 30 kW_{el}, working 24 h/day the year-round (i.e., with electricity consumption of ~20,000 kWh/month). The company is located in the city of Cuenca, and herein it is referred to as Company A. The number of companies/consumers in the initial list was ~41,800. Next, the resulting list was analyzed and filtered again to remove companies and/or institutions (both public and private) in which, although their electricity consumption was >20,000 kWh/month, no fuels are required for their operation, except diesel for transportation and LPG (Liquid Petroleum Gas) for cooking at a small scale. This is the case of:

(1) Elementary schools, high schools, colleges/universities, government buildings and offices at a national or municipal level where, as previously mentioned, due to climate conditions in Ecuador, there is no necessity of cogeneration intending, for example, indoor heating (which is common in temperate places) or water heating.

(2) Construction and civil engineering companies (e.g., roads construction companies) that report high electricity consumption (for example for reducing the particle size of rocks).

It was also observed that the possibilities of cogeneration in a few companies that process polymers/plastics (e.g., High Density Polyethylene-HDPE, Polypropylene-PP, Polyvinyl Chloride-PVC) for producing plastic toys, plastic bags and/or plastic furniture for both domestic and industrial use (with electricity consumption > 20,000 kWh/month) should be verified in situ. Thus, these companies were kept in the list. The amount of companies after this filtering process was approximately 2000.

2.1.1.2. Fuel Consumption Baseline

The fuel consumption baseline started by analyzing the possibilities of cogeneration in the representative Company A (Section 2.1.1.1), which uses heat (produced by burning diesel) for its manufacturing process. The fuel consumption of this company served as the basis to start filtering the data provided by the ARCH. The company uses a typical small boiler (186 kW$_{th}$) that produces saturated steam at 140–150 °C, working ~6 h/day, 5 days/week, and employing up to 7570 L/month (i.e., 90,840 L/year) of diesel. A preliminary computation (following works of [43,44,47] and energy balances) showed that, if the company was interested in adopting cogeneration, the size of the cogeneration plant would be close to 300 kW$_{el}$. This cogeneration unit could operate, for instance, on a diesel or a gas engine (depending on the fuel available) and use the waste heat for producing the steam for the process (in a HRSG). However, according to a study conducted in the industrial sector in Mexico (with weather conditions somehow similar to those in Ecuador), the projects on cogeneration that offer better prospective, from an economic viewpoint, are those larger than 500 kW$_{el}$ [72]. Therefore, the minimum capacity of the cogeneration plants in the Ecuadorian industry, in all cases and at this level of the study, should be 500 kW$_{el}$, which corresponds to a cogeneration plant that demands ~90,800 L/year of diesel (or any diesel equivalent fuel) Consequently, the fuel consumption data filtering process started by considering a baseline of diesel or fuel oil consumption of 90,800 L/year (76.19 t/year).

The information on fossil fuel consumption provided by the ARCH included data on type of fuel, amount, company's name, location and information on the main products of the company. This information was used to identify the location of each industrial plant. The types of fuels consumed in the country are as follows: fuel oil, diesel fuel (for both industry and transportation), gasoline (both regular and premium), liquefied petroleum gas (LPG), and NG in a smaller amount (all fuels were converted to diesel equivalent fuel). The initial list included ~500,000 companies and institutions. An initial filtering process removed from the list companies that a) reported LPG consumption, since in the country LPG is not used for industrial processes, except some hotels, hospitals, and shopping malls that have centralized LPG supply in relatively small amounts, and b) companies that sell diesel and gasoline for transportation (i.e., gas stations). The resulting list was filtered again by removing institutions that reported large amounts of diesel consumption for transportation only (e.g., municipal governments; ministries from the Ecuadorian government; and civil engineering companies that use diesel for transport/operation of heavy machinery for the construction of roads, bridges, and large buildings in the country). After a quantitative analysis, similar to that conducted for company A, it was found that the cogeneration capacity in companies consuming <151,400 L/year of fuel-oil or diesel will be <500 kW$_{el}$. Thus, the final fuel consumption baseline for selecting the companies where cogeneration could potentially be adopted was 151,400 L/year of diesel and/or fuel oil (both with approximately similar high heating value—HHV). Therefore, the list was reduced to ~1000 companies.

2.1.1.3. Final List of Industrial Companies That Could Adopt Cogeneration

The resulting lists (after filtering the ARCONEL and the ARCH data) were put together to prepare a final list of industrial companies (including hotels and hospitals) at a national level. Although the majority of the companies from the filtered ARCH list were also present in the filtered ARCONEL list, some companies were present in one list only since they reported high electricity consumption but low fuel consumption (e.g., plastics processing and ice making companies) and vice versa (e.g., fishing companies). After a case by case analysis, the final list was comprised of 555 companies (See Figure 4). All the 555 companies from the list, except 2 (from the oil palm industry, which are located in the Amazonian region), are located in the coast (~57%) and in the Andean highlands (~43%) regions. Among this list, there were sixteen companies working on shrimp growing/processing and eight ice making plants. These companies reported both high electricity and diesel consumption, but the chances of cogeneration were apparently negligible, since it was identified that the fuels were used for water pumping using internal combustion (diesel) engines in places where no electricity grids were available for shrimp pools operation and/or for land transport (using trucks). Thus, we decided to keep these companies in the final list to confirm the possibilities of cogeneration after visiting some of those plants.

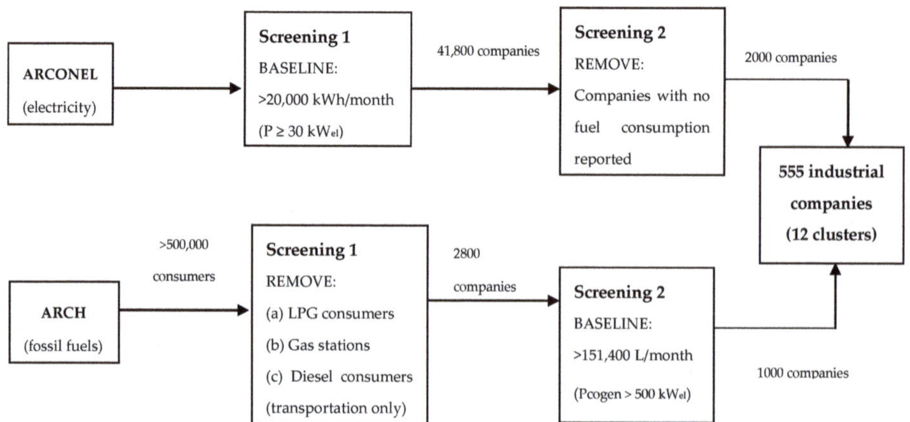

Figure 4. Flow diagram showing the selection of the companies where cogeneration is proposed.

2.1.2. Classification of Companies by Clusters and Validation of Data

The 555 companies in the final list were classified by clusters, which helped to organize visits to confirm the energy consumption data and to identify and record the corresponding industrial processes, including the identification of hot/cold fluids and their characteristics. The companies were grouped into twelve categories or clusters of industries, following the International Standard Industrial Classification of All Economic Activities (ISIC) [73,74]. Airports, shopping malls, and oil refineries were included in the cluster "others". Table 3 shows the list of clusters and the number of companies in each cluster. The information provided by the ARCONEL and the ARCH was validated by visiting 162 companies (~30% of the total), as detailed in Table 3. The selection of the companies to visit considered the amount of companies per cluster, the sizes, location, and the types of manufacturing processes to guarantee that all types of industries were visited. Interview survey formats (asking about energy consumption, types and amounts of fuels, industrial process, types and conditions of industrial fluids, if cogeneration has been adopted in the plant and the corresponding conditions, and other aspects to determine cogeneration potential) were used to collect the information provided by the industrial companies.

Table 3. Classification of industrial companies into clusters, types of industries in each cluster, amount of industrial plants visited, and types of predominant cold/hot fluids identified.

No.	Classification ISIC	Cluster Name	Number of Companies	Number (and %) of Companies Visited	Predominant Work Fluid(s)
1	C13	Textile industry	56	15 (27%)	Steam, Hot gases
2	C23	Construction materials (cement, ceramics/tiles)	23	17 (74%)	Hot gases, Steam
3	C10	Food industry (grain mills, fruit processing/juice, dairy, seafood, etc.)	132	36 (27%)	Steam, Hot water, Cold water
4	C11	Alcoholic and no-alcoholic beverages	35	18 (51%)	Steam, Hot water, Cold water
5	C16	Wood and wood composites	5	2 (40%)	Steam, Hot gases
6	C17	Pulp and paper	22	6 (27%)	Steam, Hot gases
7	C24 y C25	Metal processing industry	29	10 (34%)	Hot gases
8	C20	Agroindustry (includes oil palm industry)	58	14 (24%)	Steam, Hot water
9	Q86	Hospitals	47	17 (36%)	Steam, Hot water, A/C ***
10	I55	Hotels	17	9 (53%)	Steam, Hot water, A/C
11	-	Others (chemical products, tires, glass, shopping malls, airports *, refineries **)	63	15 (24%)	Steam, Hot water, A/C, Hot gases
12	C22	Plastics	68	3 (4.8%)	Hot water
		TOTAL:	555	162 (~30%)	

* Three airports were included in the study: Guayaquil, Quito, and Cuenca. The rest of airports in the country operate only sporadically and are not candidates for cogeneration. ** The three main oil refineries in the country [6] were included. *** Air conditioning.

2.1.3. Selection of Cogeneration Technologies

The following considerations were made for selecting the cogeneration technology that fits into the industrial plants' requirements:

(1) The proposed cogeneration/trigeneration system must fit into the current plant's requirements of heat (e.g., steam and/or hot water necessities) or cold fluids (including A/C) to guarantee cogeneration plants with high capacity factors. Therefore, the plant requirement of thermal energy with heating and/or cooling effect defined the cogeneration/trigeneration capacity of the plant.

(2) The prime mover selected will allow one to cover the electricity requirements totally or partially. In the case of deficit of electricity, and as long as the thermal energy production is met, it is preferred to import electricity from the national grid. If the cogeneration system produces electricity surplus, then it can be sold to the national grid. No sell or purchase of hot/cold fluids (i.e., transport of these fluids from or to the plant) were considered.

(3) The type of fuel (e.g., biomass, biogas, NG, diesel, heavy oil) proposed for cogeneration should be readily available in the place the cogeneration plant will be located. Therefore, fuel availability is a key component for deciding on the technology proposed.

(4) The yearly average thermal energy requirements (not the peak requirements) were used for sizing the cogeneration/trigeneration plant.

(5) No indoor heating and/or district heating are required. This is expected due to geographical location [75].

(6) The selection of the prime movers considered the limitations imposed by geographical conditions, specifically altitude. For the case study, industrial plants in the Ecuadorian Andes highlands are located at approximately 2500 m above the sea level (m.a.s.l.); thus, in these places, it is preferred to use diesel engines, gas engines, or boiler and steam turbines instead of gas turbines to guarantee adequate levels of efficiency of the cogeneration plant [76–78].

(7) The selection of the prime movers also considered possible partial loads requirements (i.e., the ability to vary thermal and electrical output depending on hourly requirements, or the necessity for frequent stopping and starting). Consequently, diesel and/or gas engines are preferable for cogeneration instead of gas turbines or steam turbines coupled with boilers in companies that do not operate 24/7. Diesel and gas engines, additionally, are able to run with renewable fuels (biodiesel and biogas, respectively), which are expected to be available in the country in the future [79] (See Section 3.2).

(8) Trigeneration can be projected only in industrial plants where air conditioning and/or process cooling fluids (above the water freezing temperature) for the industrial process are required. In this case, both air conditioning and/or cold fluids will be produced by using residual heat from the prime mover. The trigeneration system will mostly work on LiBr (lithium bromide) absorption equipment for air conditioning in the Coastal region and, in some cases, hotels, hospitals, and airports in the Andean highlands. Ammonia (NH_3) absorption systems are proposed only when fluids with low temperatures are required for the industrial process (e.g., for pasteurization in the beverages, food, and dairy industries). Freezing is not part of the proposed trigeneration systems.

2.1.4. Computation of the Potential of Cogeneration of Ecuador

The potential of cogeneration of Ecuador was determined in two steps. First, the sum of the potential of cogeneration of all industries by each cluster was conducted. Then, the potential of each cluster was added to obtain the potential at a national level. Regarding cogeneration sizing at the industry level, the computations were first conducted for the industrial plants that were visited (see Section 2.1.2), and computations were carried out for the rest of the plants, using the information on the fuels and electricity consumption, as well as its location, working conditions, and size in a case by case basis. The main steps for computing the potential of cogeneration of a specific company were as follows (see Appendix B for equations used):

1. Identify the location of the industrial plant and the availability of electricity grids to ensure interconnection to import/export electricity when electricity deficit/surplus exists.
2. Collect/verify data on electrical and thermal loads and types of fuels used. This information was compared with the data from the ARCONEL and the ARCH (Section 2.1.1).
3. Gather data on the company's process: types of products, heat requirements (e.g., steam or hot gases) and other fluids used (e.g., cold fluids, air conditioning, hot water).
4. Identify types of fuels that are or could be available in the company (or plant) location place.
5. Select the appropriate cogeneration prime mover and the corresponding fuel.
6. Compute the cogeneration plant capacity, based on the necessities of thermal energy. Table 4 presents equipment parameters used for the computations.
7. Standardize the size of the equipment suggested for a specific company by using catalogues from companies that provide equipment for cogeneration/trigeneration (e.g., boilers, diesel engines, gas engines, steam turbines, HRSGs, and absorption chillers).
8. Compute the amount of fuel that the cogeneration/trigeneration plant will require (Appendix B).
9. Compute the amount of electricity that will be produced by the prime movers in the operating conditions of the cogeneration plant and how much of this electricity will be available for exporting to the national grid (if surplus electricity is available).

Table 4. Parameters corresponding to the equipment used in the computations.

Equipment and Type	Efficiency	Comments
Diesel engine	Up to 40% electric efficiency [78])	Expected heat recovery: up to 86% from the total heat released by the engine (i.e., heat from exhaust gases and heat from jacket coolant), depending on the size of the engine.
Gas engine (working with biogas)	Up to 45% electric efficiency [78]	Expected heat recovery: up to 88% from the total heat released by the engine (i.e., heat from exhaust gases and heat from jacket coolant), depending on the size of the engine.
Steam turbine (back pressure)	~55% [78]	
Heat recovery steam generator (HRSG)	82% [80]	
Absorption chillers (single effect in all cases) *	Coefficient of performance, COP = 0.7	LiBr absorption chillers for air conditioning and for producing cold fluids, except for low temperature fluids (close to 4 °C, where NH_3 absorption chillers are suggested).
Biomass boilers	75–80%	Depends on the capacity of the boiler.

* Single effect chillers are more convenient for diesel (and gas) engines [71].

2.1.5. Assessment of Impacts of Cogeneration in Ecuador

2.1.5.1. Environmental Impacts

The computation of the environmental impacts of cogeneration considered two types of impacts: (a) the GHG emissions resulting from the fuel burned in each cogeneration plant, and (b) the avoided GHG emissions resulting from the possible replacement of large thermal power plants in the country (that use fossil-derived fuels for electricity production) by cogeneration plants in the industry. It is expected that the availability of cogeneration plants could remove the necessity of installing a thermal power plant (that uses oil-derived fuels to run) with capacity equal to that corresponding to the total cogeneration potential. Both results were added to obtain the net GHG emissions.

(a) Emissions in cogeneration plants

The fuels required for cogeneration depend on the prime mover selected. Cogeneration in Ecuador will use diesel, biogas, and lignocellulosic biomass, which are the fuels available currently in the country (See Section 3.2). The GHG emissions were estimated for each type of fuel. The computations followed the concept of conservation of carbon, from the fuel combusted into CO_2, according to the guidelines from the International Energy Agency [81]. For biogas, GHG emissions also considered the release of methane to the environment that can be avoided by using effluents in palm oil mills to produce biogas via anaerobic digestion [28].

(b)   Emissions avoided by replacing thermal power plants

This computation consisted of determining how much fossil-derived fuels could save the country due to the substitution of existing or expected thermal power plants for electricity production (which could be a necessity to offset hydropower generation capacity in the country, especially during the dry season of the year) by cogeneration in industrial plants. To make easier the computations, it was assumed that the efficiency of large thermal power plants is ~35% [80] (although the efficiency of some existing thermal power plants in Ecuador is lower). The expected efficiency of the cogeneration plants taken as a reference was calculated in five representative companies (including a hospital and a hotel, where trigeneration is possible). Results showed efficiencies >70% in all cases. Thus, the difference in efficiency in a scenario without cogeneration and a scenario with cogeneration was conservatively taken as 30%.

2.1.5.2. Economic Impacts

Economic analysis was carried out to understand the convenience of cogeneration in the country from an economic point of view. The analysis consisted of (a) estimating the costs avoided if cogeneration is used instead of large thermal power plants that operate on fossil fuels, and (b) computing the cost of generating electricity in cogeneration plants if the whole potential of cogeneration calculated is installed. Table 5 summarizes the parameters employed for conducting the economic analysis. Some of these parameters are in agreement with the work of [42]. The prices of fuels and electricity are similar in all regions of the country.

Table 5. Parameters used for the economic analysis.

Parameter	Details
Cost of both diesel and gas engines for cogeneration	USD 1,000,000/MW$_{el}$ * installed
Cost of equipment for Rankine cycle (boiler + steam turbine)	USD 3,000,000/MW$_{el}$ installed, from which, approximately 15% corresponds to the cost of the steam turbine and the rest to the boiler and auxiliary equipment and accessories [78,82,83]
Cost of HRSG	USD 300,000 per MW$_{el}$ of cogeneration capacity installed (this value is above that in [80], Ch. 24).
Cost of LiBr absorption chiller	USD 500/TR ** installed [80,84].
Cost of NH$_3$+H$_2$O absorption chiller	USD 700/TR installed [84].
Operation and maintenance costs	Value varies from 2% of the investment during the first years of the projects to up to 7% after year 10. Values are in the range of those reported by [85], although a little higher after year 5 due to the necessity of importing parts.
Expected capacity factor	95% to 60%, depending on the type of industry (see Table 6).
discount rate (includes financial cost and financial risk)	12% (rate currently used for electricity projects in Ecuador).
Reinvestment	25% of the initial investment will be required on year 10.
Projects lifetime	15 years.
Plant location and land requirements	No land will be bought for cogeneration plants since the plant will be installed at existing companies' facilities.
Substation and transmission facilities	Cost is included in the cost of prime movers.
Insurance	0.5% of the investment per year
Cost of diesel and natural gas	USD 2.12/gallon (USD ~0.57 US/L) and USD 0.45/kg, respectively (without subsidies) [86].
Cost of biomass ***	USD 20/t, which is in the range of or above the costs of residues from the agroindustry (e.g., oil palm residues) in the Ecuadorian coast region (resulting from a field study).
Workforce salaries	Each cogeneration plant will require one employee per MW$_{el}$ installed per every 8 h of operation, with salaries of USD 1250/month (in the conditions of Ecuador), plus one supervisor and one person in charge of maintenance.

* Includes project management and design engineering as well as construction and start-up. This is a referential cost due to discrepancy of values in the literature. The authors of [78] show higher values, but [87] and [85] report values in the range of USD 1000/kW. However, the cost of a gas engine (1 MW) operating at a landfill in Cuenca was USD 450/kW. The value considered in this work could be adequate due to economy of scale when contracting and installing several cogeneration plants. ** TR refers to ton of refrigeration (equivalent to 3.52 kW). *** Electricity to be sold to the national electricity grid after operation of the plant and service loads are met *** To operate cogeneration plants based on Rankine cycle.

Table 6. Summary of prime movers selected for cogeneration/trigeneration in Ecuador, range of sizes, and expected capacity factor for each type of industry.

Type of Industry (Cluster)	Location of Company	Prime Mover Suggested	Range of Sizes	Expected Average Capacity Factor
Food industry: Dairy	Coastal region and Andean highlands	Internal combustion engine (diesel engine)	0.75 to 2 MW_{el}, using one or more engines	75%
Food industry	Coastal region and Andean highlands	Internal combustion engine (diesel engine) or steam turbine (biomass fired boiler)	0.5 to 5 MW_{el}, using 1 or more engines	80%
Textile industry	Andean highlands	Internal combustion engine (diesel engine)	1 to 5 MW_{el}, using normally more than 1 engine	80%
Agroindustry (except oil palm industry)	Coastal region	Internal combustion engine (biogas or diesel engine) (4)	0.5 to 3 MW_{el}, using normally more than 1 engine	80%
Agroindustry: Oil palm industry	Coastal region	Internal combustion engine (gas engine) (1)	1 to 5 MW_{el}, using normally more than 1 engine	85%
Beverage industry	Coastal region and Andean highlands	Internal combustion engine (diesel engine)	0.5 to 5 MW_{el}, using 1 or more engines	80%
Wood and wood composites industry	Andean highlands	Boiler (biomass fired) and steam turbine (Rankine cycle) (2)	2 to 7 MW_{el}	85%
Cement and ceramic tiles	Coastal region and Andean highlands	Organic Rankine cycle	Up to 3 MW_{el}	80%
Pulp and paper	Coastal region and Andean highlands	Internal combustion engine (diesel engine) or biomass fired boiler (steam turbine) (3)	0.5 to 3 MW_{el}	90%
Metals	Coastal region and Andean highlands	Organic Rankine cycle	0.9 to 1.25 MW_{el}	80%
Hospitals	Coastal region and Andean highlands	Internal combustion engine (diesel engine)	0.5 to 5 MW_{el}, using normally more than 1 engine	60%
Hotels	Coastal region and Andean highlands	Internal combustion engine (diesel engine)	0.5 to 3.75 MW_{el}, using normally more than 1 engine	60%
Other: Airports	Coastal region and Andean highlands	Internal combustion engine (diesel engine)	0.6 to 3 MW_{el}, using normally more than 1 engine	65%
Other: Shopping malls	Coastal region and Andean highlands	Internal combustion engine (diesel engine)	2 MW_{el}	65%
Other: Tires	Andean highlands	Rankine cycle	~1.2 MW_{el}	95%

(1) Using only gas engines running with biogas. (2) Using biomass from the same plant. (3) Depending on the size of the company. (4) Further study is required to analyze the possibility of using biomass.

2.1.5.3. Social Impacts

According to [88] (p. 43), social impacts are the 'consequences of social relations (interactions) weaved in the context of an activity (production, consumption or disposal) and/or engendered by it and/or by preventive or reinforcing actions taken by stakeholders (ex. enforcing safety measures in a facility)'. A social life cycle analysis (SLCA) should consider the potential social impacts on local communities, workers, and consumers [89]. However, the literature shows that the social implications of projects related, for instance, with the use of lignocellulosic natural resources for energy [90] or wood-based products [91] are hard to estimate due to the difficulty of correlating cause–effect chains with regards to production activities and their potential social effects. Therefore, the computation of the social impacts of adopting cogeneration in a whole country is even more difficult. For this reason, in this work, the social impacts of cogeneration are focused on a preliminary estimation of such impacts on the creation of new jobs in the places where cogeneration plants could be installed. Such jobs are required, generally, for operating the cogeneration plants. Each plant will require at least five people: three for operation, one for maintenance, and one for management/supervision.

3. Results and Discussion

3.1. Current Electricity Demand and Fuel Consumption in the Industrial Sector of Ecuador

The electricity demand (from de National Interconnected System—SNI) and the fuel consumption in the 555 companies are 409,199 MWh/month and 61.73×10^6 L/month (51,773 t/month) of diesel equivalent, respectively. Figure 5 shows the electricity demand and fuel consumption by each type of cluster of companies (See Table 3). It is seen that the electricity consumption (Figure 5a) is higher in the clusters of food and construction materials industries, with 19% and 17% of the total, respectively. The fuel consumption, as seen in Figure 5b, is higher, again, in the cluster of companies of construction materials and in the cluster of food industries, with 17% and 16% of the total, respectively. The large amount of companies in the food industry cluster and the presence of energy intensive industries in the construction materials cluster (e.g., cement and ceramic tiles) explain these results.

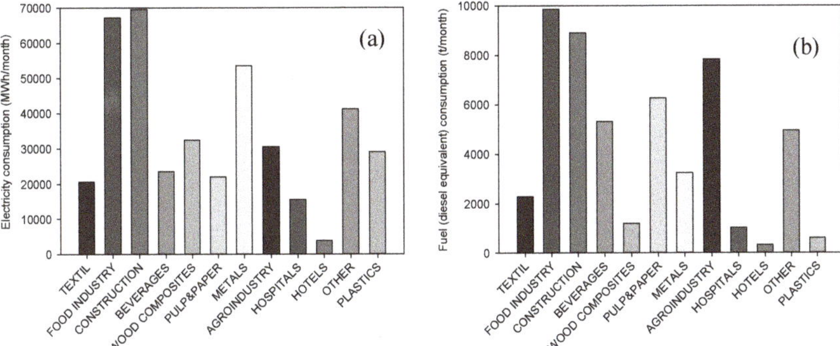

Figure 5. (**a**) Electricity and (**b**) fuel (diesel equivalent) consumption by industrial clusters in the list of the companies analyzed.

3.2. Cogeneration Technology by Type of Industry

Table 6 presents the technologies suggested for cogeneration schemes in each type of industry in Ecuador. The table also shows the geographic location of each cluster of industries. Internal combustion engines (diesel and gas engines) are the most prominent prime movers suggested due to their advantages, as discussed in Section 2.1.3. In addition, these engines offer the possibility of working with biodiesel and biogas, in substitution of diesel and NG, respectively, which is of interest in Ecuador. Currently the country produces only ~30 t/year of biodiesel from *Jatropha curcas* to operate diesel engines in thermal power plants the Galapagos Islands [92]. The program to produce biodiesel from this plant is in its infancy, but it is expected that the biodiesel production capacity will increase in coming years. The use of gas engines deserves further study since it is expected that the agroindustrial sector in Ecuador will start producing biogas using their residues via anaerobic digestion. However, this topic is out of the scope of this paper.

3.3. Potential of Cogeneration/Trigeneration

The estimated potential of cogeneration in Ecuador is 598 MW$_{el}$, which, as mentioned in Section 2.1, consists of the potential of cogeneration of industries with expected installed cogeneration capacity above 0.5 MW$_{el}$. The value excludes the existing cogeneration capacity shown in Table 1. This potential is ~7% of the current electricity generation installed capacity in Ecuador and could produce up to 17% of the total electricity consumed in 2017 in the country. This last value is, interestingly, in the range of percentages of the cogeneration share (respect to the total electricity produced) in countries such as Germany (17%), Brazil (18%), Spain (12%), or the United States (12%) [58,93–95]. Even though in the case of Ecuador this amount refers to potential cogeneration (i.e., not installed cogeneration capacity), such value is important because of the possibility of using cogeneration during the driest season of the year, when hydropower generation is negatively affected by weather conditions (See Section 1). For this reason, cogeneration has been seen in the country as an important strategy for electricity production in the near future, and new laws and regulations are under study to promote cogeneration/trigeneration.

Figure 6 summarizes the potential of cogeneration in Ecuador by type of prime mover selected. Diesel engines are the predominant prime movers suggested for cogeneration (Section 3.2). These engines can run with biodiesel (mixed with diesel) when available. Figure 7 presents the potential of cogeneration by cluster, showing that the textile, food, and agroindustry industries are the clusters with higher potential. Moreover, the potential of trigeneration in the country is 212 MW$_{el}$. Approximately 17% of the 555 companies identified in Section 2.1 could adopt trigeneration, especially in the food and beverages industries, as well as in hotels and hospitals (Figure 8).

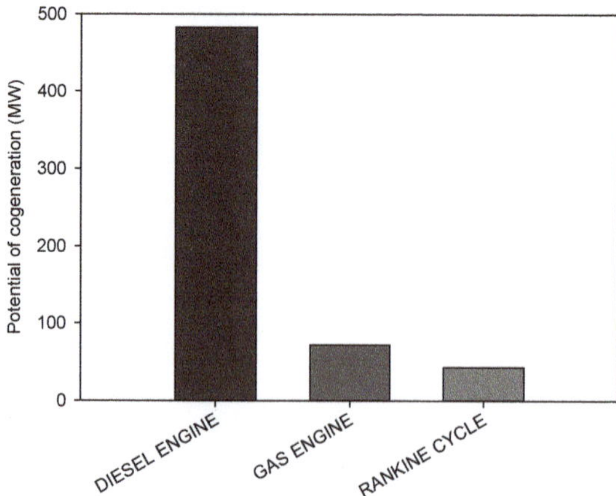

Figure 6. Potential of cogeneration in Ecuador by type of prime mover suggested (MW$_{el}$).

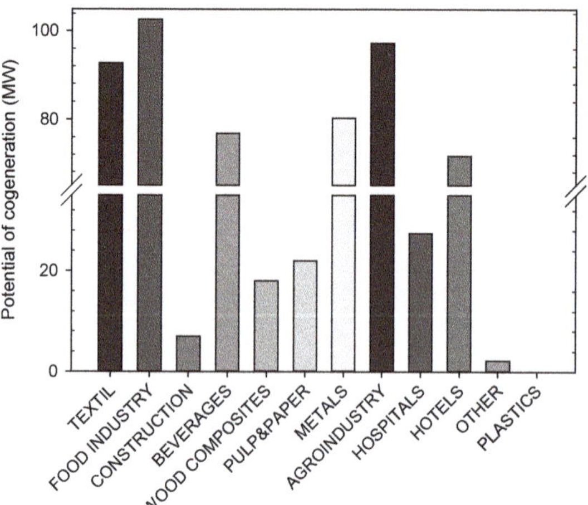

Figure 7. Potential of cogeneration by cluster of industries (MW$_{el}$).

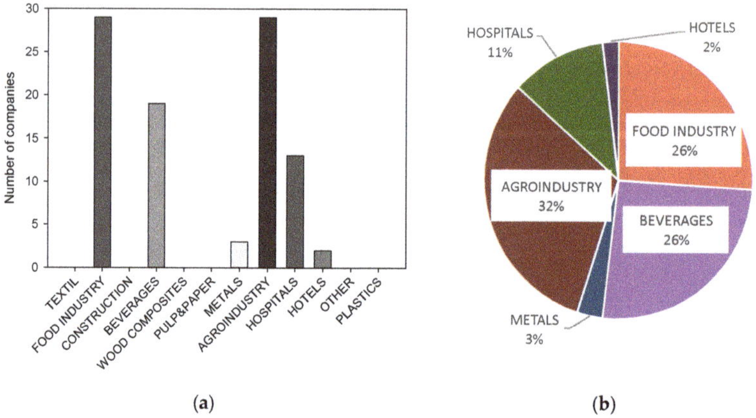

Figure 8. (**a**) Amount of companies that could adopt trigeneration, and (**b**) contribution (in %) of each cluster to trigeneration (based on a trigeneration potential of 212 MW$_{el}$).

3.4. Impacts of Cogeneration in Ecuador

3.4.1. Fuel Consumption, Improvement of Energy Efficiency, and GHG Emissions Reduction

The adoption of cogeneration in Ecuador will require different types of fuels. Due to the lack of NG in the country (the preferred fuel for cogeneration in most countries from tempered regions), in the conditions of this study and considering current fuel availability in Ecuador (See Section 3.2), diesel has been selected. Diesel could comprise approximately 81% of the fuel requirements for cogeneration (if the whole potential of 598 MW$_{el}$ is installed), while biogas and biomass could, together, cover approximately 17% (as shown in Table 7). Biomass fuel is constituted by solid residues generated by the agroindustry (e.g., oil palm and rice), which are abundant biomass resources in the coast region. Although the potential of biomass for cogeneration can be higher than this value, its use deserves more analysis due to the difficulty of hauling and burning this fuel in industrial plants located in urban areas far away from biomass sources. The potential use of NG for cogeneration is very low (~2%). Because of NG is an important fuel for cogeneration in most countries (due to availability, competitive prices, and cleanliness during burning), Ecuador urgently needs to look for NG as an alternative (at least partially) to diesel. For this purpose, two options are being analyzed in the country: (a) importing NG from neighbor countries such as Peru, which, in addition to its high potential production [45], could also import it from Bolivia, as part of the so-called Latin America Energy Integration [96–98], and (b) exploring the Gulf of Guayaquil for more NG, since there is no certainty about the NG reserves in this part of the country.

Table 7. Types and quantities of fuels required for cogeneration in Ecuador and potential contribution to greenhouse gas (GHG) generation/reduction.

Type of Fuel	Potential of Cogeneration (MW$_{el}$) and Share in the Total (%)	Amount of Fuel	Expected Electricity Generation (GWh/year)	Potential GHG Emissions (tCO$_2$/year)
Diesel	482.9 (81%)	368,950,000 kg/year	4231.3	+1,150,500 (a)
Biogas	60.0 (10%)	100,126,800 kg/year	525.7	−704,147 (b)
Biomass	43.0 (7%)	71,781,943 kg/year	376.9	−227,770
Natural Gas	11.9 (2%)	19,772,571 kg/year	102.9	+62,210
TOTAL:	598		5236.8	280,793 (Total 1)
			Emissions in the SNI	−576,800
			Net GHG (reduction)	−296,007 (Total 2)

Table 7 also shows the electricity that could be produced by type of fuel (column four) and the corresponding potential contribution to GHG emissions (Table 7, column five). The negative sign in the Table indicates avoided GHG emissions, which results from (1) burning biomass and biogas instead of oil-derived fuels to produce electricity (in cogeneration plants), and (2) the avoided methane formation from liquid effluents from the oil palm industry. Currently, although the majority of the 35 oil palm companies in the country (See Section 1) are aware about the necessity of using liquid effluents for biogas production, these effluents are discharged to pools for stabilization prior to final disposal due to the lack of incentives/regulations from the State to use them for energy.

The adoption of cogeneration could promote a reduction 18.55 million L/month (15,556 t/month) of diesel (and/or heavy fuel oil) and avoid up to 576,800 tCO$_2$/year. This value results from considering that the country would need to install and operate a 600 MW$_{el}$ power plant (or several plants with equivalent total capacity) to offset the reduction of hydropower during the dry season and that, instead of installing such thermal power plant, cogeneration in the industry will be adopted. The positive impact of cogeneration in the industrial sector's energy efficiency of the country is proportional to the amount of fuels saved. Thus, in the conditions of this study, the increase in energy efficiency, if the whole cogeneration potential was installed, could reach between 35% and 40%.

The net GHG emissions (i.e., total 1 in Table 7 minus 576,800) could be −296,007 tCO$_2$/year (total 2), showing that installing cogeneration/trigeneration in the industry can be an important strategy to avoid GHG emissions in Ecuador. Figure 9 shows that the clusters in which fuel savings could be higher are the food industry, the beverage industry, and the agroindustry. Further study is necessary for analyzing the environmental positive impacts of changing diesel and natural gas by biodiesel and biogas, respectively. However, Table 7 shows that potential GHG emissions are reduced even using diesel and NG, as a consequence of higher efficiency on burning these fuels in cogeneration plants.

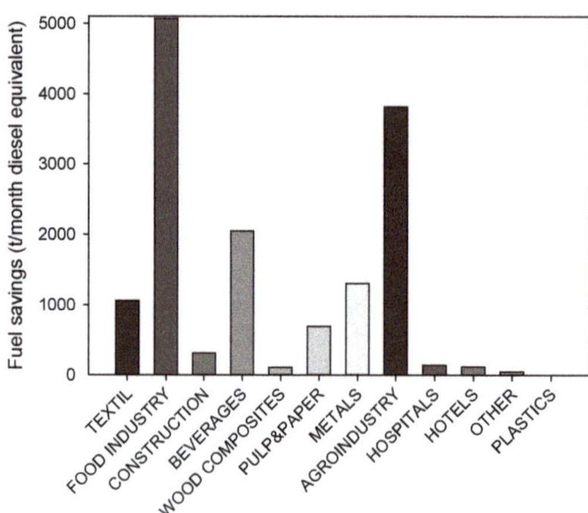

Figure 9. Expected fuel savings (by cluster) resulting from the possible adoption of cogeneration.

3.4.2. Economic Analysis

The economic analysis showed that an important consequence for Ecuador is that, if cogeneration is installed instead of a large thermal power plant to offset the future lack of hydroelectricity, the country could save up to USD 125 million per year by avoiding the use of oil-derived fuels for electricity generation. The cost of the electricity produced in cogeneration plants will depend on the type of cogeneration scheme and the type of fuel used, as seen in Table 8. The cost for electricity produced in cogeneration plants (considering the cost of fuels shown in Table 5, but excluding NG), will vary from USD 0.09/kWh to USD0.17/kWh for electricity produced in the oil palm industry (using lignocellulosic biomass) and in hospitals (using diesel), respectively. Table 8 also shows that some types of cogeneration plants, even using diesel, could produce electricity at costs lower than USD 0.17/kWh. For instance, the hotels industry and the textile industry could produce electricity at USD 0.12/kWh and USD 0.13/kWh (using diesel as fuel), respectively. Although these values are higher than the cost of generating electricity in hydropower plants in Ecuador (up to 0.08 USD/kWh), cogeneration in these conditions is still of interest for Ecuador due to the necessity of diversification of electricity generation and the opportunity of having installed capacity for electricity generation during the dry season of the year. Because of insufficient electricity generation (especially before 2016), Ecuador has often required to import electricity from both Colombia and Peru at prices up to USD 0.28/kWh or to produce electricity using thermal power plants at even higher costs (up to USD 0.50/kWh in old thermal power plants).

An analysis of sensitivity was carried out to understand the effect of using NG (when available in the future) instead of diesel for cogeneration in the country. Results showed that NG could promote a substantial reduction of the costs of electricity production in cogeneration plants. For instance, the dairy industry could produce electricity at around USD 0.06/kWh, hotels at USD 0.08/kWh, and hospitals at USD 0.05/kWh (Table 8). These results reinforce the notion that the country must look for options for buying NG overseas, especially in neighboring countries (see Section 3.4.1). The production and use of biofuels for cogeneration requires further analysis.

Table 8. Examples of costs of electricity generated in some types of clusters of industries in the conditions of the study (including the potential use of NG).

Type of Cluster of Industries	Type of Fuel Suggested	Type of Plant	Expected Cost of Electricity Generated (USD/kWh)
Oil palm industry	Biomass	Cogeneration	0.09
Oil palm industry	Biogas	Cogeneration	0.02
Dairy industry	Diesel	Trigeneration	0.14
Dairy industry	NG	Trigeneration	0.06
Textile industry	Diesel	Cogeneration	0.13
Textile industry	Biomass	Cogeneration	0.10
Hotels	Diesel	Trigeneration	0.12
Hotels	NG	Trigeneration	0.08
Hospitals	Diesel	Trigeneration	0.17
Hospitals	NG	Trigeneration	0.05

3.4.3. Social Impacts of Cogeneration

The adoption of cogeneration/trigeneration in Ecuador could promote more than 2600 new jobs. As mentioned in Section 2.1.5.3, these direct jobs are required for operating, managing, and maintaining the cogeneration plants. There is evidence showing positive impacts of energy efficiency measures on GDP, employment, economic structure, and welfare [99]. In addition, there is an important element that was not included in the economic analysis: the benefit to the state of avoiding the release of CO_2 by installing cogeneration plants, which is related to the "social cost of carbon" or marginal damage caused by an additional ton of carbon dioxide emissions [2–4,100]. Therefore, these and other benefits that are not considered at this level of the study (e.g., the impact on rural areas where some cogeneration will be installed, the benefits on health due to better air quality or the creation of indirect jobs) deserve further study.

4. Conclusions

In tropical climate countries, the potential of cogeneration (and as such, its calculation) of the industrial sector is dependent on particular climate conditions, consumption behavior, cogeneration schemes, and fuel availability. Tropical countries such as Ecuador do not necessitate indoor heating (an important energy requirement in tempered climate countries), although air conditioning is prominently used. Thus, large cogeneration projects are more suitable in the industrial sector and in places where hot and cold fluids are used (e.g., hospitals, hotels, airports, and shopping malls). This study has shown that the adoption of cogeneration at a large scale promotes environmental, economic, and social benefits to countries by reducing GHG emissions, promoting fuel savings and energy efficiency, and by creating new jobs, respectively. In the case of Ecuador, the potential of cogeneration in the industrial sector (including hospitals, hotels, shopping malls and two airports) is approximately 600 MW$_{el}$, which is around 7% of the total electricity generation installed capacity in the country. If this cogeneration potential is implemented, the energy efficiency in the Ecuadorian industry could be increased by 35–40%. This potential could save up to 18.6×106 L/month of oil-derived fuels, avoiding up to 576,800 tCO$_2$/year, and creating more than 2600 direct jobs. Lack of NG for cogeneration is seen as a problem that needs to be addressed in the future to reduce the cost of electricity generation in cogeneration plants. The use of diesel and gas engines (the main types of prime movers in the conditions of the industry in Ecuador) presents opportunities to easily move from fossil-derived fuels to renewable fuels, i.e., to use biodiesel and biogas in substitution of diesel and NG, respectively. This topic deserves further analysis, especially in identifying options for producing biofuels. Further studies should also address the logistics of integration of cogeneration with other electricity generation sources such as hydropower, or the logistics of biomass for cogeneration, to mention two aspects. Distributed generation through cogeneration offers opportunities to diversify local (small scale) electricity generation to optimize the use of the national grid and offset one of the problems of the Ecuadorian electricity sector: its high dependency on hydropower that has large seasonal variations due to water flow reductions.

Author Contributions: Conceptualization, M.R.P.-S., J.L.E., J.J.-A., P.R.-G. and P.R.; methodology, M.R.P.-S., J.L.E., J.J.-A., F.M.-A., P.A.-R.; validation, M.R.P.-S., J.L.E., J.J.-A., P.A.-R.; data curation, M.R.P.-S., J.L.E. and T.G.-P.; writing—original draft preparation, M.R.P.-S. and J.L.E.; writing—review and editing, M.R.P.-S., J.L.E., J.J.-A., F.M.-A. and T.G.-P.; supervision, J.L.E. and M.R.P.-S.; project administration, J.J.-A.; funding acquisition, J.J.-A., P.R.-G. and P.R. All authors have read and agreed to the published version of the manuscript.

Funding: This research was funded by the former Ministry of Electricity and Renewable Energy—MEER (currently Ministry of Energy and Natural Non-renewable Resources) and the Corporación Eléctrica del Ecuador—CELEC EP (Contract No. 042-2016).

Acknowledgments: To the former Ministry of Electricity and Renewable Energy—MEER (currently Ministry of Energy and Natural Non-renewable Resources) and the Corporación Eléctrica del Ecuador—CELEC EP, institutions that have authorized this publication. Thanks to Stalin Vaca, Diego Suárez, Jaime Alvarado, Ricardo Álvarez, Robinson Machuca, Guillermo Pérez, Tamara Serrano, Rommel Vargas, and Fernanda Orellana for their support in conducting field research; to all the Ecuadorian industrial facilities that contributed information to conduct this study; to the ARCH and the ARCONEL for providing information on energy consumption in Ecuador; to Paul Martinez and Jorge Ortiz (CELEC EP) for their constructive comments on the results of the study; and to Raul Pelaez-Garcia for revising the English in the text.

Conflicts of Interest: The authors declare no conflict of interest.

Appendix A. Schematics of Proposed Cogeneration Systems

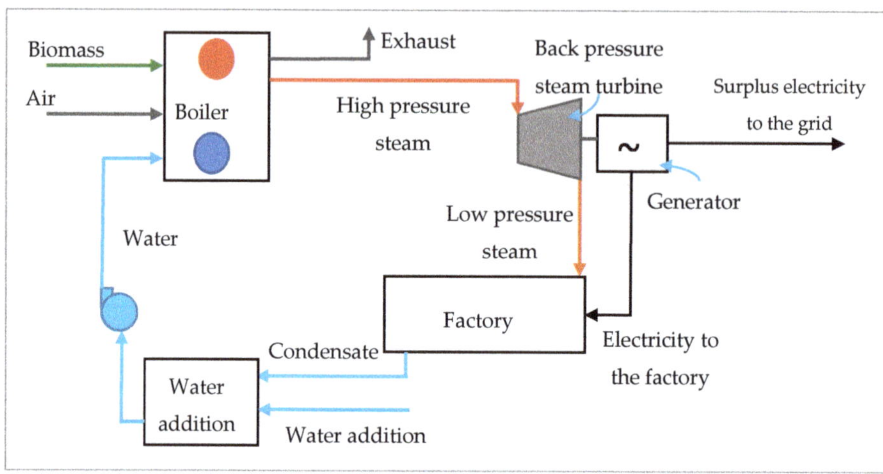

Figure A1. Schematic of cogeneration system based on Rankine cycle for companies that can use biomass as fuel (e.g., sugarcane, pulp and paper, oil palm industries) and back pressure steam turbines. Adapted from [18,19,23,28,32,49,51].

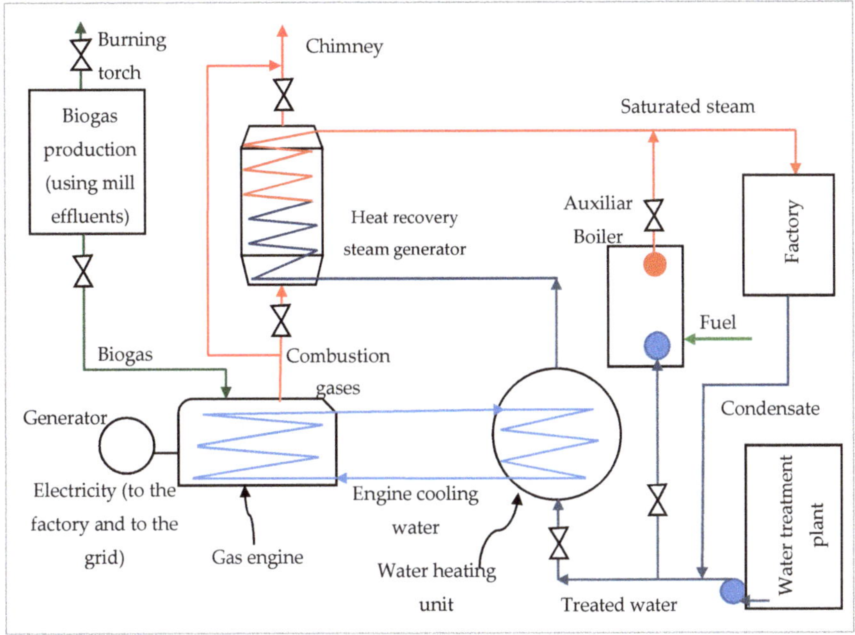

Figure A2. Schematic of a cogeneration system using gas engines for the oil palm industry.

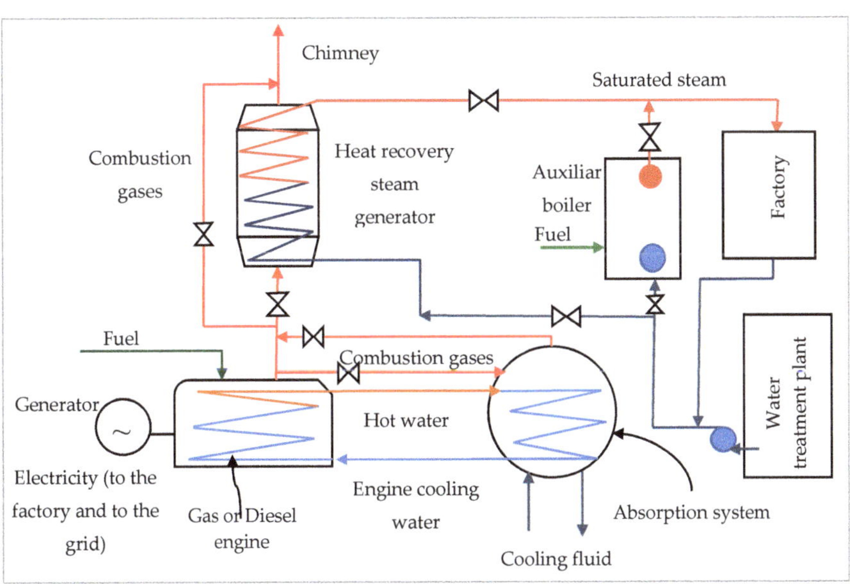

Figure A3. Schematic of a proposed trigeneration system using gas engines or diesel engines for the beverage industry, dairy industry, and food industry

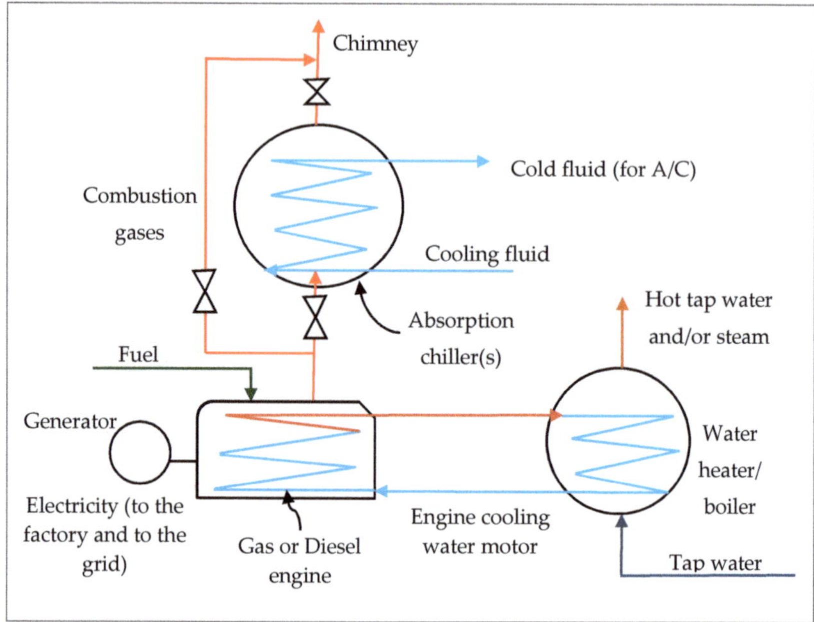

Figure A4. Schematic of a proposed trigeneration system using gas engines or diesel engines for service industries (e.g., hotels, hospitals). Adapted from [46].

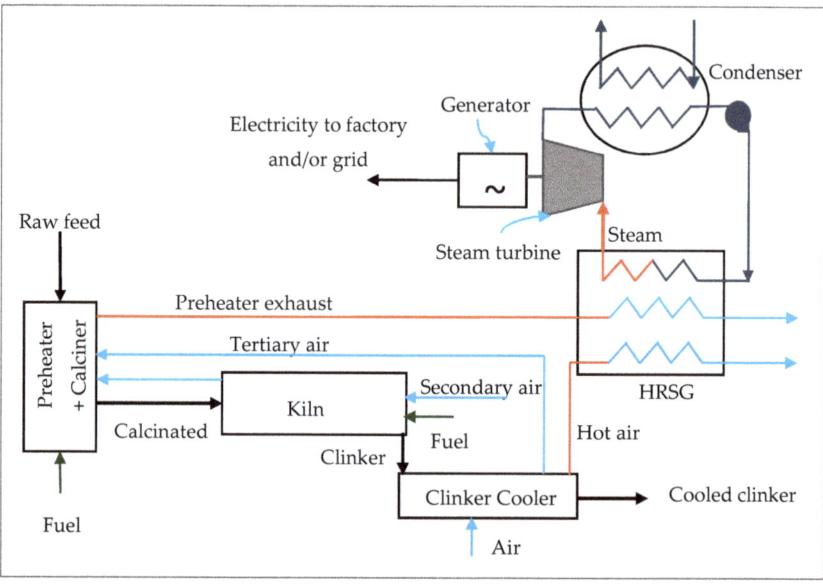

Figure A5. Schematic of a proposed bottom cogeneration system used in cement industry. HRSG—heat recovery steam generator (adapted from [55,101]).

Appendix B. Main Equations Used for the Computation of Cogeneration Systems (Units Are Presented in Brackets)

(a) Cogeneration plant efficiency (CHP_{eff})

CHP_{eff} = (power output + useful heat recovered)/energy in fuel

(b) Energy in fuel (Q_{fuel})

$$Q_{fuel} = m_{fuel} * LHV_{fuel} \ [kW]$$

m_{fuel}—fuel rate [kg/s]

LHV_{fuel}—fuel lower heating value [kJ/kg]

(c) Energy in steam (Q_{steam})

$$Q_{steam} = m_{steam} * (h_{steam} - h_{water}) \ [kW]$$

m_{steam}—flow rate (production) of steam [kg/s]

h_{steam}—enthalpy of steam at the boiler exit [kJ/kg]

h_{water}—enthalpy of water at the entrance of boiler [kJ/kg]

(d) Efficiency of boiler [η_{boiler}]

$$\eta_{boiler} = Q_{steam}/Q_{fuel}$$

(e) Energy in combustion gases (Q_{cgas}) that is used, for instance, in a heat recovery steam generator (HRSG)

$$Q_{cgas} = m_{cgas} * (h_{hotgas} - h_{coldgas}) \ [kW]$$

m_{cgas}—flow rate of combustion gases [kg/s] (e.g., gases from gas engine)

h_{hotgas}—enthalpy of combustion gases at the entrance of heat recovery unit [kJ/kg]

$h_{coldgas}$—enthalpy of combustion gases after passing through the heat recovery equipment [kJ/kg]

(f) Electric energy efficiency of prime movers (motors) (η_{EE})

$$\eta_{EE} = W_{elec}/Q_{fuel}$$

W_{elec}—electric power (useful energy output) [kW]

(g) Heat recovery unit (HRU) efficiency (η_{HRU}) for water heating

$$\eta_{HRU} = Q_{HRUactual}/Q_{HRUtheor}$$

$Q_{HRUactual}$—actual heat transfer rate [kJ/s]

$Q_{HRUtheor}$—maximum possible heat transfer rate [kJ/s]

$$Q_{HRUactual} = m_{waterHRU} * (h_{waterHRUent} - h_{waterHRUexit})$$

$m_{waterHRU}$—water flow rate in the HRU [kg/s]

$h_{waterHRUexit}$—enthalpy of water at the entrance of the HRU [kJ/kg]

$h_{waterHRUent}$—enthalpy of water at the exit of the HRU [kJ/kg]

(h) Heat recovery steam generator efficiency (η_{HRSG}) (for steam production)

$$\eta_{HRSG} = Q_{HRSGactual}/Q_{HRSGtheor}$$

$Q_{HRSGactual}$—actual heat transfer rate [kJ/s]

$Q_{HRSGheor}$—maximum possible heat transfer rate [kJ/s]

$$Q_{HRSGactual} = m_{steamHRSG} * (h_{steamHRSGent} - h_{waterHRSGexit})$$

$m_{waterHRU}$—steam (or water) flow rate in the HRSG [kg/s]

$h_{waterHRUexit}$—enthalpy of water at the entrance of the HRSG [kJ/kg]

$h_{steamHRUent}$—enthalpy of steam at the exit of the HRSG [kJ/kg]

(i) Efficiency of absorption chiller (COP_{Achill})

$COP_{Achill} = Q_{evap}/Q_{in}$

Q_{evap}—rate at which water is cooled by the evaporator [kJ/s]

Q_{in}—heat input (rate of heat loss from exhaust gas or steam that are used by the absorption unit) [kJ/s]

(j) Electricity produced by steam turbine-generator (E_{gen}) [kWh/month]

$E_{gen} = Q_{steam}*\eta_{turb}*\eta_{gen}*T_{oper}*p_f$ [kWh/month]

η_{turb}—steam turbine efficiency

η_{gen}—generator efficiency

T_{oper}—time generator operates [h/month]

(k) Present worth (present value) (C_t) of C monetary units

$C_t = C/(1+i)^t$;

i—discount rate; t—number of time periods

(l) Net present value (NPV)

$NPV = (C_1+C_2+C_3+ \ldots + C_n)$

$C_1, C_2, C_3, \ldots , C_n$ – Present worth of anticipated cash flows

References

1. BP STATS. *Statistical Review of World Energy Statistical Review of World*, 68th ed.; BP: London, UK, 2019.
2. Ackerman, F.; Stanton, E.A. Climate risks and carbon prices: Revising the social cost of carbon. *Economics* **2012**. [CrossRef]
3. Moore, F.C.; Diaz, D.B. Erratum: Temperature impacts on economic growth warrant stringent mitigation policy. *Nat. Clim. Chang.* **2015**, *5*, 280. [CrossRef]
4. Dell, M.; Jones, B.F.; Olken, B.A. Temperature shocks and economic growth: Evidence from the last half century. *Am. Econ. J. Macroecon.* **2012**. [CrossRef]
5. Peláez-Samaniego, M.R.; Garcia-Perez, M.; Cortez, L.A.B.; Oscullo, J.; Olmedo, G. Energy sector in Ecuador: Current status. *Energy Policy* **2007**. [CrossRef]
6. Pelaez-Samaniego, M.R.; Riveros-Godoy, G.; Torres-Contreras, S.; Garcia-Perez, T.; Albornoz-Vintimilla, E. Production and use of electrolytic hydrogen in Ecuador towards a low carbon economy. *Energy* **2014**, *64*, 626–631. [CrossRef]
7. Ponce-Jara, M.A.; Castro, M.; Pelaez-Samaniego, M.R.; Espinoza-Abad, J.L.; Ruiz, E. Electricity sector in Ecuador: An overview of the 2007–2017 decade. *Energy Policy* **2018**, *113*. [CrossRef]
8. BID; Ministerio de Electricidad y Energía Renovable; MEER. *Plan Nacional de Eficiencia Energética 2016–2035*; Ministerio de Electricidad y Energía Renovable: Quito, Ecuador, 2017.
9. Insituto de Investigación Geológico y Energético (IIGE). *Balance Energético Nacional 2018*; Ministerio de Energía y Recursos Naturales no Renovables: Quito, Ecuador, 2019.
10. MAE. Resumen del Inventario Nacional de Gases de Efecto Invernadero del Ecuador. Available online: https://info.undp.org/docs/pdc/Documents/ECU/06%20Resumen%20Ejecutivo%20INGEI%20de%20Ecuador.%20Serie%20Temporal%201994-2012.pdf (accessed on 20 May 2019).
11. MEER; BID. *Plan Nacional de Eficiencia Energética*; Ministerio de Electricidad y Energía Renovable: Quito, Ecuador, 2016.

12. Posso, F.; Sánchez, J.; Espinoza, J.L.; Siguencia, J. Preliminary estimation of electrolytic hydrogen production potential from renewable energies in Ecuador. *Int. J. Hydrogen Energy* **2016**, *41*, 2326–2344. [CrossRef]

13. ARCONEL. Balance Nacional de Energía Eléctrica. Available online: https://www.regulacionelectrica.gob.ec/balance-nacional/ (accessed on 20 December 2019).

14. CENACE. Energía Neta Producida por las Centrales de Generación. Available online: http://www.cenace.org.ec/docs/Produci{\protect\edefT1{T5}\let\enc@update\relaxó}n-de-Energ{\protect\edefT1{T5}\let\enc@update\relaxí}a-(GWh)-mensual.htm (accessed on 12 March 2020).

15. MEER; ARCONEL. *Plan Maestro de Electrificación 2016–2025*; Ministerio de Electricidad y Energía Renovable: Quito, Ecuador, 2017.

16. Nzotcha, U.; Kenfack, J. Contribution of the wood-processing industry for sustainable power generation: Viability of biomass-fuelled cogeneration in Sub-Saharan Africa. *Biomass Bioenergy* **2019**. [CrossRef]

17. Arshad, M.; Ahmed, S. Cogeneration through bagasse: A renewable strategy to meet the future energy needs. *Renew. Sustain. Energy Rev.* **2016**. [CrossRef]

18. Deshmukh, R.; Jacobson, A.; Chamberlin, C.; Kammen, D. Thermal gasification or direct combustion? Comparison of advanced cogeneration systems in the sugarcane industry. *Biomass Bioenergy* **2013**. [CrossRef]

19. Dias, M.O.S.; Modesto, M.; Ensinas, A.V.; Nebra, S.A.; Filho, R.M.; Rossell, C.E.V. Improving bioethanol production from sugarcane: Evaluation of distillation, thermal integration and cogeneration systems. *Energy* **2011**. [CrossRef]

20. Salem Szklo, A.; Soares, J.B.; Tolmasquim, M.T. Energy consumption indicators and CHP technical potential in the Brazilian hospital sector. *Energy Convers. Manag.* **2004**. [CrossRef]

21. Restuti, D.; Michaelowa, A. The economic potential of bagasse cogeneration as CDM projects in Indonesia. *Energy Policy* **2007**. [CrossRef]

22. To, L.S.; Seebaluck, V.; Leach, M. Future energy transitions for bagasse cogeneration: Lessons from multi-level and policy innovations in Mauritius. *Energy Res. Soc. Sci.* **2018**. [CrossRef]

23. Gongora, A.; Villafranco, D. Sugarcane bagasse cogeneration in Belize: A review. *Renew. Sustain. Energy Rev.* **2018**. [CrossRef]

24. Deepchand, K. Commercial scale cogeneration of bagasse energy in Mauritius. *Energy Sustain. Dev.* **2001**. [CrossRef]

25. Rincón, L.E.; Becerra, L.A.; Moncada, J.; Cardona, C.A. Techno-economic analysis of the use of fired cogeneration systems based on sugar cane bagasse in south eastern and mid-western regions of Mexico. *Waste Biomass Valoriz.* **2014**. [CrossRef]

26. Booneimsri, P.; Kubaha, K.; Chullabodhi, C. Increasing power generation with enhanced cogeneration using waste energy in palm oil mills. *Energy Sci. Eng.* **2018**. [CrossRef]

27. Nasution, M.A.; Herawan, T.; Rivani, M. Analysis of palm biomass as electricity from palm oil mills in north sumatera. *Energy Procedia* **2014**, *47*, 166–172. [CrossRef]

28. Garcia-Nunez, J.A.; Rodriguez, D.T.; Fontanilla, C.A.; Ramirez, N.E.; Silva Lora, E.E.; Frear, C.S.; Stockle, C.; Amonette, J.; Garcia-Perez, M. Evaluation of alternatives for the evolution of palm oil mills into biorefineries. *Biomass Bioenergy* **2016**. [CrossRef]

29. Simo, A.; Siyam Siwe, S. Availability and conversion to energy potentials of wood-based industry residues in Cameroon. *Renew. Energy* **2000**. [CrossRef]

30. Mujeebu, M.A.; Jayaraj, S.; Ashok, S.; Abdullah, M.Z.; Khalil, M. Feasibility study of cogeneration in a plywood industry with power export to grid. *Appl. Energy* **2009**. [CrossRef]

31. Duval, Y. Environmental impact of modern biomass cogeneration in Southeast Asia. *Biomass Bioenergy* **2001**. [CrossRef]

32. Coelho, S.T.; Velázquez, S.G.; Zylbersztajn, D. Cogeneration in Brazilian Pulp and Paper Industry from Biomass-Origin to Reduce CO_2 Emissions. In *Developments in Thermochemical Biomass Conversion*; Bridgwater, A.V., Boocock, D.G.B., Eds.; Springer: Dordrecht, The Netherlands, 1997.

33. Coronado, C.R.; Yoshioka, J.T.; Silveira, J.L. Electricity, hot water and cold water production from biomass. Energetic and economical analysis of the compact system of cogeneration run with woodgas from a small downdraft gasifier. *Renew. Energy* **2011**. [CrossRef]

34. Udomsri, S.; Martin, A.R.; Martin, V. Thermally driven cooling coupled with municipal solid waste-fired power plant: Application of combined heat, cooling and power in tropical urban areas. *Appl. Energy* **2011**. [CrossRef]

35. Mahlia, T.M.I.; Chan, P.L. Life cycle cost analysis of fuel cell based cogeneration system for residential application in Malaysia. *Renew. Sustain. Energy Rev.* **2011**, *15*, 416–426. [CrossRef]
36. Silva, H.C.N.; Dutra, J.C.C.; Costa, J.A.P.; Ochoa, A.A.V.; dos Santos, C.A.C.; Araújo, M.M.D. Modeling and simulation of cogeneration systems for buildings on a university campus in Northeast Brazil—A case study. *Energy Convers. Manag.* **2019**. [CrossRef]
37. IDEA. *Análisis del Potencial Cogeneración de Alta Eficiencia en España 2010–2015–2020*; Instituto para la Diversificación y Ahorro de la Energía: Madrid, Spain, 2016.
38. NSW. Office of Environment and Heritage NSW, Cogeneration Feasibility Guide; Sidney, 2014. Available online: https://www.environment.nsw.gov.au/resources/business/140685-cogeneration-feasibility-guide.pdf (accessed on 20 January 2018).
39. Alexis, G.K.; Liakos, P. A case study of a cogeneration system for a hospital in Greece. Economic and environmental impacts. *Appl. Therm. Eng.* **2013**. [CrossRef]
40. Renedo, C.J.; Ortiz, A.; Mañana, M.; Silió, D.; Pérez, S. Study of different cogeneration alternatives for a Spanish hospital center. *Energy Build.* **2006**. [CrossRef]
41. Fabrizio, E. Feasibility of polygeneration in energy supply systems for health-care facilities under the Italian climate and boundary conditions. *Energy Sustain. Dev.* **2011**. [CrossRef]
42. Armanasco, F.; Colombo, L.P.M.; Lucchini, A.; Rossetti, A. Techno-economic evaluation of commercial cogeneration plants for small and medium size companies in the Italian industrial and service sector. *Appl. Therm. Eng.* **2012**. [CrossRef]
43. Sugiartha, N.; Chaer, I.; Marriott, D.; Tassou, S.A. Combined heating refrigeration and power system in food industry. *J. Energy Inst.* **2008**. [CrossRef]
44. Bianco, V.; De Rosa, M.; Scarpa, F.; Tagliafico, L.A. Implementation of a cogeneration plant for a food processing facility. A case study. *Appl. Therm. Eng.* **2016**. [CrossRef]
45. Gonzales Palomino, R.; Nebra, S.A. The potential of natural gas use including cogeneration in large-sized industry and commercial sector in Peru. *Energy Policy* **2012**. [CrossRef]
46. Arteconi, A.; Brandoni, C.; Polonara, F. Distributed generation and trigeneration: Energy saving opportunities in Italian supermarket sector. *Appl. Therm. Eng.* **2009**. [CrossRef]
47. Fantozzi, F.; Ferico, S.D.; Desideri, U. Study of a cogeneration plant for agro-food industry. *Appl. Therm. Eng.* **2000**. [CrossRef]
48. Szklo, A.S.; Tolmasquim, M.T. Strategic cogeneration-Fresh horizons for the development of cogeneration in Brazil. *Appl. Energy* **2001**. [CrossRef]
49. Bechara, R.; Gomez, A.; Saint-Antonin, V.; Schweitzer, J.M.; Maréchal, F. Methodology for the optimal design of an integrated sugarcane distillery and cogeneration process for ethanol and power production. *Energy* **2016**. [CrossRef]
50. Jenjariyakosoln, S.; Gheewala, S.H.; Sajjakulnukit, B.; Garivait, S. Energy and GHG emission reduction potential of power generation from sugarcane residues in Thailand. *Energy Sustain. Dev.* **2014**. [CrossRef]
51. Husain, Z.; Zainal, Z.A.; Abdullah, M.Z. Analysis of biomass-residue-based cogeneration system in palm oil mills. *Biomass Bioenergy* **2003**. [CrossRef]
52. Oliveira, R.C.d.; Silva, R.D.d.S.e; Tostes, M.E.D.L. A methodology for analysis of cogeneration projects using oil palm biomass wastes as an energy source in the Amazon. *DYNA* **2015**. [CrossRef]
53. Wu, Q.; Qiang, T.C.; Zeng, G.; Zhang, H.; Huang, Y.; Wang, Y. Sustainable and renewable energy from biomass wastes in palm oil industry: A case study in Malaysia. *Int. J. Hydrogen Energy* **2017**. [CrossRef]
54. Arrieta, F.R.P.; Teixeira, F.N.; Yáñez, E.; Lora, E.; Castillo, E. Cogeneration potential in the Columbian palm oil industry: Three case studies. *Biomass Bioenergy* **2007**, *31*, 503–511. [CrossRef]
55. Pradeep Varma, G.V.; Srinivas, T. Design and analysis of a cogeneration plant using heat recovery of a cement factory. *Case Stud. Therm. Eng.* **2015**, *5*, 24–31. [CrossRef]
56. Cardona, E.; Piacentino, A. A methodology for sizing a trigeneration plant in mediterranean areas. *Appl. Therm. Eng.* **2003**, *23*, 1665–1680. [CrossRef]
57. Costa, M.H.A.; Balestieri, J.A.P. Comparative study of cogeneration systems in a chemical industry. *Appl. Therm. Eng.* **2001**. [CrossRef]
58. Dillingham, G. *Combined Heat and Power (CHP) in Breweries—Better Beer at Lower Costs*; U.S. Department of Energy: Washington, DC, USA, 2016; pp. 1–37.

59. Freschi, F.; Giaccone, L.; Lazzeroni, P.; Repetto, M. Economic and environmental analysis of a trigeneration system for food-industry: A case study. *Appl. Energy* **2013**. [CrossRef]

60. Chicco, G.; Mancarella, P. Assessment of the greenhouse gas emissions from cogeneration and trigeneration systems. Part I: Models and indicators. *Energy* **2008**. [CrossRef]

61. Darabadi Zareh, A.; Khoshbakhti Saray, R.; Mirmasoumi, S.; Bahlouli, K. Extensive thermodynamic and economic analysis of the cogeneration of heat and power system fueled by the blend of natural gas and biogas. *Energy Convers. Manag.* **2018**. [CrossRef]

62. Su, B.; Han, W.; Chen, Y.; Wang, Z.; Qu, W.; Jin, H. Performance optimization of a solar assisted CCHP based on biogas reforming. *Energy Convers. Manag.* **2018**. [CrossRef]

63. CEN-CENELC. *Manual for Determination of Combined Heat and Power (CHP)*; CEN: Brussels, Belgium, 2004.

64. Cho, W.; Lee, K.S. A simple sizing method for combined heat and power units. *Energy* **2014**. [CrossRef]

65. Howard, B.; Saba, A.; Gerrard, M.; Modi, V. Combined heat and power's potential to meet New York city's sustainability goals. *Energy Policy* **2014**. [CrossRef]

66. Gambini, M.; Vellini, M. High efficiency cogeneration: Performance assessment of industrial cogeneration power plants. *Energy Procedia* **2014**. [CrossRef]

67. Kavvadias, K.C.; Tosios, A.P.; Maroulis, Z.B. Design of a combined heating, cooling and power system: Sizing, operation strategy selection and parametric analysis. *Energy Convers. Manag.* **2010**. [CrossRef]

68. Liu, M.; Shi, Y.; Fang, F. Combined cooling, heating and power systems: A survey. *Renew. Sustain. Energy Rev.* **2014**. [CrossRef]

69. Al Moussawi, H.; Fardoun, F.; Louahlia-Gualous, H. Review of tri-generation technologies: Design evaluation, optimization, decision-making, and selection approach. *Energy Convers. Manag.* **2016**. [CrossRef]

70. Sanaye, S.; Meybodi, M.A.; Shokrollahi, S. Selecting the prime movers and nominal powers in combined heat and power systems. *Appl. Therm. Eng.* **2008**. [CrossRef]

71. Shelar, M.N.; Bagade, S.D.; Kulkarni, G.N. Energy and Exergy Analysis of Diesel Engine Powered Trigeneration Systems. *Energy Procedia* **2016**. [CrossRef]

72. GIZ; BMZ. *Micro y Pequeña Cogeneración y Trigeneración en México*; GIZ Mexico: Mexico City, Mexico, 2013.

73. United Nations. *United Nations Statistical Division International Standard Industrial Classification of All Economic Activities (ISIC)*; Rev.4; United Nations: New York, NY, USA, 2008; ISBN 9789211615180.

74. INE. *Clasificación Nacional de Actividades Económicas*; Instituto Nacional de Estadistica: Madrid, Spain, 2012.

75. Carvalho, M.; Serra, L.M.; Lozano, M.A. Geographic evaluation of trigeneration systems in the tertiary sector. Effect of climatic and electricity supply conditions. *Energy* **2011**. [CrossRef]

76. Chaker, M.; Meher-Homji, C.B.; Mee, T.; Nicolson, A. Inlet Fogging of Gas Turbine Engines-Detailed Climatic Analysis of Gas Turbine Evaporative Cooling Potential. In Proceedings of the ASME Turbo Expo, New Orleans, LA, USA, 4–7 June 2001.

77. Santoianni, D. Power plant performance under extreme ambient conditions. *WÄRTSILÄ Tech. J.* **2015**, *1*, 22–27.

78. USEPA. Catalogue of CHP Technologies. U.S. Environmental Protection Agency Combined Heat and Power Partnership. Available online: https://www.epa.gov/sites/production/files/2015-07/documents/catalog_of_chp_technologies.pdf (accessed on 14 December 2017).

79. MAGAP. Ecuador Marca su Rumbo en la Industria de los Agrocombustibles. Available online: https://www.agricultura.gob.ec/ecuador-marca-su-rumbo-en-la-industria-de-los-agrocombustibles/ (accessed on 15 May 2020).

80. Hyman, L.B.; Meckel, M. *Sustainable On-Site CHP Systems*; Meckler, M., Hyan, L., Eds.; McGraw Hill: New York, NY, USA, 2010.

81. IEA. *CO2 Emissions from Fuel Combustion. Statistics*; OCD Publishing: Paris, France, 2017. [CrossRef]

82. Black & Vetch. *Cost and Performance Data for Power Generation Technologies*; Black & Vetch: Overland Park, KS, USA, 2012.

83. IRENA. Renewable Power Generation Costs in 2014. Available online: www.irena.org/publications (accessed on 20 June 2017).

84. Midwest Application CHP Center; Avalon Consulting. *Combined Heat and Power (CHP) Resource Guide*, 2nd ed.; University of Illinois at Chicago–Energy Resource Center: Chicago, IL, USA, 2005.

85. GIZ. *Cogeneration & Trigeneration How to Produce Energy Efficiently*; GIZ: Bonn, Germany, 2016.

86. Petroecuador Subsidio Proyectado Por Producto Del 11 De Septiembre Al 10 De Octubre De 2020. Available online: https://www.eppetroecuador.ec/wp-content/uploads/downloads/2020/09/PRODUCTOS-SUBSIDIADOS-SEPTIEMBRE-2020-COMERCIAL-11-AL-10.pdf (accessed on 18 September 2020).

87. Arne, O.; Schlag, N.; Patel, K.; Kwok, G. *Capital Cost Review of Generation Technologies*; Energy and Environmental Economics Inc.: San Francisco, CA, USA, 2014.

88. UNEP; SETAC. *Guidelines for Social Life Cycle Assesment of Products*; Benoît, C., Mazijn, B., Eds.; UNEP/SETAC: Druk in de weer, Belgium, 2009; ISBN 978-92-807-3021-0.

89. Rafiaani, P.; Kuppens, T.; Dael, M.V.; Azadi, H.; Lebailly, P.; Passel, S.V. Social sustainability assessments in the biobased economy: Towards a systemic approach. *Renew. Sustain. Energy Rev.* **2018**. [CrossRef]

90. Contreras-Lisperguer, R.; Batuecas, E.; Mayo, C.; Díaz, R.; Pérez, F.J.; Springer, C. Sustainability assessment of electricity cogeneration from sugarcane bagasse in Jamaica. *J. Clean. Prod.* **2018**. [CrossRef]

91. Siebert, A.; Bezama, A.; O'Keeffe, S.; Thrän, D. Social life cycle assessment indices and indicators to monitor the social implications of wood-based products. *J. Clean. Prod.* **2018**. [CrossRef]

92. Salgado Torres, R. *Plan Maestro de Electricidad 2019–2027*; Ministerio de Electricidad y Energía Renovable: Quito, Ecuador, 2019.

93. COGEN Europe. *Cogeneration Country Fact Sheet GERMANY*; COGEN Europe: Brussels, Belgium, 2014.

94. Mercados, E. *Análisis de la Industria de Cogeneración en España*; COGEN Espana: Madrid, Spain, 2010.

95. DOE; EPA. *Combined Heat and Power: A Clean Energy Solution*; United States Environmental Protection Agency: Washington, DC, USA, 2012.

96. Raineri, R.; Dyner, I.; Goñi, J.; Castro, N.; Olaya, Y.; Franco, C. Latin America Energy Integration: An Outstanding Dilemma. In *Evolution of Global Electricity Markets: New Paradigms, New Challenges, New Approaches*; Elsevier: Amsterdam, The Netherlands, 2013; ISBN 9780123978912.

97. Corneille, F. Energy Integration in Latin America: The Connecting the Americas 2022 Initiative. Available online: http://www.ecpamericas.org/Blog/Default.aspx?id=134 (accessed on 3 January 2018).

98. The Atlantic Council Regional Energy Integration: The Growing Role of LNG in Latin America. Available online: http://www.atlanticcouncil.org/events/webcasts/regional-energy-integration-the-growing-role-of-lng-in-latin-america (accessed on 22 January 2018).

99. Bataille, C.; Melton, N. Energy efficiency and economic growth: A retrospective CGE analysis for Canada from 2002 to 2012. *Energy Econ.* **2017**. [CrossRef]

100. Tol, R.S. *The Private Benefit of Carbon and Its Social Cost*; Working Paper Series; Department of Economics, University of Sussex Business School: Sussex, UK, 2017.

101. Khurana, S.; Banerjee, R.; Gaitonde, U. Energy balance and cogeneration for a cement plant. *Appl. Therm. Eng.* **2002**, *22*, 485–494. [CrossRef]

Article

Conventional and Advanced Exergy and Exergoeconomic Analysis of a Spray Drying System: A Case Study of an Instant Coffee Factory in Ecuador

Diana L. Tinoco-Caicedo [1,2,*], Alexis Lozano-Medina [2] and Ana M. Blanco-Marigorta [2]

[1] Facultad de Ciencias Naturales y Matemáticas, Escuela Superior Politécnica del Litoral, 090903 Guayaquil, Ecuador
[2] Department of Process Engineering, Universidad de las Palmas de Gran Canaria, 35017 Las Palmas de Gran Canaria, Spain; alexis.lozano@ulpgc.es (A.L.-M.); anamaria.blanco@ulpgc.es (A.M.B.-M.)
* Correspondence: dtinoco@espol.edu.ec

Received: 14 September 2020; Accepted: 23 October 2020; Published: 27 October 2020

Abstract: Instant coffee is produced worldwide by spray drying coffee extract on an industrial scale. This production process is energy intensive, 70% of the operational costs are due to energy requirements. This study aims to identify the potential for energy and cost improvements by performing a conventional and advanced exergy and exergoeconomic analysis to an industrial-scale spray drying process for the production of instant coffee, using actual operational data. The study analyzed the steam generation unit, the air and coffee extract preheater, the drying section, and the final post treatment process. The performance parameters such as exergetic efficiency, exergoeconomic factor, and avoidable investment cost rate for each individual component were determined. The overall energy and exergy efficiencies of the spray drying system are 67.6% and 30.6%, respectively. The highest rate of exergy destruction is located in the boiler, which amounts to 543 kW. However, the advanced exergoeconomic analysis shows that the highest exergy destruction cost rates are located in the spray dryer and the air heat exchanger (106.9 $/h and 60.5 $/h, respectively), of which 47.7% and 3.8%, respectively, are avoidable. Accordingly, any process improvement should focus on the exergoeconomic optimization of the spray dryer.

Keywords: advanced exergoeconomic analysis; spray dryer; exergy destruction cost rate

---

1. Introduction

Instant coffee is one of the most commonly consumed drinks worldwide; around 118 billion dollars of it were sold in the global market in 2019. The worldwide market for instant coffee has high growth expectations: projected to grow by 11.6% in the next 5 years [1]. Coffee has a high concentration of antioxidants [2], vitamins B, and minerals [3]. It benefits physical performance and stimulates the central nervous system [4]. Coffee is sold as whole bean, ground coffee, instant coffee, coffee pods, and capsules. Among these, instant coffee is quickly becoming popular all over the world because of cheaper transportation and convenience in preparation, which increases its demand among urban consumers [5]. Many industrial-scale plants have been established around the word to produce this kind of coffee.

The production process of instant coffee powder begins with roasting the coffee beans and grinding them. Later, they pass through a liquid solid extraction. The extracted liquid is then concentrated and, finally, it is spray dried. The drying process reduces the amount of water in the coffee and allows its shelf-life to be increased. This operation requires the most energy resources [6], and is also considered highly exergy-destructive [7]. Spray dryers are considered to be limiting units within a productive process, and one of the operations with the highest exergy improvement potential [8]. A previous study has demonstrated that the exergy efficiency of spray dryers is lower than that of other drying technologies such as tray dryers, continuous dryers, heat pump assisted dryers, fluidized bed dryers, solar dryers, freeze dryers, vacuum dryers, and flash dryers [7].

Exergy analysis has become an important tool for the assessment of different energy-intensive industrial processes, such as spray drying [9]. These analyses have allowed for the identification of the components with the highest exergy losses, the avoidable exergy losses, and the operational conditions, which most affect the irreversibility of the systems. Erbay et al. [10] used a pilot-scale spray dryer on white cheese slurry to demonstrate experimentally that parameters like atomization pressure and drying air temperature can affect the exergetic efficiency of the spray dryer. Another study of the same scale for the drying of cherry puree showed that drying agents could reduce the exergy destruction rate of the process [11]. Some studies were done at a laboratory scale. One lab-scale study, evaluated the exergetic efficiency of spray drying of photochromic dyes and obtained efficiency below 4% [12]. Further, Aghbashlo et al. [13] studied the influence of parameters such as air and feed flow rate in the exergy destruction rate of the spray drying of microencapsulation of fish oil. Only two studies have been done on industrial-scale spray dryers, and both took place in a powdered milk factory. The first analyzed each step of the production process and concluded that the spray dryer was one of the most exergy destructive components (2196 kW) [14]. In the second study, Camci et al. [15] analyzed a spray drying system with solar collectors for preheating the drying air in a closed loop, resulting in an increase of the exergetic efficiency to 22.6%.

However, although the exergetic analysis identifies the location and magnitude of the thermal energy losses, it has limitations given that it can not quantify the cost of those losses. Furthermore, an exergy analysis is not conclusive about which components should have investment priority in order to reduce the exergy losses [16]. In order to complete an exergy analysis, an exergoeconomic analysis can be applied, which combines exergy and economic principles at the component level to identify the real cost sources in a thermal system [17]. Since the thermodynamic considerations of exergoeconomics are based on the exergy concept, the term exergoeconomics can also be used to describe the combination of exergy analysis and economics [18]. Exergoeconomic analysis has been applied in different industrial processes in order to minimize the economic losses due to irreversibility, and, consequently, provide the added benefit of reducing production costs of the entire complex energy system. Few conventional exergoeconomic analyses of different drying technologies on both the pilot and industrial scale have been found in the literature; they focused on the production of pasta [19], tea leaves [20], powdered cheese [21], and powdered milk [22,23]. Of these, only the last two refer to spray drying technology at an industrial scale. These exergoeconomic analysis performed were useful for the evaluation of the economic viability of the proposed improvements to the spray dryer in a powdered milk factory. Erbay et al. [21] also performed an exergoeconomic analysis on a pilot-scale spray dryer for cheese powder and concluded that some investments should be made in order to reduce the operational cost rates by increasing exergetic efficiency of the process.

Although the exergy and exergoeconomic analyses allow for the quantification of the exergetic and cost losses, they do not provide sufficient information about which losses are avoidable; this information is essential for industrial plants to make decisions about improvement potential. Advanced exergoeconomic analysis is a proposed tool that has been applied to different industrial processes in order to quantify the avoidable and unavoidable economic losses and determine the potential for improvement [24]. However, there have not been any studies that apply an advanced exergoeconomic analysis in spray-drying technology in order to quantify this kind of exergy destruction.

The aim of the present work is to carry out a conventional and advanced exergy and exergoeconomic analysis on the spray drying process of instant coffee at a factory in Guayaquil, Ecuador in order to quantify total operating cost rates at a component level and split into avoidable and unavoidable parts. There are two main novelties in this study: first, real data from an instant coffee plant in operation have been used; second, an advanced exergoeconomic analysis on the spray-drying system of an instant coffee plant has been applied for the first time. This analysis will be a valuable decision-making tool for the factory for future improvements focused on operational cost reduction, and sustainability increase.

2. Materials and Methods

2.1. System Description

The instant coffee was dried in an industrial scale spray drying system. Figure 1 illustrates a schematic diagram of the process. The coffee extract (44% m/m of soluble coffee) comes from a storage tank that had a temperature of 12 °C. A flow rate of 528 kg/h of coffee extract (stream 2) was pumped by a low-pressure pump (LP) and mixed with 7.4 kg/h of carbon dioxide (stream 1). Then it was pumped by a high-pressure pump (HP) into a heat exchanger unit (HXE) where steam increased its temperature to 32 °C. The coffee extract (stream 6) was sprayed by a nozzle into the drying unit (SD), which is at vacuum pressure. A flow rate of 9922 kg/h of ambient air (stream 7) was heated by the main heat exchanger (MHX) using steam until it reached the temperature of 180 °C. A flow rate of 4002 kg/h of ambient air (stream 10) with an absolute humidity of 0.02 kg water/kg dry air was dehumidified to 8×10^{-3} kg water/kg dry air by a cooler (CHX) and then a fraction of it (stream 11) was heated and distributed in order to maintain a fluidized bed in the bottom of the spray dryer. The dried instant coffee produced with a humidity of about 3% m/m (stream 23) was then collected on a belt (BT), where two streams of dehumidified air at 85 °C (stream 16) and 27 °C (stream 20) were used to gradually cool the coffee and prevent it from agglomerating. Then the instant coffee (stream 25) was passed through vibratory screen (S) in order to obtain the required particle size. The fraction of instant coffee with the smallest particle size (stream 28) was recirculated to the process using dried air at 27 °C (stream 22) while the biggest particle size of instant coffee (stream 27) was considered waste. The humidified air (stream 29) that exits the spray dryer was passed through a cyclone separator (FF) to remove solid coffee particles. These solid particles (stream 32) were recirculated into the process and the humidified air (stream 31) was released to the environment.

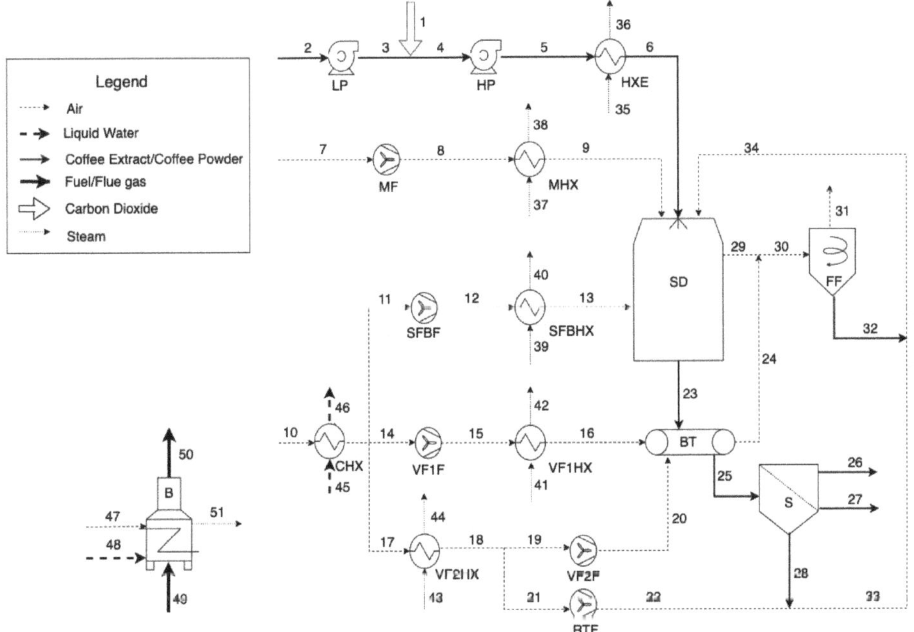

Figure 1. Process flow diagram of the spray dryer system.

To develop the process modeling, the following assumptions were made:

- The process was at a steady state condition.
- The coffee extract was modeled as a solution with a constant concentration of soluble solids from *Coffea arabica* beans.
- The heat losses from the components were neglected.
- The pressure losses in the pipes, heat exchangers, bag filter, and spray dryer were neglected.
- The properties of the incoming air were considered as constants.

2.2. Exergy Analysis

The analysis of the spray drying system was performed by using the engineering equation solver (EES) software for the formulation of mass, energy, and exergy balances for each component. In their general form, they are, respectively:

$$\sum_{in} \dot{m}_{in} - \sum_{out} \dot{m}_{out} = 0 \tag{1}$$

$$\sum_{in} h_{in}\dot{m}_{in} - \sum_{out} h_{out}\dot{m}_{out} + \dot{W}_k + \dot{Q}_k = 0 \tag{2}$$

$$\sum_{k} \dot{E}_{q,k} + \dot{W}_k + \sum_{in} \dot{E}_{in} - \sum_{out} \dot{E}_{out} - \dot{E}_{D,k} = 0 \tag{3}$$

The exergy rate, specific exergy, physical exergy, kinetics exergy, and potential exergy were calculated using Equations (4)–(8). Table 1 shows the expressions of both fuel and product exergy of each component.

$$\dot{E} = \dot{m} * e \tag{4}$$

$$e = e^{PH} + e^{CH} + e^{KN} + e^{PT} \tag{5}$$

$$e^{PH} = (h - h_0) - T_0(s - s_0) \tag{6}$$

$$e^{PT} = gz \tag{7}$$

$$e^{KN} = \frac{v^2}{2} \tag{8}$$

Table 1. Composition of the different states.

State	Description	Soluble Solids (kg/kg)	Water (kg/kg)	Dried Air (kg/kg)
2	Coffee extract	0.440	0.560	-
23	Soluble Coffee powder	0.970	0.030	-
24	Mixture BT	0.001	0.009	0.990
29	Mixture SD	0.004	0.040	0.955
33	Mixture S	0.038	0.008	0.954
34	Mixture FF	0.117	0.001	0.882

The velocities of different streams were estimated by the Bernoulli relationship, Equation (9), where γ is the specific heat ratio and ρ is the density of the stream.

$$\frac{\Delta v^2}{2} + \left(\frac{\gamma}{\gamma - 1}\right) * \frac{P}{\rho} = \left(\frac{\gamma}{\gamma - 1}\right) * \frac{P_0}{\rho_0} \tag{9}$$

For the streams that had soluble coffee solids as part of their compositions, Equations (10) and (11) were used to determine the thermodynamic properties such as entropy and enthalpy. The c_p value was obtained from Burmester et al. [25]. The dead state conditions have been taken as $T_0 = 27.5\,°C$ and $P_0 = 101.13$ kPa.

$$h - h_0 = c_p(T - T_0) \tag{10}$$

$$s - s_0 = c_p \ln\left(\frac{T}{T_0}\right) - R\ln\left(\frac{P}{P_0}\right) \tag{11}$$

The composition for the different states of the system is shown in Table 1. This information was used to calculate the different thermodynamic properties.

For the calculation of chemical exergy of each state point that has soluble coffee solids and water, Equation (12) [17] was used. The concentration of water and coffee in equilibrium with the environment (x_i^e) was chosen as the dead state of reference. Those values were obtained from previous studies on Arabica coffee by Yao et al. [26]. For the calculation of the chemical exergy of each state point that has soluble coffee solids, water, and air, Equation (13) [17] was used, where x_i is the mole fraction of the different substances.

$$e_{CE}^{CH} = -RT_0 \sum x_i \ln\left(\frac{x_i^e}{x_i}\right) \tag{12}$$

$$e_{mix}^{CH} = \sum x_i e_i^{ch} + RT_0 \sum x_i \ln(x_i) \tag{13}$$

The chemical exergy of air for the different moisture content in air was calculated using an expression from Wepfer et al. [27], according to Equation (14), where w_0 and w are mole fraction of water vapor at environmental conditions and operational conditions, respectively.

$$e_{air}^{CH} = 0.2857\, c_{p,air} T_0 ln\left[\left[\frac{1 + 1.6078 w_0}{1 + 1.6078 w}\right]^{(1+1.6078w)} \left[\frac{w}{w_0}\right]^{1.6078w}\right] \tag{14}$$

The exergy balance can also be formulated as Equation (15).

$$\dot{E}_{F,k} - \dot{E}_{P,k} = \dot{E}_{D,k} - \dot{E}_{L,k} \tag{15}$$

where $\dot{E}_{F,k}$ corresponds to the fuel exergy, $\dot{E}_{P,k}$ is the product exergy, $\dot{E}_{D,k}$ is the destroyed, exergy and $\dot{E}_{L,k}$ is the exergy loss. The exergy of the fuel and the exergy of the product for each single component were formulated following Lazzareto and Tsatsaronis rules [28] and they are shown in Table 2.

For the total system the exergetic efficiency was calculated as the sum of the product exergy rates divided by the sum of the fuel exergy rates.

Other interesting parameters involved in an exergy analysis were the relative exergy destruction $(y_{D,k}^*)$, which represents the relationship between the destroyed exergy of a component and the total destroyed exergy of the system, as shown in Equation (16) [17]. The exergy destruction ratio $(y_{D,k})$, which relates the destroyed exergy of a component with the total fuel exergy of the system, is shown in Equation (17). The exergetic efficiency $(n_{ex,k})$, which represents the amount of exergy that is useful in relation to the fuel exergy in the component, is shown in Equation (18).

$$y_{D,k}^* = \frac{\dot{E}_{D,k}}{\dot{E}_{D,tot}} \tag{16}$$

$$y_{D,k} = \frac{\dot{E}_{D,k}}{\dot{E}_{F,tot}} \tag{17}$$

$$n_{ex,k} = \frac{\dot{E}_{P,k}}{\dot{E}_{F,k}} \tag{18}$$

Table 2. Definitions of fuel and product exergy for each component.

Component	$\dot{E}_{P,k}$	$\dot{E}_{F,k}$
LP	$\dot{E}_3 - \dot{E}_2$	$\dot{W}_{LP}$
HP	$\dot{E}_5 - \dot{E}_4$	$\dot{W}_{HP}$
HXE	$\dot{E}_6 - \dot{E}_5$	$\dot{E}_{35} - \dot{E}_{36}$
MHX	$\dot{E}_9 - \dot{E}_8$	$\dot{E}_{37} - \dot{E}_{38}$
SFBHX	$\dot{E}_{13} - \dot{E}_{12}$	$\dot{E}_{39} - \dot{E}_{40}$
VF1HX	$\dot{E}_{16} - \dot{E}_{15}$	$\dot{E}_{41} - \dot{E}_{42}$
VF2HX	$\dot{E}_{18} - \dot{E}_{17}$	$\dot{E}_{43} - \dot{E}_{44}$
MF	$\dot{E}_8 - \dot{E}_7$	$\dot{W}_{MF}$
SFBF	$\dot{E}_{12} - \dot{E}_{11}$	$\dot{W}_{SFBF}$
VF1F	$\dot{E}_{15} - \dot{E}_{14}$	$\dot{W}_{VF1F}$
VF2F	$\dot{E}_{20} - \dot{E}_{19}$	$\dot{W}_{VF2F}$
RFF	$\dot{E}_{22} - \dot{E}_{21}$	$\dot{W}_{RFF}$
FF	$\dot{E}_{32} + \dot{E}_{31} - \dot{E}_{30}$	$\dot{W}_{FF}$
SD	$\dot{E}_{23} - \dot{E}_6 - \dot{E}_{34}$	$\dot{E}_{13} + \dot{E}_9 - \dot{E}_{29}$
BT	$\dot{E}_{25} - \dot{E}_{23}$	$\dot{E}_{20} + \dot{E}_{16} - \dot{E}_{24}$
S	$\dot{E}_{26} + \dot{E}_{28} - \dot{E}_{25}$	$\dot{E}_{27} + \dot{W}_S$
B	$\dot{E}_{51} - \dot{E}_{48}$	$\left(\dot{E}_{49} + \dot{E}_{47}\right) - \dot{E}_{50}$
CHX	$\dot{E}_{14} + \dot{E}_{17} + \dot{E}_{11} - \dot{E}_{10}$	$\dot{E}_{45} - \dot{E}_{46}$

2.3. Advanced Exergy Analysis

In order to obtain the real potential of improvement of each component, the avoidable and unavoidable parts of the exergy destruction were calculated. The unavoidable part of the exergy destruction $(\dot{E}_{D,k}^{UN})$ would be the exergy that will inevitably be destroyed, due to technological limitations, no matter how much capital is invested, and can be calculated by using Equation (19) [29], where $(\dot{E}_D/\dot{E}_P)_k^{UN}$ is the relationship between the exergy destruction and exergy product rates estimated using the unavoidable conditions for each component.

$$\dot{E}_{D,k}^{UN} = \dot{E}_{P,k}\left(\frac{\dot{E}_D}{\dot{E}_P}\right)_k^{UN} \tag{19}$$

Values of the unavoidable and real operation conditions of the components are summarized in Table 3, and were assumed according to previous studies [14,30]. For the spray dryer, the minimum air flow required to supply the energy for water evaporation was calculated as an avoidable condition [31].

Table 3. Assumptions that are considered for real conditions (RC), unavoidable thermodynamic inefficiency conditions (RTI), and unavoidable investment cost conditions (UIC).

Component	RC	RTI	UIC
Heat Exchangers	$\Delta T_{min,\ HXE} = 51$ $\Delta T_{min,\ MHX} = 12$ $\Delta T_{min,\ SFBHX} = 69$ $\Delta T_{min,\ VF1HX} = 80$ $\Delta T_{min,\ VF2HX} = 139$ $\Delta T_{min,\ CHX} = 9$	$\Delta T_{min,\ HXE} = 30$ $\Delta T_{min,\ MHX} = 10$ $\Delta T_{min,\ SFBHX,\ VFIHX} = 20$ $\Delta T_{min,\ VF2HX} = 80$ $\Delta T_{min,\ CHX} = 4$	$\Delta T_{min,\ HXE} = 60$ $\Delta T_{min,\ MHX} = 20$ $\Delta T_{min,\ SFBHX} = 80$ $\Delta T_{min,\ VF1HX} = 90$ $\Delta T_{min,\ VF2HX} = 145$ $\Delta T_{min,\ CHX} = 15$
Pumps	$\eta_{is} = 60\%$	$\eta_{is} = 86\%$	$\eta_{is} = 65\%$
F	$\eta_{is} = 60\%$	$\eta_{is} = 90\%$	$0.85\ \dot{Z}_k^{real}$
S	$\eta_{elec} = 78\%$	$\eta_{elec} = 90\%$	$\eta_{elec} = 78\%$
BT	$\eta_{elec} = 60\%$	$\eta_{elec} = 85\%$	$\eta_{elec} = 60\%$
B	$\eta_{con} = 90\%$	$\eta_{con} = 95\%$	$0.66\ \dot{Z}_k^{real}$
SD	AP-Ratio $= 18.8$	AP-Ratio $= 8.6$	$0.90\ \dot{Z}_k^{real}$

2.4. Exergoeconomic Analysis

The exergoeconomic analysis consists of the formulation of a cost balance and its auxiliary equations at a component level, for each component of the process. The general cost balance [17] is shown in Equation (20) where c_{out} and c_{in} represent the costs of the outflows and inflows respectively, $c_{w,k}$ represents the cost rate related with the work and $\dot{Z}_k$ represents the investment cost of each component. Table 2 shows the cost balance of each component present in the system.

$$\sum_k c_{q,k}\dot{E}_{q,k} + c_{w,k}\dot{W}_k + \sum_{in} c_{in}\dot{E}_{in} - \sum_{out} c_{out}\dot{E}_{out} - c_{D,k}\dot{E}_{D,k} + \dot{Z}_k = 0 \tag{20}$$

The cost balance can be written in terms of the fuel and product formulation [28] as is shown in Equations (21) and (22).

$$\dot{C}_{P,k} = \dot{C}_{F,k} + \dot{Z}_k - \dot{C}_{D,k} \tag{21}$$

$$c_{P,k}\dot{E}_{P,k} = c_{F,k}\dot{E}_{F,k} + \dot{Z}_k - \dot{C}_{D,k} \tag{22}$$

where $\dot{C}_{P,k}$ is the product cost rate, $\dot{C}_{F,k}$ is the fuel cost rate, and $\dot{C}_{D,k}$ is the cost rate associated with the destroyed exergy for each component.

The exergy destroyed in the k-th component has an associated cost rate $\dot{C}_{D,k}$ that can be calculated in terms of the cost of the additional fuel ($c_{F,k}$) that needs to be supplied to this component to cover the exergy destruction and to generate the same exergy flow rate of the product, when $\dot{E}_{P,k}$ stay constant (Equation (23)) [17]. Table 4 shows the cost balance of each component present in the system.

$$\dot{C}_{D,k} = c_{F,k}\dot{E}_{D,k} \tag{23}$$

Table 4. Cost balance equations and auxiliary equations for exergy costs of the system.

Component	Fuel Cost Expression	Product Cost Expression	Auxiliary Equations
LP	$\dot{C}_3 + \dot{W}_{LP}$	$\dot{C}_2 + \dot{Z}_{LP}$	-
HP	$\dot{C}_5 + \dot{W}_{HP}$	$\dot{C}_4 + \dot{Z}_{HP}$	$c_4 = c_3 + c_1$
HXE	$\dot{C}_6 + \dot{C}_{36}$	$\dot{C}_5 + \dot{C}_{35}$	$c_{36} = c_{35} = c_{51}$
MHX	$\dot{C}_9 + \dot{C}_{38}$	$\dot{C}_8 + \dot{C}_{37}$	$c_{38} = c_{37} = c_{51}$
SFBHX	$\dot{C}_{13} + \dot{C}_{40}$	$\dot{C}_{12} + \dot{C}_{39}$	$c_{40} = c_{39} = c_{51}$
VF1HX	$\dot{C}_{16} + \dot{C}_{42}$	$\dot{C}_{15} + \dot{C}_{41}$	$c_{42} = c_{41} = c_{51}$
VF2HX	$\dot{C}_{18} + \dot{C}_{44}$	$\dot{C}_{17} + \dot{C}_{43}$	$c_{44} = c_{43} = c_{51}$
MF	$\dot{C}_8 + \dot{W}_{MF}$	$\dot{C}_7$	$c_7 = 0$
SFBF	$\dot{C}_{12} + \dot{W}_{SFBF}$	$\dot{C}_{11}$	-
VF1F	$\dot{C}_{15} + \dot{W}_{VF1F}$	$\dot{C}_{14}$	-
VF2F	$\dot{C}_{20} + \dot{W}_{VF2F}$	$\dot{C}_{19}$	$c_{19} = c_{18}$
RFF	$\dot{C}_{22} + \dot{W}_{RFF}$	$\dot{C}_{21}$	$c_{21} = c_{18}$
FF	$\dot{C}_{31} + \dot{C}_{32} + \dot{W}_{FF}$	$\dot{C}_{30}$	$c_{31} = c_{32}$
SD	$\dot{C}_{29} + \dot{C}_{23}$	$\dot{C}_6 + \dot{C}_9 + \dot{C}_{13} + \dot{C}_{34}$	$c_{29} = c_9$
BT	$\dot{C}_{24} + \dot{C}_{25}$	$\dot{C}_{16} + \dot{C}_{23}$	$c_{24} = c_{16}$
S	$\dot{W}_S$	$\dot{C}_{26} + \dot{C}_{27} - \dot{C}_{25} - \dot{C}_{28}$	$c_{28} = c_{30};\ c_{29} = c_{31}$
B	$\dot{C}_{50} + \dot{C}_{51}$	$\dot{C}_{47} + \dot{C}_{48} + \dot{C}_{49}$	$c_{47} = 0;\ c_{49} = c_{50}$
CHX	$\dot{C}_{11} + \dot{C}_{14} + \dot{C}_{17} + \dot{C}_{46}$	$\dot{C}_{10} + \dot{C}_{45}$	$c_{10} = 0;\ c_{45} = c_{46}$ $c_{11} = c_{14} = c_{17}$

There are some non-energetic costs used in the calculations of the cost balance of each component In the boiler, the fuel used to generate vapor was fuel oil 6. The price of the liquid fuel (stream 49) was

$1.07 per gallon [32]. The potable water (stream 48) had a cost of $0.53 per cubic meter [33]. The price of carbon dioxide (stream 1) injected into the coffee extract was $24.22 per kg.

The variable $\dot{Z}_k$ was calculated as the sum of capital investment ($\dot{Z}_k^{CI}$) and operation and maintenance costs ($\dot{Z}_k^{OM}$) for each component, as is shown in Equation (24) [17].

$$\dot{Z}_k = \dot{Z}_k^{OM} + \dot{Z}_k^{CI} \tag{24}$$

The capital investment for each component can be calculated by using Equation (25) [17]:

$$\dot{Z}_k^{CI} = \frac{PEC_k * CRF}{\tau} \tag{25}$$

where PEC_k is the purchase price of the kth component and τ is the number of annual operating hours (24 h per day, 365 days per year). It was assumed that the ordinary annuities transaction occurs at the end of each time interval, thus the CRF (capital recovery factor) could be obtained using Equation (26) [17], where i_{eff} is the interest rate (10%), and n is the lifetime of the system (20 years).

$$CRF = \frac{i_{eff} * \left(1 + i_{eff}\right)^n}{\left(1 + i_{eff}\right)^n - 1} \tag{26}$$

The rate of operation and maintenance costs ($\dot{Z}_k^{OM}$) can be calculated by using Equation (27). The operation and maintenance cost (OMC_k) of each component is determined by using Equation (28), which is a close approximation used by Bejan et al [17]. The constant-escalation levelization factor (CELF) was determined by using Equation (29), which depends on the factor k_{OMC} defined by Equation (30) [17]. For the nominal escalation rate (r_{OM}), it was assumed that all costs except fuel costs and the values of by-products change annually with the constant average inflation rate of 4% [17].

$$\dot{Z}_k^{OM} = \frac{OMC_k * CELF_{OM}}{\tau} \tag{27}$$

$$OMC_k = 0.2 * PEC_k \tag{28}$$

$$CELF_{OM} = \frac{k_{OMC} * (1 - k_{OMC}^n) * CRF}{(1 - k_{OMC})} \tag{29}$$

$$k_{OMC} = \frac{1 + r_{OM}}{1 + i_{eff}} \tag{30}$$

For a better interpretation of the results, the exergoeconomic factor (f_k) and relative cost difference (r_k) were determined. The first factor represents the relationship between the investment cost and the total operating cost rate, while the r_k represents the increase of the specific exergy cost in a component divided by the specific exergy cost of the fuel.

$$f_k = \frac{\dot{Z}_k}{\dot{Z}_k + \dot{C}_{D,k}} \tag{31}$$

$$r_k = \frac{c_{P,k} - c_{F,k}}{c_{F,k}} \tag{32}$$

2.5. Advanced Exergoeconomic Analysis

The unavoidable ($\dot{C}_{D,k}^{UN}$) and avoidable cost ($\dot{C}_{D,k}^{AV}$) associated with exergy destruction were calculated using Equations (33) and (34). The unavoidable ($\dot{Z}_k^{UN}$) and avoidable investment cost rates

$(\dot{Z}_k^{AV})$ were calculated by using Equations (35) and (36). The relation between the investment cost rate and the exergy product rate $(\dot{Z}_k/\dot{E}_P)_k^{UN}$ was estimated by using the unavoidable cost conditions presented in Table 4. For the heat exchangers, a Pro/II ®simulator was used to estimate the new heat transfer area based on the minimum temperature difference.

$$\dot{C}_{D,k}^{UN} = c_{F,k}\dot{E}_{D,k}^{UN} \tag{33}$$

$$\dot{C}_{D,k}^{AV} = \dot{C}_{D,k} - \dot{C}_{D,k}^{AV} \tag{34}$$

$$\dot{Z}_k^{UN} = \dot{E}_{P,k}\left(\frac{\dot{Z}_k}{\dot{E}_P}\right)_k^{UN} \tag{35}$$

$$\dot{Z}_k^{AV} = \dot{Z}_k - \dot{Z}_k^{UN} \tag{36}$$

3. Results and Discussions

3.1. Conventional Exergy Analysis

The parameters of the exergetic analysis were calculated for each state throughout the entire studied system. Table 5 shows the flow rate ($\dot{m}$), temperature (T), pressure (P), specific chemical exergy (e^{CH}), specific physical exergy (e^{PH}), specific kinetic exergy (e^{KN}), and exergy rate ($\dot{E}$) of each stream.

Table 5. Thermodynamic values of the streams.

State	$\dot{m}$ (kg/h)	T (°C)	P (kPa)	e^{CH} (kJ/kg)	e^{PH} (kJ/kg)	e^{KN} (kJ/kg)	$\dot{E}$ (kJ/h)
1	7.4	12	101	322	0.22	0.0	2383
2	528	14	101	2.25	10.8	0.0	6891
3	528	15	750	2.25	9.70	0.5	6593
4	528	16	750	1.56	8.84	0.5	5776
5	528	18	5400	1.56	4.73	4.0	5470
6	528	39	5400	1.56	13.4	4.0	10,045
7	9922	28	101	0.00	0.00	0.0	0
8	9922	28	105	0.01	0.00	1.0	10,286
9	9922	178	105	0.01	29.9	1.0	307,205
10	4002	28	101	0.00	0.00	0.0	0
11	1626	15	101	0.002	0.27	0.0	436
12	1626	15	105	0.012	0.27	1.0	2126
13	1626	96	105	0.012	6.97	1.0	13,031
14	1100	15	101	0.002	0.27	0.0	295
15	1100	15	105	0.012	0.27	1.0	1438
16	1100	85	105	0.012	5.02	1.0	6665
17	1276	15	101	0.002	0.27	0.0	342
18	1276	26	101	0.002	0.00	0.0	6
19	1101	26	101	0.002	0.00	0.0	6
20	1101	27	105	0.012	0.00	1.0	1146
21	175	26	101	0.002	0.00	0.0	1
22	175	27	105	0.012	0.00	1.0	182
23	209	80	101	5.80	8.24	6.0	4202
24	2203	58	101	0.002	1.49	0.0	3298
25	207	35	101	5.80	0.18	1.0	1450
27	0.04	30	101	4.18	0.07	0.0	0.04
26	200	30	101	5.80	0.02	0.0	1163
28	6.96	30	101	5.80	0.02	0.0	40
29	12,065	96	100	0.001	7.45	2.1	114,790
30	14,268	94	100	0.003	6.94	0.0	99,094

Table 5. *Cont.*

State	$\dot{m}$ (kg/h)	T (°C)	P (kPa)	e^{CH} (kJ/kg)	e^{PH} (kJ/kg)	e^{KN} (kJ/kg)	$\dot{E}$ (kJ/h)
31	14,252	94	105	0.003	6.94	0.9	111,685
32	16	94	100	1647	11.2	0.0	26,942
33	182	30	101	1647	0.01	0.0	26,763
34	198	40	101	1647	0.26	0.9	27,009
35	20	90	70	480	418	0.0	18,231
36	20	90	70	2.50	23.9	0.0	537
37	806	190	1250	480	499	0.0	789,231
38	806	190	1250	2.50	29.0	0.0	25,387
39	80	165	700	480	753	0.0	98,620
40	80	165	700	2.50	104	0.0	8507
41	43	165	700	480	753	0.0	53,008
42	43	165	700	2.50	104	0.0	4581
43	10	165	700	480	753	0.0	12,328
44	10	165	700	2.50	104	0.0	1063
45	25,438	2	500	2.50	5.13	0.0	194,111
46	25,438	6	500	2.50	3.71	0.0	157,878
47	959	190	1250	480	499	0.0	938,861
48	959	104	1250	2.50	39.3	0.0	40,075
49	2217	28	101	0.00	0.00	0.0	0
50	77	28	101	43,293	0.00	0.0	3,332,277
51	2294	650	101	26.0	331	0.0	817,815

The exergy rate of the fuel ($\dot{E}_F$) and the product ($\dot{E}_P$), the exergetic (n_{ex}) and energetic (n_{en}) efficiencies, and the exergy destruction ratios ($y^*_{D,k}$ and $y_{D,k}$) were calculated for each component in the system. The results are summarized in Table 6. The components with the highest exergy fuel rates were the B, the MHX, and the SD. The MHX is the component with the highest exergetic efficiency (38.9%), followed by the boiler (37%). There is a big difference between the exergetic and the energetic efficiencies of the majority of the components, and consequently the overall system also exhibited the same behavior. Therefore, despite the energy efficiency of the system (the conservation of the quantity of energy) being 67.8%, the overall exergy efficiency (the quality of that energy) was only 33.3%. Similar results were obtained in a study on the spray drying process in an industrial scale ceramic factory, in which the energetic efficiency was found to be between 43% and 87% [34], and the exergetic efficiency was between 12% and 64% [35]. However, in a pilot-scale study of spray drying of cherry puree the energetic and exergetic efficiencies were only 3.2% and 0.7%, respectively [11]. This, along with laboratory-scale studies [10,12,36], demonstrates that pilot-scale and laboratory-scale studies do not accurately represent the energetic and exergetic performances of the industrial-scale spray drying process.

Figure 2 shows the fuel and product exergy rate of the overall system, and the destroyed exergy rate of each component. The results show that the components that had electric energy as the main fuel exergy source such as the vibrating screen, belt, and fans had the lowest impact on the exergetic destruction. This occurs because the electric energy was used for mechanical operations, instead of being used as a heat source. The exergy destruction ratio (y_D) was lower than 5% for these components. These results were similar to other studies that determined an exergy destruction ratio lower than 2% for the compressors and pumps in a CCHP system [37]. Furthermore, in a yogurt plant the devices that required electric energy accounted for less than 5% of the total exergy destruction [38].

Table 6. Results of the exergy analysis of all the components of the spray drying system.

Component	$\dot{E}_F$ (kJ/h)	$\dot{E}_P$ (kJ/h)	n_{ex} (%)	n_{en} (%)	$y^*_{D,k}$	$y_{D,k}$
SD	205,446	32,852	16.0	93.9	0.058	0.040
LP	7920	298	3.8	27.2	0.003	0.002
HP	19,800	307	1.5	34.2	0.007	0.005
HXE	17,694	4576	25.9	76.4	0.005	0.003
MHX	763,844	296,918	38.9	79.4	0.174	0.116
SFBHX	90,114	10,906	12.1	81.4	0.030	0.020
VF1HX	48,427	5227	10.8	88.6	0.016	0.011
VF2HX	11,264	336	3.0	69.2	0.004	0.003
MF	66,600	10,286	15.4	46.3	0.021	0.014
SFBF	19,800	1690	8.5	24.0	0.007	0.005
VF1F	14,400	1143	7.9	22.4	0.005	0.003
VF2F	14,400	1140	7.9	23.3	0.005	0.003
RFF	1980	181	9.2	29.4	0.001	0.001
FF	108,000	39,534	36.6	51.5	0.026	0.017
CHX	36,233	1072	3.0	27.8	0.013	0.009
B	2,514,427	898,786	35.7	73.3	0.611	0.374
BT	7920	546	6.9	69.9	0.003	0.002
S	3600	247	6.9	n/a	0.001	0.001

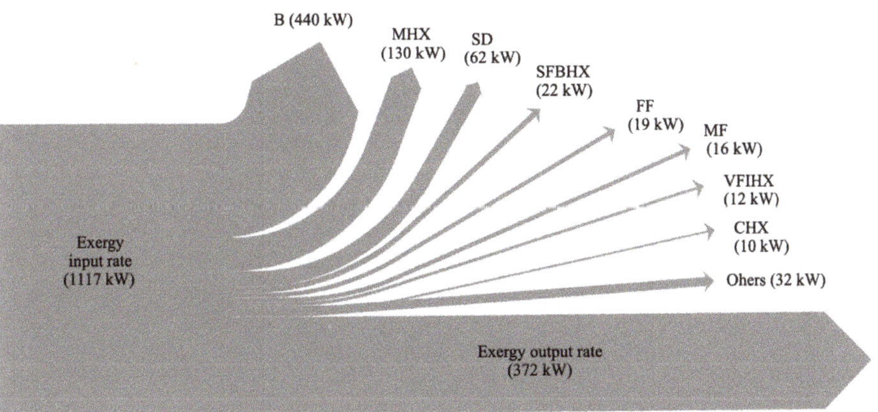

Figure 2. Grassmann's diagram of the spray drying process.

Conversely, the boiler destroyed 39.4% of the overall fuel exergy rate. This percentage was similar to other plants where the boiler was used as an auxiliary supply of steam. For instance, in a factory, which produces ghee, the boiler has the highest exergy destruction ratio 39% [39]. This is because the main purpose of this component is to convert a high-quality energy (chemical energy of fuel oil) to a low-quality energy (heat).

The MHX also has a high exergy destruction rate, despite having one of the highest exergetic efficiencies. The air heater used in this process was a steam-heated type, which is one of the most used in food industry, it had an exergy efficiency of 38.9% and a high specific exergy destruction of 287 kJ per kg of heated air, with a minimum temperature difference of 12 °C. There are other types of air heaters that could reduce the exergy destruction rate and the minimum temperature difference such as a system with a heat exchanger that uses geothermic fluid. A previous study showed that this kind of heat exchanger has an exergy efficiency of 42% and specific destruction exergy of 57.5 kJ per kilogram of heated air with a minimum temperature difference of 5 °C [40]. Another type of air heater was one that uses electric energy as the source of heat. A previous study on the spray drying of photochromic dyes determined that the exergy efficiency of this kind of heater was 16.4% [12], this has the lowest exergy efficiency because it is transforming high quality energy (electric energy) to low quality energy (heat).

The SD also affects the performance of the overall system, since it has one of the highest rates of exergy destruction at 595 kJ/kg of evaporated water. Previous studies by Bühler et al. [31] found that the spray dryer is a highly exergy-destructive component in a powdered milk factory. Similarly in a large dairy factory producing primarily milk powder, they obtained an exergy destruction rate of 1345 kJ/kg of evaporated water [14]. In a ceramic plant, the exergy destruction rate was 1111.4 kJ/kg of evaporated water [35].

3.2. Advanced Exergy Analysis

In order to determine the avoidable and unavoidable fractions of the exergy destruction rate, it was split at a component level by considering the unavoidable thermodynamic inefficiency conditions listed in Table 3. Figure 3 shows that the components with the highest avoidable exergy destruction rates. Even though the MHX had one of the highest exergy destruction rates, more than 96% of the MHX destroyed exergy was unavoidable, this is because the real operational conditions were close to the unavoidable ones.

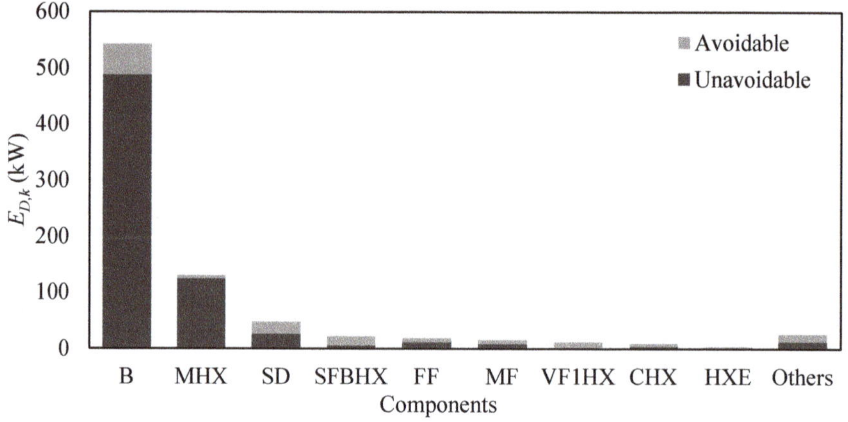

Figure 3. Irreversibility rate distribution of the main components of the system.

Conversely, the B and the SD were responsible for 38% (54 kW) and 15% (21 kW) of the total avoidable exergy destruction rate, respectively. Vuckovic et al. [30] and Bühler [14] found similar results for the boiler in an industrial energy supply plant (16.4%) and the spray dryer for a milk processing factory (16.5%), respectively.

Structural changes in spray drying systems have been studied as an alternative to reduce avoidable exergy destruction rates. Walmsley et al. [22] concluded that a closed drying air loop for the recovery of heat waste in a spray drying system for the production of powdered milk could achieve a reduction of 14.4% of steam used. This reduction would consequently reduce the avoidable exergy destruction rate for the system. In addition, Camci et al. [15] determined that the exergy destruction rate could decrease by 11% when solar collectors for preheating the drying air were used.

3.3. Conventional Exergoeconomic Analysis

The conventional exergoeconomic analysis was carried out at a level component and it is presented in Table 7 different indicators such as the specific fuel cost (c_F), the destruction exergy cost rate ($\dot{C}_D$), the exergoeconomic factor (f_k), the relative cost difference (r_k), and the total operating cost rate ($\dot{C}_D + \dot{Z}_k$) in descending order.

Table 7. Results of the thermoeconomic analysis.

Component	c_F ($/kJ)	$\dot{C}_D$ ($/h)	$\dot{Z}_k + \dot{C}_D$ ($/h)	r_k	f_k (%)
SD	6.2×10^{-4}	106.8	109.6	0.02	2.50
MHX	1.3×10^{-4}	60.5	61.6	0.01	1.73
B	6.7×10^{-6}	13.1	14.4	0.07	9.03
SFBHX	1.0×10^{-4}	8.2	8.3	0.02	2.06
VF1HX	1.0×10^{-4}	4.4	4.6	0.02	2.61
BT	5.7×10^{-4}	4.2	4.5	0.06	5.71
CHX	7.0×10^{-5}	2.4	3.0	0.20	17.43
HXE	1.4×10^{-4}	1.9	1.9	0.01	1.37
VF2HX	1.0×10^{-4}	1.1	1.2	0.03	3.17
FF	2.6×10^{-5}	1.8	1.9	0.05	7.83
MF	2.6×10^{-5}	1.5	1.6	0.08	9.00
HP	2.6×10^{-5}	0.5	1.1	1.14	53.73
RTF	2.6×10^{-5}	0.05	0.2	3.30	78.42
SFBF	2.6×10^{-5}	0.5	0.6	0.33	26.52
VF2F	2.6×10^{-5}	0.3	0.5	0.45	33.02
VF1F	2.6×10^{-5}	0.3	0.5	0.45	33.03
LP	2.6×10^{-5}	0.2	0.4	1.21	55.73
S	2.6×10^{-5}	0.1	0.4	8.16	78.49

The results show that the two highest total operating cost rates ($\dot{Z}_k + \dot{C}_D$) were from the SD followed by the MHX, meaning that the influence of these components on the total costs associated with the overall system was significant. Interesting results are presented, because although the B had a higher avoidable exergy destruction rate than the SD and MHX, the specific cost rate was higher in the SD than in the B, thus making the SD the component that had the greatest influence on the total operating cost rate. In contrast, the fans, the pumps, and the vibrating stream were the three components that contributed least to the total operating cost rate. Similar results were obtained by an exergoeconomic analysis in a corn dryer, where the drying chamber represented more than 98% of the total operational costs [41].

Furthermore, although the percentage relative cost differences for components such as the B (7%), SD (2%), and MHX (1%) were found to be low, their exergy destruction cost rates were high. The MHX and the SD had exergoeconomic factors of 1.6% and 3.3%, respectively, which means that the exergetic efficiency of these components must increase in order to reduce the overall system cost. Similar results were found in other drying technologies such as gas engine-driven heat pump dryer and a ground-source heat pump food dryer, which had exergoeconomic factors of 25% [42] and 14.6% [43], respectively. Another previous study on a pilot-scale spray dryer for the production of cheese powder, concluded similarly that in order to reduce the operational cost in spray drying systems, the exergy efficiency in the drying chamber should be increased even though this would require an increment in the capital investment [21].

3.4. Advanced Exergoeconomic Analysis

In order to determine the system's potential of improvement for the reduction of the overall operational cost, an advanced exergoeconomic analysis was performed. In Figure 4, the avoidable ($\dot{C}_{D,k}^{AV}$) and unavoidable ($\dot{C}_{D,k}^{UN}$) cost of exergy destruction, and the avoidable ($\dot{Z}_k^{AV}$) and unavoidable ($\dot{Z}_k^{UN}$) investment cost rates of the different components of the system are presented.

As it is shown in Figure 4 the combined avoidable investment cost rates of the B, the SD and the MHX, represents only 10.2% of the overall investment cost rate and less than 1% of the overall operational cost rate. These results show that the improvement potential for the investment cost rate of the SD and the MHX was low.

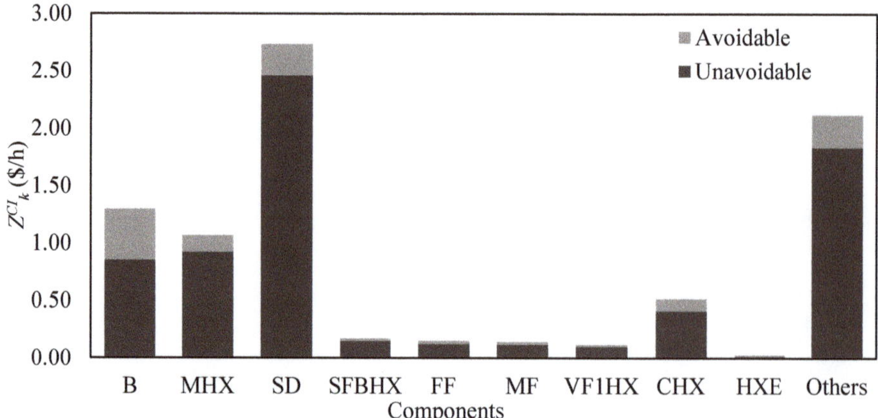

Figure 4. Avoidable and unavoidable investment cost rate of the components of the system.

On the other hand, the avoidable exergy destruction cost rate for the overall system represents 30% of the operational cost and 31% of the overall destruction cost. Only three advanced exergoeconomic analyses have been done in drying systems, but all of them were performed on heat pump dryers [44,45]. These previous studies reported that 46% and 74% of the overall destruction cost were avoidable. This indicates that spray drying process could have lower improvement potential than the heat pump drying process.

In Figure 5, the avoidable and unavoidable exergy destruction cost rates are presented at a component level. It is shown that the B and MHX had high unavoidable exergy destruction cost rate, combined they represented 49% of the total unavoidable exergy destruction. A previous advanced exergoeconomic analysis in a power plant showed similar results for the boiler: around 90% of the destruction cost rate was unavoidable [46].

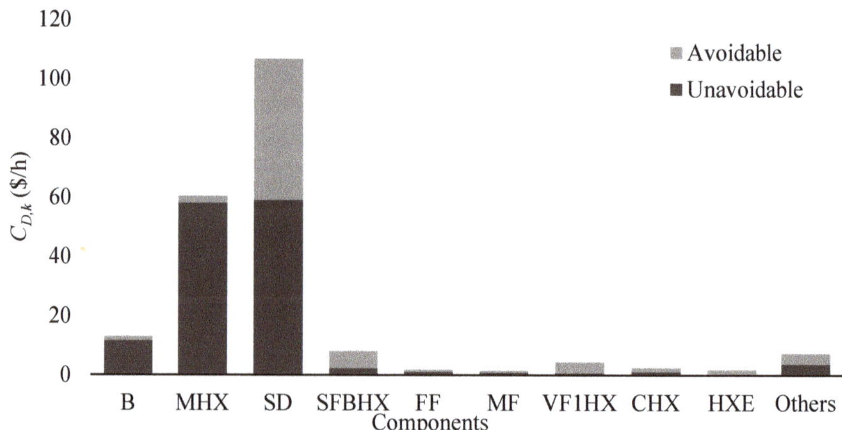

Figure 5. Avoidable and unavoidable exergy destruction cost rate of the components of the system.

Other components such as fans, pumps, and the vibrating screen had also low avoidable cost rates associated with exergy destruction (accounting for less than 1% of the total avoidable cost), which means that any improvement in these components will not significantly reduce the total operating cost. This result is also shown in other food drying systems where the components that require electric energy have avoidable costs that represent less than 1% of the total cost [45].

Conversely, although the B has the highest avoidable exergy destruction rate, the spray dryer has the highest avoidable exergy destruction cost rate ($47.7/h), which represents 73% of the overall avoidable destruction cost rate of the process. A previous study on a pump food dryer similarly concluded that 68.6% of the destruction cost rates were avoidable in the drying chamber [47]. These results imply that the SD had the highest level of improvement potential. A reduction of the exergy destruction rate in the spray dryer could reduce the total cost of the overall system by 22%.

4. Conclusions

According to the aim of this study, we developed conventional and advanced exergy and exergoeconomic analyses of a spray drying system of instant coffee for the first time, using real operational data. The components of the system where analyzed individually. The advanced analysis was found to be useful for quantifying the flow costs in the process and also for identifying which components have the greatest potential for improvement in order to make the overall system more cost effective.

According to the analysis and discussion, the following conclusions were obtained:

- The overall energy and exergy efficiencies of the spray drying system were calculated as 71% and 33% respectively, where the B had the highest exergy destruction rate, but most of it (90%) was unavoidable exergy destruction.
- The conventional exergoeconomic analysis allows for the quantification of the overall operational cost rate ($207.9/h); more than 70% of that cost rate was due to the SD and the MHX.
- The exergoeconomic factor allowed for the identification of the SD and MHX as the sources with the highest cost rate. More than 97% of the operating cost rate of the SD and the MHX were due to a high exergy destruction rate; of all the components in the studied system, these components were the most exergy destructive. The cost rates of the exergy destruction for the SD and the MHX were 106.9 $/h and 60.5$/h, respectively.
- The advanced exergoeconomic analysis revealed that 33% of the exergy destruction cost rate of the overall system was avoidable. Additionally, it established that 70% of the avoidable exergy destruction cost rate was located in the SD, demonstrating that this was the component with the highest improvement potential.

Finally, based on the results obtained in this analysis, the following recommendations were made for the plant: It would be useful to reduce the exergetic destruction cost rate of the SD and the MHX, by performing a parametric study and implementing structural changes within an exergoeconomic optimization in order to obtain f_k values as close to 50% as possible [48]. Further studies are necessary to analyze the interdependence of the SD and the rest of the system's components, in order to determine the percentage of avoidable costs that can be attributed to the irreversibilities of each component's operation.

Author Contributions: Conceptualization, investigation and data curation, D.L.T.-C.; methodology, A.M.B.-M.; software and validation, formal analysis, D.L.T.-C. and A.M.B.-M.; writing—original draft preparation, D.L.T.-C.; writing—review and editing, A.M.B.-M.; supervision and project administration, A.L.-M. and A.M.B.-M. All authors have read and agreed to the published version of the manuscript.

Funding: This research received no external funding.

Conflicts of Interest: The authors declare no conflict of interest.

Nomenclature

$\dot{C}$	cost rate associated with an exergy stream (\$/h)
y	destruction rate
$\dot{E}$	exergy rate (kJ/h)
f	exergy rate (kJ/h)
i	interest rate
c_p	heat capacity (kJ/K*kg)
$\dot{Q}$	heat flow rate (kJ/h)
R	ideal gas constant (kJ/kmol*K)
$\dot{Z}$	investment cost rate (\$/h)
$\dot{m}$	mass flow rate (kg/h)
n	life time of the system
P	pressure (kPa)
r	relative cost difference
y^*	relative irreversibility
h	specific enthalpy (kJ/kg)
s	specific entropy (kJ/kg)
e	specific exergy rate (kJ/kg)
T	temperature (°C)
c	unit exergy cost (\$/kJ)
w	mole fraction of water vapor
$\dot{W}$	power (kJ/h)
x	mole fraction

Greek letters

Δ	difference
γ	specific heat ratio
η	efficiency
ρ	air density (kg/m^3)
τ	annual operating hours (h)

Abbreviations

B	boiler
BT	belt
CHX	cooler heat exchanger
HXE	extract heat exchanger
RFF	fine returns fan
SFBF	fluidized bed fan
SFBHX	fluidized bed heat exchanger
HP	high pressure pump
LP	low pressure pump
MF	main fan
MHX	main heat exchanger
N	nozzle
PEC	purchased equipment cost
SD	spray dryer
FF	vacuum pump
VF1F	vf1 fan
VF1HX	vf1 heat exchanger
VF2F	vf2 fan
VF2HX	vf2 heat exchanger
S	vibrating screen

Subscripts

con	conversion
D	exergy destruction
elec	electric
en	energy
ex	exergy
F	fuel exergy
in	inflow
is	isentropic
k	kth component
mech	mechanical
min	minimum
mix	mixture
out	outflow
P	product exergy
L	loss
tot	overall system
o	thermodynamic environment

Superscripts

AV	avoidable
CH	chemical
CI	capital investment
KN	kinetic
OM	operating and maintenance
PH	physical
PT	potential
UN	unavoidable

References

1. Statista Consumer Market Outlook for Instant Coffee. 2020. Available online: https://www.statista.com/outlook/30010200/100/instant-coffee/worldwide (accessed on 20 May 2020).
2. Svilaas, A.; Sakhi, A.K.; Andersen, L.F.; Svilaas, T.; Ström, E.C.; Jacobs, D.R.; Ose, L.; Blomhoff, R. Intakes of Antioxidants in Coffee, Wine, and Vegetables Are Correlated with Plasma Carotenoids in Humans. *J. Nutr.* **2004**, *134*, 562–567. [CrossRef]
3. Gebhardt, S.; Lemar, L.; Haytowitz, D.; Pehrsson, P.; Nickle, M.; Showell, B.; Holden, J. *USDA National Nutrient Database for Standard Reference, Release 21*; United States Department of Agriculture Agricultural Research Service: Washington, DC, USA, 2008.
4. Lucas, M.; Mirzaei, F.; Pan, A.; Okereke, O.I.; Willett, W.C.; O'Reilly, É.J.; Koenen, K.; Ascherio, A. Coffee, Caffeine, and Risk of Depression Among Women. *Arch. Intern. Med.* **2011**, *171*, 1571–1578. [CrossRef]
5. Market, A. Global Industry Analysis, Size, Share, Growth, Trends and Forecast 2019–2025. 2019. Available online: https://www.transparencymarketresearch.com/logistics-market.html (accessed on 20 May 2020).
6. Bhandari, B. *Handbook of Industrial Drying*; Mujumdar, A.S., Ed.; CRC Press: Boca Raton, FL, USA, 2015; ISBN 978-1-4665-9665-8.
7. Aghbashlo, M.; Mobli, H.; Rafiee, S.; Madadlou, A. A review on exergy analysis of drying processes and systems. *Renew. Sustain. Energy Rev.* **2013**, *22*, 1–22. [CrossRef]
8. Johnson, W.P.; Langrish, A.T. Interpreting exergy analysis as applied to spray drying systems. *Int. J. Exergy* **2020**, *31*, 120–149. [CrossRef]
9. Johnson, P.W.; Langrish, T.A.G. Exergy analysis of a spray dryer: Methods and interpretations. *Dry. Technol.* **2017**, *36*, 578–596. [CrossRef]
10. Erbay, Z.; Koca, N. Energetic, Exergetic, and Exergoeconomic Analyses of Spray-Drying Process during White Cheese Powder Production. *Dry. Technol.* **2012**, *30*, 435–444. [CrossRef]
11. Saygı, G.; Erbay, Z.; Koca, N.; Pazir, F. Energy and exergy analyses of spray drying of a fruit puree (cornelian cherry puree). *Int. J. Exergy* **2015**, *16*, 315. [CrossRef]

12. Çay, A.; Kumbasar, E.P.A.; Morsunbul, S. Exergy analysis of encapsulation of photochromic dye by spray drying. *IOP Conf. Ser. Mater. Sci. Eng.* **2017**, *254*, 22003. [CrossRef]

13. Aghbashlo, M.; Mobli, H.; Madadlou, A.; Rafiee, S. Influence of spray dryer parameters on exergetic performance of microencapsulation processs. *Int. J. Exergy* **2012**, *10*, 267. [CrossRef]

14. Bühler, F.; Nguyen, T.-V.; Jensen, J.K.; Holm, F.M.; Elmegaard, B. Energy, exergy and advanced exergy analysis of a milk processing factory. *Energy* **2018**, *162*, 576–592. [CrossRef]

15. Camci, M. Thermodynamic analysis of a novel integration of a spray dryer and solar collectors: A case study of a milk powder drying system. *Dry. Technol.* **2019**, *38*, 350–360. [CrossRef]

16. Tsatsaronis, G. Definitions and nomenclature in exergy analysis and exergoeconomics. *Energy* **2007**, *32*, 249–253. [CrossRef]

17. Bejan, A.; Tsatsaronis, G.; Michael, M. *Thermal Design and Optimization*; John Wiley & Sons: New York, NY, USA, 1996; Volume 21, ISBN 0471584673.

18. Tsatsaronis, G. Thermoeconomic analysis and optimization of energy systems. *Prog. Energy Combust. Sci.* **1993**, *19*, 227–257. [CrossRef]

19. Ozgener, L. Exergoeconomic analysis of small industrial pasta drying systems. *Proc. Inst. Mech. Eng. Part A J. Power Energy* **2007**, *221*, 899–906. [CrossRef]

20. Ozturk, M.; Dincer, I. Exergoeconomic analysis of a solar assisted tea drying system. *Dry. Technol.* **2019**, *38*, 655–662. [CrossRef]

21. Erbay, Z.; Koca, N. Exergoeconomic performance assessment of a pilot-scale spray dryer using the specific exergy costing method. *Biosyst. Eng.* **2014**, *122*, 127–138. [CrossRef]

22. Walmsley, T.G.; Walmsley, M.R.; Atkins, M.J.; Neale, J.R.; Tarighaleslami, A.H. Thermo-economic optimisation of industrial milk spray dryer exhaust to inlet air heat recovery. *Energy* **2015**, *90*, 95–104. [CrossRef]

23. Walmsley, T.G.; Walmsley, M.R.W.; Atkins, M.J.; Neale, J.R. Thermo-Economic assessment tool for industrial milk spray dryer exhaust heat recovery systems with particulate fouling. *Chem. Eng. Trans.* **2014**, *39*, 1459–1464. [CrossRef]

24. Petrakopoulou, F.; Tsatsaronis, G.; Morosuk, T.; Carassai, A. Advanced Exergoeconomic Analysis Applied to a Complex Energy Conversion System. *J. Eng. Gas Turbines Power* **2011**, *134*, 031801. [CrossRef]

25. Burmester, K.; Fehr, H.; Eggers, R. A Comprehensive Study on Thermophysical Material Properties for an Innovative Coffee Drying Process. *Dry. Technol.* **2011**, *29*, 1562–1570. [CrossRef]

26. Clément, Y.B.Y.; Benjamin, Y.N.; Roger, K.B.; Clement, A.D.; Kablan, T. Moisture Adsorption Isotherms Characteristic of Coffee (Arabusta) Powder at Various Fitting Models. *Int. J. Curr. Res. Biosci. Plant Biol.* **2018**, *5*, 26–35. [CrossRef]

27. Wepfer, W.J.; Gaggioli, R.A.; Obert, E.F. Proper evaluation of available energy for HVAC. *ASHRAE Trans.* **1979**, *85*, 214–230.

28. Lazzaretto, A.; Tsatsaronis, G. SPECO: A systematic and general methodology for calculating efficiencies and costs in thermal systems. *Energy* **2006**, *31*, 1257–1289. [CrossRef]

29. Tsatsaronis, G.; Park, M.-H. On avoidable and unavoidable exergy destructions and investment costs in thermal systems. *Energy Convers. Manag.* **2002**, *43*, 1259–1270. [CrossRef]

30. Vučković, G.D.; Stojiljković, M.M.; Vukić, M.V.; Stefanović, G.M.; Dedeić, E.M. Advanced exergy analysis and exergoeconomic performance evaluation of thermal processes in an existing industrial plant. *Energy Convers. Manag.* **2014**, *85*, 655–662. [CrossRef]

31. Cziesla, F.; Tsatsaronis, G.; Gao, Z. Avoidable thermodynamic inefficiencies and costs in an externally fired combined cycle power plant. *Energy* **2006**, *31*, 1472–1489. [CrossRef]

32. Petroecuador, E.P. Precios de Venta a Nivel de Terminal para las Comercializadoras Calificadas y Autorizadas a Nivel Nacional. Ep Petroecuador Gerencia de Comericalización Nacional. 2019. Available online: http://bit.ly/2XeM8Zq (accessed on 20 March 2020).

33. Interagua. *Informe Anual 2018–2019*; International Water Services (Guayaquil) Interagua C. Ltda: Guayaquil, Ecuador, 2019.

34. Apak. Energy exergy analyses, Turkey (Supervisor: Prof. Dr. Ramazan KƐOSE). In *A Ceramic Factory, Mechanical Engineering Branche*; Engineering Faculty, Dumlupınar University: Kütahya, Turkey, 2007; p. 110. (In Turkish)

35. Utlu, Z.; Hepbasli, A.; Hepbaşlı, A. Exergoeconomic analysis of energy utilization of drying process in a ceramic production. *Appl. Therm. Eng.* **2014**, *70*, 748–762. [CrossRef]

36. Kurozawa, L.E.; Park, K.J.; Hubinger, M.D. Spray Drying of Chicken Meat Protein Hydrolysate: Influence of Process Conditions on Powder Property and Dryer Performance. *Dry. Technol.* **2011**, *29*, 163–173. [CrossRef]

37. Mohammadi, A.; Ahmadi, M.H.; Bidi, M.; Joda, F.; Valero, A.; Uson, S. Exergy analysis of a Combined Cooling, Heating and Power system integrated with wind turbine and compressed air energy storage system. *Energy Convers. Manag.* **2017**, *131*, 69–78. [CrossRef]

38. Jokandan, M.J.; Aghbashlo, M.; Mohtasebi, S.S. Comprehensive exergy analysis of an industrial-scale yogurt production plant. *Energy* **2015**, *93*, 1832–1851. [CrossRef]

39. Singh, G.; Singh, P.; Tyagi, V.; Barnwal, P.; Pandey, A. Exergy and thermo-economic analysis of ghee production plant in dairy industry. *Energy* **2019**, *167*, 602–618. [CrossRef]

40. Yildirim, N.; Genc, S. Energy and exergy analysis of a milk powder production system. *Energy Convers. Manag.* **2017**, *149*, 698–705. [CrossRef]

41. Li, B.; Li, C.; Huang, J.; Li, C. Exergoeconomic Analysis of Corn Drying in a Novel Industrial Drying System. *Entropy* **2020**, *22*, 689. [CrossRef]

42. Gungor, A.; Erbay, Z.; Hepbasli, A. Exergoeconomic (Thermoeconomic) Analysis and Performance Assessment of a Gas Engine–Driven Heat Pump Drying System Based on Experimental Data. *Dry. Technol.* **2012**, *30*, 52–62. [CrossRef]

43. Erbay, Z.; Hepbasli, A. Exergoeconomic evaluation of a ground-source heat pump food dryer at varying dead state temperatures. *J. Clean. Prod.* **2017**, *142*, 1425–1435. [CrossRef]

44. Erbay, Z.; Hepbasli, A. Advanced exergoeconomic evaluation of a heat pump food dryer. *Biosyst. Eng.* **2014**, *124*, 29–39. [CrossRef]

45. Gungor, A.; Tsatsaronis, G.; Gunerhan, H.; Hepbasli, A. Advanced exergoeconomic analysis of a gas engine heat pump (GEHP) for food drying processes. *Energy Convers. Manag.* **2015**, *91*, 132–139. [CrossRef]

46. Wang, L.; Fu, P.; Yang, Z.; Lin, T.-E.; Yang, Y.; Tsatsaronis, G. Advanced Exergoeconomic Evaluation of Large-Scale Coal-Fired Power Plant. *J. Energy Eng.* **2020**, *146*, 04019032. [CrossRef]

47. Erbay, Z.; Hepbasli, A. Assessment of cost sources and improvement potentials of a ground-source heat pump food drying system through advanced exergoeconomic analysis method. *Energy* **2017**, *127*, 502–515. [CrossRef]

48. Hamdy, S.; Morosuk, T.; Tsatsaronis, G. Exergoeconomic optimization of an adiabatic cryogenics-based energy storage system. *Energy* **2019**, *183*, 812–824. [CrossRef]

Publisher's Note: MDPI stays neutral with regard to jurisdictional claims in published maps and institutional affiliations.

Review

A Review of Energy Management Assessment Models for Industrial Energy Efficiency

A S M Monjurul Hasan and Andrea Trianni *

School of Information, Systems, and Modelling, Faculty of Engineering and IT, University of Technology Sydney, 81 Broadway, Ultimo 2007, Australia; asmmonjurul.hasan@uts.edu.au
* Correspondence: andrea.trianni@uts.edu.au

Received: 25 September 2020; Accepted: 29 October 2020; Published: 1 November 2020

Abstract: The necessity to ensure energy efficiency in the industries is of significant importance to attain reduction of energy consumption and greenhouse gases emissions. Energy management is one of the effective features that ensure energy efficiency in the industries. Energy management models are the infancy in the industrial energy domain with practical guidelines towards implementation in the organizations. Despite the increased interest in energy efficiency, a gap exists concerning energy management literature and present application practices. This paper aims to methodologically review the energy management assessment models that facilitate the assessment of industrial energy management. In this context, the minimum requirements model, maturity model, energy management matrix model, and energy efficiency measures characterization framework are discussed with implications. The study concludes with interesting propositions for academia and industrial think tanks delineating few further research opportunities.

Keywords: energy management; industrial energy efficiency; energy management practices; assessment model

1. Introduction

The industrial sector, being one of the largest entities for consuming energy [1], is responsible for 30% of total carbon emission [2]. Further, the up-rising of energy expenses, stringent environmental restrictions, and fossil fuel depletion have shaped increased demand to the reduction of energy consumption and its associated costs in the industries [3]. In this context, ensuring energy efficiency is one of the significant mainstays of the industrial processes that must be addressed as a priority. Energy efficiency gains through the implementation of energy management practices can provide multiple benefits to an organization ensuring the optimum usage of energy resources maintaining the desired energy productivity level and reduce the energy costs [4–6].

The energy management programs are being developed to endorse energy efficiency in the industries for facilitating energy savings, reduction of greenhouse gas emissions, and productivity benefits [7,8]. However, industrial energy efficiency still remains unattained [9,10]; with low implementation rates of energy efficiency measures [11,12] because of certain barriers [13,14], although research has shown its immense potential. There are multiple studies conducted at local, regional, national, and multinational focusing the barriers to adopt energy efficiency in the industries [15–19]. On the contrary, the drivers are also found towards energy-efficient technology adoption by several studies [20–22]. The energy efficiency gap has been conferred, keeping the relevance on technical aspects and appliances [23], whilst it has also incorporated behavioral issues [24].

Energy management and energy services are mostly studied through theoretically or conceptually, whilst energy management practices are studied in an empiric way [25]. Academic studies have been conducted regionally and beyond by many researchers about energy management practices and their characterization [15,26–28]. Energy management practices, as well as energy services, are perceived as significant explanations, and few efforts are paid to depict them including the assessment model to facilitate industrial think tank focusing particular set of actions for improved energy management [24]. It is notable to mention that research mainly acquainted the idea of an "extended energy efficiency gap", expressing the gap abide by technical as well as managerial components. In addition, a vast portion of unexplored market potential namely "energy service gap" exists because of high operating cost at the industrial application phase [29], even though energy services speak for a favorable market-centric resolution for improved energy efficiency [30]. So far, the avenues between integration of energy management with production systems are unexplored. Further, energy management into industrial decision-making process is not discussed thoroughly till now. Therefore, it is imperative to explore the domain of energy management to support industrial decision-makers pointing to the specified actions which are required to minimize the energy management lagging aspects, still keeping mind the multi-dimensional context and complexity of industrial energy management systems [31,32].

Given the introductory context, the paper aims to methodologically review the energy management assessment models that facilitate the assessment of industrial energy management. Notably, this study does not consider energy generation part and confines its focus to energy management framework only to help the industrial decision-making process covering energy consumption aspects in the industries.

This study is novel considering the fact that there has not been any study focusing on energy management 4.0 in the industrial decision-making process and comprehends the energy management framework to the best of authors' knowledge. In this study, we have worked to synthesize this gap in the greater interest of academia. By doing the review, we want to highlight future research avenues having nexus with energy management and industrial energy efficiency. Interestingly, all of the present research gaps fall into the big area, which is energy management 4.0 in industrial decision making. On another note, this study would help the industrial managers and engineers by figuring out improvement options in their energy management activities and supply chain system. In addition, the available options for policymakers to address energy management regulations are also incorporated in this study.

The rest of this paper is designed as follows: an introduction to the energy management concept is presented in Section 1. Section 2 describes the methodology. Section 3 provides the descriptive results of reviewing the literature on energy management assessment model. Subsequently, this paper concludes with explaining and incorporating the results in Section 4. Section 5 presents the concluding remarks.

2. Methodology

A systematic and rigorous review process was conducted in this paper. The primary focus of such reviews is to point out the related available studies established on pre-formulated research queries to synthesize the conclusion based on the evidence [33]. It is notable to mention that the systematic review features substantial leverages contrast to conventional narrative approaches of literature work. The conventional review does not apprehend formal methodological approaches, whilst the systematic review incites to minimize research biases through the adoption of search strategies, preordained inquiry string, and inclusion and elimination criterion [33]. Moreover, the comprehensive documentation nature of review enhances the clarity of review as well as facilitates subsequent replication [34].

In this paper, the relevant literature search methodology comprised of scientific literature sources, mainly the "Scopus" and "Web of Science" as both of the sources are well accepted in academia for their research quality and reliability [35]. We checked the online databases indexed in "Scopus" and "Web of Science" to identify the articles based on our keyword. In this research, the selected keywords to sort out the literature are "Energy Management", "Industry", "Energy Management Model", "Energy Management Practices", and "Energy Efficiency". Nonetheless, there was no specific starting timeframe for searching the literature in the database, though attempts were made to consider the recent researches. Table 1 presents the selection basis of the literature review.

Table 1. Selection basis of the literature review.

Heading	Remark
Research domain	Energy; Engineering; Management
Search string	Industrial Energy Management; Energy management Practice; Energy Management Framework
Publication Type	The academic journals, conference proceedings, and book chapters. Working papers are not considered due to their review process state and reliability issue [36]. The included publications are Elsevier, Springer, IEEE Xplore, MDPI, Taylor & Francis, John Wiley & Sons, and Emerald.
Availability	Available online
Area	Industry
Relevance	Articles articulate energy management; energy efficiency proceedings at the institutional perspective
Time	Focus on the recent researches

Each of the selected articles has been checked manually for content analysis in stage 2, the "screening" process. During the screening process, expulsion criteria that are followed in this research are presented in Table 2. Articles were discarded in this stage based on the criterion EXC 1, EXC 2. In stage 3, a backward review was conducted to reconsider relevant articles based on our selected keywords. The following stage consists of the exclusion of articles based on the criterion EXC 2, EXC 3, EXC 4, EXC 5, EXC 6, and EXC 7. Finally, the last step of methodology replicates the content analysis of selected articles. The entire methodological steps are illustrated in Figure 1.

In the phase of analyzing the content, it was essential to distinguish between energy management and energy management assessment framework/model. Therefore, the situation was very critical and decisive to the inclusion of such specification in this study. Nonetheless, discarding any concept related to energy management and its framework additional resolutions and aspects were also introduced that were not considered in the initial phase.

Table 2. Exclusion criterion of the literature.

Exclusion Heading	Remark
EXC 1	The article published not in English
EXC 2	The article uses "Energy management" term only in title and does not incorporate in any energy management framework or model in an elaborated form
EXC 3	The article uses "Energy management" only as a part of the future research direction or future perspective
EXC 4	The article uses "Energy Management" just as a cited term
EXC 5	Articles deals only with drivers, barriers to energy management practices in the industries
EXC 6	Availability of full texts
EXC 7	Working papers

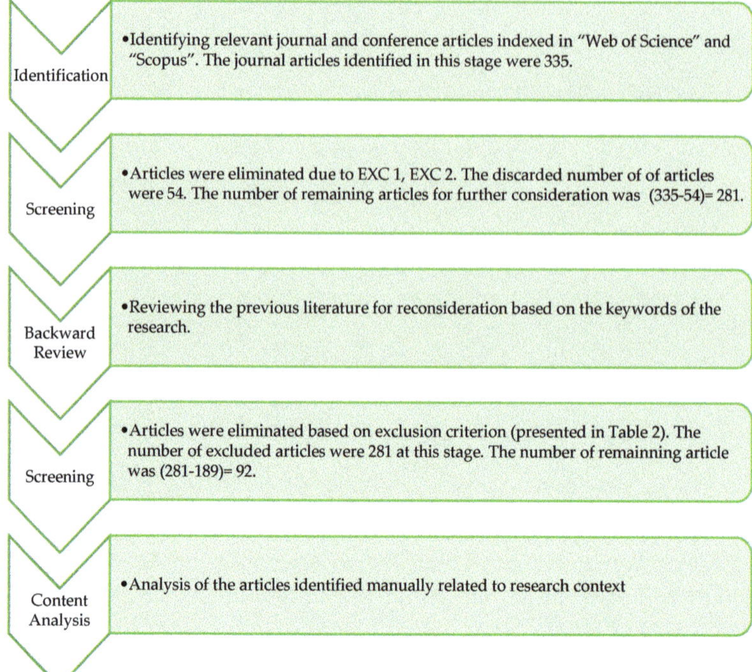

Figure 1. The methodological steps followed in the research.

3. Results and Analysis

3.1. Energy Management Definition

Defining energy management is significant when it comes to the point at energy management modelling or energy system practices implementation. Energy management concept is specified by many studies that incorporate multiple arenas. The prime areas covered by multiple studies to define energy management are energy consumption, strategic aspect, the involvement of managerial perspective, and people relevancy [25].

The German Federal Environment Agency defined energy management as the inclusion of planned and execution of actions to ensure predefined performance by a minimum amount of energy input [37]. B.L. Capehart has characterized the term energy management as the proficient and effective usage of energy towards maximization of profits and increasing reasonable positions [38]. O'Callaghan et al., defined the energy management as the application of resources in regards of supply, conversion, and utilization which integrates monitoring, measurement, archiving, critical examination and analyzation, control, and rerouting of energy as well as material flows through the systems for ensuring minimal energy usage and achieve meaningful goals [39]. To define energy management, Bunse et al. focused on the inclusion of control, supervision, and improvement activities towards energy efficiency [6]. On the contrary, Ates et al. strengthened on the combination of techniques, activities, and managerial processes that leads to reduce energy cost and anthropogenic emissions [40]. One of the studies by Abdelaziz et al. promoted energy management focusing on energy optimization strategy that incorporates compelling the energy demand [41]. A comprehensive definition of energy management has been proposed by Schulze et al. that incorporates all necessary energy management elements and energy management practices in the industries [32].

In academic literature, energy management is portrayed as a holistic combination of applying resources, conversion, and application of energy [16,20,25,32]. The system involves checking, auditing, recording, scrutinizing, and more importantly controlling the energy flows to ensure the minimum consumptions of energy but to achieve maximum energy productivity [16,42]. Academicians have pointed some of the minimal prerequisites for implementation and operation of energy management in the industries [27,40,41,43,44]. Table 3 illustrates the requirements toward energy management with specifications whether the requirements are considered full, partly, or not under consideration.

Table 3. Minimal prerequisites for energy management in the industries. This table is adopted from Schulze et al. [32].

Minimum Prerequisite	Abdelaziz et al. [41]	Christoffersen et al. [44]	Thollander and Ottosson [27]	McKane et al. [43]	Ates and Durakbasa [40]
Long-term strategic plan; inclusion of energy policy; energy saving targets.	✓✓	✓	✓✓	✓	✓
Energy activities by dedicated responsibilities and actions	✓	✓	×	×	×
Acquaintance of energy management team led by the energy manager	✓✓	×	×	✓✓	✓
Policies and proceedings	×	✓	×	✓✓	✓
Energy audit to explore energy-saving features	✓	×	✓✓	×	×
Planning and implementation of an explicit energy-saving program	✓	✓✓	✓✓	✓✓	✓✓
Identification of key performance indicators	×	×	×	✓✓	×
Meter and monitoring of energy consumption	✓	×	✓✓	×	✓
Energy reporting	✓✓	×	×	✓✓	×
Top management commitment	×	×	✓✓	×	×
Employee engagement in energy management activities	✓	✓	✓	×	✓

Abbreviations: ✓✓ (Full Consideration); ✓ (Partial Consideration); × (Not Considered).

It becomes discernible by analyzing the minimum requirements for energy management from Table 3 that the sets of minimum requirements elucidated in the studies contrast in the number of elements as well as conformation of the individual features. In addition, it shows indistinctness on the conclusiveness of the list of minimum requirements whether it is suitable to describe a fully developed energy management. By analyzing earlier contributions on the topic, we can note the lack of a comprehensive conceptual framework about energy management. Therefore, in this study, we respond to this research gap by complying a review of academic journal publications in the area of industrial energy management and use its results to propose future research avenues to explore further.

3.2. Approaches to Energy Management Models

There are research streams which are considered in academia as well as the industries to assess the energy management models. The streams can be categorized as "Minimum requirements", "Maturity models", and "Energy management matrixes" [25]. Furthermore, there is assessment tool namely "Energy Management Measures Characterization Framework", so to shape the energy management aspects accordingly". This is practice based, therefore basing on energy management practices with characteristics.

3.2.1. Minimum Requirements

The ISO 50001 standard that deals with energy management issues is incorporated at the first stream and thus apprehends guidelines to enable energy management system [45]. Enabling the organizations towards energy efficiency is the primary purpose of ISO 50001 Energy Management System standard. The standard is reviewed and published by the ISO/TC 301 Technical Standardization Committee, Energy Management, and Energy Saving in 2018 [45]. The protocol has a high level of hierarchical structure consists of ten chapters with a homologous architecture. The ISO 50001 standard is a consistent improvement framework which consists of "Plan-Do-Check-Act" at organizational practices. Table 4 presents the phases that are comprehended at ISO 50001 Energy Management System standard. However, it does not apprehend the critical assessment of the enterprises' effectiveness for a taken initiative of particular energy management practice. In addition, the initial stream incorporates primary endeavor to evaluate energy management, maintaining the limit of analysis [40,44].

Table 4. The continual phases of ISO 50001 Energy Management System standard [45].

Phase	Remark
Plan	To apprehend the organizational context; incorporation of energy policy; incorporation of energy management team; consideration of actions towards risks and opportunities; conduct of energy review; identification of significant energy uses and establishment of energy performance indicators; energy baseline; objectives and energy targets; necessary action plans to improve energy performance in accordance with the organization's energy policy.
Do	Implementation of the action plans; operation and maintenance controls, and communication; ensuring competence in energy domain i.e., energy performance in design and procurement.
Check	Monitor; quantify; analyzation; evaluation; audit and conducting management review of energy performance as well as energy management system.
Act	Activities to address non-conformities and continuation for improving energy performance.

3.2.2. Maturity Models

This second stream solicits a systematic perspective for assessing energy management in the organization [8] that includes the analysis for the requisite steps to enact energy management system [46]. Continuous improvement options are one of the significant features of maturity model. Therefore, the maturity model is accepted and popular in academia as well as industries since the development of the capability maturity model (CMM) [47–49]. The maturity models help the institutional enterprises surmount the austerity and enhance the quality by measuring institutional maturity based on particular or multiple domains with the help of predefined rules [50,51]. However, the maturity models are single dimensioned that focus either on objects maturity or process maturity, whilst the process maturity levels are dominant than the object-based model [52]. In one of the studies, Bojana et al. presented the maturity stages of energy management at activity levels [53]. Figure 2 exhibits the levels considered in maturity models for energy and utility management.

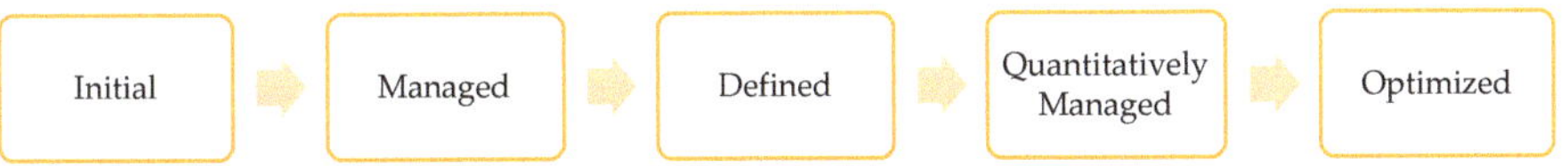

Figure 2. The levels in maturity models for energy and utility management (Source: [54]).

3.2.3. Energy Management Matrixes

The energy management matrixes are incorporated with the third stream [55,56], which confers multiple similarities with the maturity model. It offers an insight into the present approach to energy issues in a company and helps the management to improve energy efficiency by integrating feedback. It also shows the substantial improvement potential in energy efficiency that is achievable by technical activity alone. However, the application of the energy management matrix in a wider range of industrial organizations has acknowledged manifold activities towards improvement of energy management practice. In addition, it puts the hitherto isolated technologically-based attempts to improve energy efficiency in a more effective management framework, often for the first time. The high standpoint from an analytical perspective, maturity concept conversion into a sophistication level along with a self-appraisal approach based on organization's perspective are the common points of energy matrixes with maturity models. Hence, no additional benefits are provided from these models in terms of approaches and aspects considered for reasoning. However, introducing assessment models have brought an amelioration that incorporates detailed activity list considered as energy management practices, whilst critical factors have not been addressed for evaluation [56,57].

3.2.4. Energy Efficiency Measures (EEM) Characterization Framework

The EEM characteristics are delved by the fourth research stream [58]. The energy efficiency measures characterization framework is important to formulize in the context of information sharing both for the policy and decision-makers about energy efficiency measures. Thanks to improved knowledge and information on industrial energy efficiency measures. Indeed, the policymakers could have enhanced support to develop the operative policies for endorsing energy efficiency at the industries. In addition, the improved knowledge on energy efficiency measures characteristics can articulate in-depth comprehension of the bottlenecks that hindering the implementation of energy efficiency processes [59]. Indeed, this is an interesting fact for resolution and policy makers.

Fleiter et al. exhibited detailed and thorough narratives of characterizations that facilitated understandings of the endorsement process for EEMs [58]. The framework encompasses twelve diverse features of energy efficiency measures which are emanated from the field of technical, relative advantage, and informational perspective. Worrell et al. characterized and grouped the energy efficiency measures into multiple attributes such as waste, emission, operation and maintenance, productivity, working environment, among others, where the secondary benefits are listed [60]. On the contrary, Trianni et al. devised a framework to explore energy management practices [59]. An inclusive view on energy efficiency measures integrating the recent applicable perspectives is encompassed in this framework for industrial decision-makers. The framework has inferred in specifying energy alongside the environmental and financial aspects. Moreover, the impact on production system, including the application aspects and interaction with other systems of energy efficiency measures are also considered in the framework. Another noteworthy feature of the framework is the inclusion of corporate involvement, which is important for industrial decision-makers and policy delegates. Moreover, the inclusion of the attribute set related to non-energy benefits is one of the salient features that has been neglected in the earlier characterization framework. Nonetheless, analytical factors of energy management activities are not portrayed comprehensively. Lung et al. affirm about the impact of additional savings and productivity benefits stemming from energy efficiency initiatives resulting in more compellingly. The authors focused on the methodology to characterize the attributes of productivity benefits as well as ancillary savings into a payback forecasting framework [61].

Another model has been proposed in a contemporary study by Trianni et al. in the domain of characterization framework to assess industrial energy management, focusing on the benchmarking of energy management practices [25]. In this model, three elements have been considered that are energy management practice lists followed by specific baseline for benchmarking the performances and optimal threshold adoption in the assessment. The notable aspects of this model are the energy management practice adoption evaluation and more comprehensiveness output compared to the other models. More importantly, it features elaborate energy management approaches and capabilities assessment to an indistinct evaluation of energy management practices. On the contrary, Sorrell [62] and Benedetti et al. [63] have considered three-dimensional classification framework focusing to energy service contracts. The framework of Sorrell is customer perspective based and consisted of "Scope", "Depth", and "Finance" dimension. Benedetti et al. considered "Scope", "Intangibility of the Contract", and "Degree of Risk".

The synopsis of the existing management assessment models is presented in Table 5.

Table 5. Synopsis of the existing energy management assessment models. The table is an aggrandized approach of Trianni et al. [25].

Category	Model Narration	Remark	Reference
Minimum requirements	Significant features: energy policy, energy saving goals (quantitative) or aspirations regarding energy-saving projects and their implementation. Energy efficient purchases, specific allotment of energy responsibility and tasks. Functioning engagement of stakeholders, specially the employees by apprising, persuading, and educating.	This model consider the energy management as a comprehensive management system. Focused on policy, energy saving goals and specific energy saving projects. However, the model does not integrate the energy manager concept. Furthermore, there is no clear guideline about top or mid-level management support to achieve energy savings. Though, involvement of employee to energy saving related work are suggested.	[44]
	"Plan-Do-Check-Act" cycle is the basis for instructions. Preconditions: management liability; policy; legitimate concern and obligations; energy audit; energy performance index; energy baseline; energy targets, and energy management blueprint; proficiency, training and consciousness; communication; archiving; energy services acquisition; operation and control; monitoring, measurement and analysis; compliance evaluation maintaining the legal necessities; in-house audit of the energy management system; aberration, corrective as well as precautionary action; archive governance; management review.	ISO 50001 incorporates nineteen characteristics in the framework. Precisely, management commitment and energy manager are also inclined to the model. Moreover, the framework integrates the employee involvements and documentation and records for further assessment.	[45]
	Alteration of the merest requirements from the (27)'s set by adding the metering of major proceedings; inclusion of dedicated energy manager at the industry.	This model is an extended version of Christoffersen et al. [44]. The model integrates energy metering, energy policy, energy manager, saving target and saving projects focusing on energy.	[40]

Table 5. *Conts.*

Category	Model Narration	Remark	Reference
Maturity Models	Five stages: preliminary, arrange, delineate, managed in quantitative form and reformed; Novel process avenues are regulated towards progress focusing on environmental aspect; Four maturity phases: practice enactment, standardization of practice, performance management and recurring phase for improvement.	The model used Capability Maturity Model Integration (CMMI) as a reference framework and extended to environmental management context. It comprised of particular procedures for energy as well as resource management. No clear guideline about dedicated energy manager.	[54]
	Instructions to attain improved energy efficiency and amenability with energy management standards especially ISO 50001. Energy management actions are categorized into five maturity phases subsequent to the PDCA cycle.	The framework adapts manifold energy management practices based on PDCA cycle. Notably, top management support is incorporated in the framework. Energy management roles are characterized. However, no clear indication about energy manager inclusion in the process.	[64]
	Five levels: Emerging, Define, Integration, Optimization and Novelty; four sections on the basis of PDCA cycle, 16 pillars, and 63 sub-pillars. The model allows 5 attribution promulgation for each sub-pillar to evaluate the maturity.	Energy management review along with action plan are integrated to the framework. In addition, competence building feature is also included.	[65]
	Primary features for the energy consumption management keeping alignment to ISO 50001. Five phases: initial, intermittent, planning, supervisory and optimal. 5 dimensions that are portrayed as requisite for success: consciousness, information, and expertise (utmost significant); methodological proposition; energy performance management and archiving system; institutional architecture; alignment with strategy.	The tool is not incorporated with inclusion of energy manager.	[8]
	Incorporation with ISO 50001; knowledge base creation for self-assessment along with monitoring and improvement. The levels are depicted for each ISO 50001 process instilled by Eric et al. [54].	The assessment tool includes top management commitment, and energy manager appointment with other manifold energy management practices.	[53]
	Salient features are the assessment of compelling factors for energy management adoption, contribution towards a better understanding of suitable energy management configuration with the help of evaluation of maturity level.	The model considers inclusion of energy manger, precisely a dedicated energy management team. In addition, top management support is integrated with the considered attributes in the model.	[46]
	Incorporation of qualitative metrics; assessment model implies on PDCA cycle; inclusion of SWOT analysis tool, incorporation of global energy management team and external peers.	Incorporates their application specific purposes which are descriptive, prescriptive, and comparative. Features with manifold energy management practices along with energy manager.	[66]

Table 5. *Conts.*

Category	Model Narration	Remark	Reference
	Consists of three features: (1) energy efficiency features (2) energy efficiency maturity levels; (3) implementation method which is accustomed from ZED scheme especially for SMEs. Seven dimensions: management obligation, arrangement and procedure, compliance of regulation and fiscal enticements, archiving system, product and procedure innovation, in-house communication, and ethos. Consists of nineteen characteristics.	Total number of nineteen energy efficiency characteristics are integrated in the model. In addition, management commitment is segregated into two sections in the form of strategic priority and energy policy.	[67]
	Five levels of energy management matrixes to address six institutional aspects that are policy, organization, motivation, information scheme, marketing, and financing.	Top management support is fully integrated into the framework under policy section. Energy managerial role included in organizational structure.	[68]
	Five levels of energy management matrixes to assess six institutional issues that are energy management scheme; organization; staff inspiration; tracking, supervision and reporting systems; staff consciousness/training and promotion, and financing.	Energy manager feature is integrated with a proposition of organizational structure. Moreover, energy management is considered comprehensively in this framework.	[55]
Energy Management Matrixes	Five levels of energy management matrixes to assess six institutional issues which are policy or specific guidelines, coordinating, training, evaluation of performance, communication, and financing. Valuation model exploring the subsequent aspects reflected as energy management practice: policy and legislation, energy blueprint, organizational formation; regulation; acquisition strategy, financing scheme, observation, and analysis of energy consumption, setting of goal; identification of possible options; staff involvement and training; operational process; communications.	The Carbon Trust guidelines comprised of five aspects. Inclusion of dedicated energy manager is not integrated to this model. However, the model incorporates senior management commitment to enhance energy efficiency related initiatives.	[56]

Table 5. *Conts.*

Category	Model Narration	Remark	Reference
	Model exploring the succeeding features considered as energy management practice: energy director appointment, incorporation of energy team, apply of energy policy; collection of information and management, establishment of yardstick or threshold, analysis, assessing from technical perspective and energy audits; exploring and setting the scope, improvement option estimation, goal setting; define technical procedures and targets, roles and resources determination; formation of a communication plan, awareness raising, capacity building, inspire, trail and monitor; measurement of result, recapitulation of action plan; maintain internal recognition, and receiving external appreciation.	The ENERGY STAR guideline clearly emphasizes on appointment of energy director with dedicated energy team. In addition, the model looks to establish baselines for measuring energy performance.	[57]
EEMs characterization framework	Three main characteristics are considered. Each characteristic are divided into sub-divisions. The first character "Relative advantage" is attributed by internal rate of return, introductory expenses, reimbursement time, and benefits of non-energy. "Technical context" the second character is attributed by modification type, impact opportunity, gap among core processes, and Lifetime. The last character "Information context" is attributed by transaction expenses, planning and execution knowledge, Dissemination progress, and field wise applicableness.	One of the salient features of this framework is inclusion of non-energy benefits. Energy manager is not integrated into the framework.	[58]
	Economic characterization consists of payback time, application costs. Energy is attributed with resource stream and energy saving. Environmental characterization is attributed by waste minimization and emission contraction. Production is attributed by productivity, working environment, and operation and maintenance. Implementation related attributes are energy saving strategy, types of action, implementation easiness, success probability, community engagement in corporate level, distance among key processes, and audit regularity. Interaction-related characterization is attributed by indirect effects.	Corporate involvement is one of the notable attributes and considered as significant for industrial decision-makers. The need for analyzing energy efficiency measures as per different perspectives is highlighted; precisely having the aspects in grouped for providing more inclusive view on the pertinent outlooks distinguishing the energy efficiency measures.	[59]

4. Discussion

The energy management frameworks were mainly researched to adopt energy management practices at the technical levels in the industries. However, the reviewed papers emphasized the energy management system, ISO 50001, and PDCA cycle, while some studies suggested holistic approaches towards industrial energy efficiency.

The framework proposed by Christoffersen et al. was stood out on the Danish industries and emphasized on multiple factors, mostly energy policy, goals and capstone projects aimed at energy savings. Regulation, external relations, company characteristics, and organizational internal condition are the main out-layers of the model to frame the energy management. However, the company size and energy intensity are two factors that can be considered to categorize the industries to apply or analyze the model [44]. The main features proposed by Christofferen et al. align with ISO 50001: 2011 standard though this model has been replaced by ISO 50001: 2018 [58]. The earlier model encompassed energy management system implementation based on PDCA cycle and enlisted few prerequisites that include mainly management liability, policy, energy audit, energy performance indexing, energy management blueprint, documentation, and so forth. One of the major changes in the recent model is the PDCA cycle modification. "Checking" was the center in the earlier version, whilst "Leadership" became the focus of all cycle components. Figure 3 represents the revised PDCA cycle of ISO 50001:2018. In the minimum requirement segment, the model proposed by Ates et al. comprehended conventional streams towards energy management. One of the significant features is the inclusion of energy manager, whilst ISO 40001 (environmental permit) also act as an enabling feature along with ISO 50001 [40].

Figure 3. The "Plan-Do-Check-Act" cycle adopted in ISO 50001: 2018 (Source: [45]).

Looking at the minimum requirement focused model, it is observed that all the energy management initiatives are not integrated into the frameworks. Christoffersen et al. [44] considered energy management as a comprehensive management system. However, the model does not integrate the energy manager concept. Furthermore, there is no clear guideline about top or mid-level management support to achieve energy savings. Though, the involvement of employee to energy-saving related works are suggested. Nonetheless, The ISO 50001 model is a significant protocol [69] along with the proposition by Ates and Durakbasa [40], manifold aspects are still to be explored regards of operational activities in the industrial energy management domain. For instance, the principles of sustainability and integral management need to be presented at the protocol. In addition, there is very little contribution on the risk management and opportunities of energy efficiency from an integral and strategic point of view, including the planning and control of product lines, process design, projects, and business approaches [69]. Notably, the fruitful operation of the energy management system requires the integrated deployment of planned, tactical, and operational levels that align the business culture with sustainable attainment. In this context, the vision that the organization plans to form should be linked to energy efficiency strategy with organization's survival plan in the market. Additionally, it is necessary to make explicit reference to newly adapted technical features through peer to peer energy management platform for optimizing the integration of energy management system component with the variable energy demand [70,71]. Moreover, an integrated perspective to control of operational features of each process are required to explore linked to energy efficiency [69].

In the energy management maturity model segment, the model proposed by Ngai et al. based on capability maturity model integration (CMMI), an extension of capability maturity model incorporated five levels according to the behavioral exhibition of the industries [54]. The levels are determined by performance area of key processes [72]. The achievement goals of key process areas must be specified for individual level for further actions. Notably, the propositions of CMM framework has been applied at multiple process enhancement programs in order to achieve the desired quality in the production system [73]. One limitation of this model is inadequate implementation time, having only one factory for consideration. However, the authors have affirmed the acceptability of the model because of prior implementation of management practices. On the contrary, Antunes et al. emphasized the PDCA cycle to design the energy management framework [64]. Additionally, the authors implied the model with ISO 50001 and incorporated energy management practices also. Notable to mention that Finnerty et al. also designed the framework based on the PDCA cycle, keeping on focus on energy management practices [66].

The model proposed by Introna et al. is comprised of five dimensions and enables the feature of self-evaluation for the industries towards energy management practices. The dimensions are characterized by identifying the necessary elements in energy management consumption segment of the industries [8]. On the contrary, Jovanović et al. focused on ISO 50001 processes as well as PDCA phases, keeping the knowledge base in the model EMMM50001 [53]. The EMMM50001 establishes the relation to EUMMM maturity levels, maintaining ISO 50001 specifications and PDCA phases. Notably to mention that EMMM50001 links the CMMI criteria, also maintaining the ISO 50001.

It can be observed that the majority of the maturity models emphasized on similar type of characteristics and areas to evaluate the energy management in an organization by a systematic set of commendations. However, the narrated models demand more time and resources to perform as per their characterization. In addition, looking at the scientific literature, all of the frameworks studied to see the requirements for providing a continuous development path following the PDCA approach and ISO 50001. Notably, few of the maturity models incorporate the implication of dedicated energy manager and top management support. In contrast, Antunes et al. [64] affirm on top management support whilst not integrating the energy manager in the framework. The framework by Introna et al. [8] also did not complied with the energy manager. Nonetheless, Jovanović and Filipović [53] and Finnerty et al. [66] considered top management support along with the energy manager in their framework.

Gordic' et al. applied the energy management matrixes model in the Serbian car manufacturer industries and critically analyzed the existing energy management system with the model [55]. Notably to mention that the energy management matrixes models proposed by Gordic' et al., Carbon Trust and Energy Star encompass all key areas to assess the energy management practices in the model, with having few modifications at the individual version.

On the contrary, Fleiter et al. [58] and Trianni et al. [52] emphasized on a characterization based model where both of the models are incorporated with specific attributes. The characterization scheme has some implications on policy design and assessment. However, formalization of the groups with categorized attributes enables the option towards relevant aspects identifying the energy efficiency measures. In addition, Trianni et al. contend a comprehensive scenario on EEMs focusing on the relevant aspects of industrial energy management [52]. One of the critical factors, "corporate involvement" for industrial decision-makers is also implied, hence allowing additional feature and an increase in the applicability of the model. In another proposed framework, Trianni et al. incorporated energy management practice-based approach. However, the authors acknowledge more compatible space for the SMEs within the model, as SMEs are sought to be entitled to more attention, considering their cumulative energy consumption percentile [74]. In a recent study, Tina et al. persuade the significance of SMEs and the policy implications in the peripheral of the industrial energy sector [74]. Referring to the SMEs, Prashar [67] proposes an energy efficiency maturity assessment framework that emphasizes SMEs. Notably, the author argues that the common energy efficiency framework approach does not facilitate fully to the SMEs; hence, a customized maturity framework is significant. The author considered steel re-rolling mill sector of India as the contextual sphere for the proposed framework.

Few of the studies on characterization the energy efficiency measure focuses on financial features. Notably to mention that these models do not frame the energy efficiency measures comprehensively, rather offer some framework without characterizing the energy efficiency measures in-depth. In one of the studies by Pye and McKane, they state that quantification of the accumulated benefits of energy efficiency scheme supports the enterprises perceive the monetary opportunities of EEMs financing [5]. The energy savings features act not as the singular primary driver for the industrial decision process; hence, the authors persuade that energy savings be viewed as a factor of the holistic approach towards energy efficiency programs. Skumatz studied the methods to find out the attributes of EEMs and established the scheme to measure both of the positive and negative secondary benefits stemming from industrial energy efficiency schemes [75,76]. On the contrary, Mills and Rosenfeld provided a framework to understand multiple benefits of energy efficiency initiatives and grouped the attributes into the better interior environment, noise lessening, savings of labor and time, improved supervision of procedure, convenience, water savings and waste reduction, and benefits due to downsizing of equipment [77].

The majority of studies on energy efficiency measures, benefits, and initial characterization frameworks propose few interesting reflections. However, a lack of consistency on the attributes grouping within existing categories from the methodological perspective is observed. It is found that the same attributes are grouped in different categories by different researchers. Moreover, the attributes are categorized and then aggregated again within other segments by different researcher. For instance, "reduced noise" and "improved indoor environment" are framed in two different categories in [77], whereas "reduced noise level" as categorized in "working environment" segment. On the other note, the decision-making process is a grey area keeping mind about the stakeholders. Nonetheless, the earlier characterization framework did not incorporate the energy efficiency measure implications in a comprehensive way. To be precise, the inclusion of non-energy benefits is not incorporated into the characterization framework. Notably, the inclusion of non-energy features in the modeling factors would double the cost-effective potential for energy-efficiency enhancement, likened to an analysis eliminating those benefit [60]. However, few attributes (e.g., improved air quality, better worker safety, reduction of noise level, and improved working situation) are there in the characterization framework, which are difficult to quantify [76]. Therefore, speculation is required to articulate the benefits into a comparable cost figure, and hence the assessment turns out to be rather subjective [60].

The study by Ngai et al. [54] features energy management with particular process areas in the manufacturing industries. In this study, few guidelines are offered to conduct analysis for organizational maturity improvement in terms of energy along with the management of utility resources. However, the integration of energy management into production process has not been complied comprehensively. This is a significant avenue that needs be to address with utmost attention in future studies considering the technical implications offered by Industry 4.0. Notable to mention here, is that increasing the efficiency at the production processes is one of the salient features of Industry 4.0 [78]. The deployment of smart machinery offers diverse benefits which primarily includes manufacturing productivity and waste reduction [79]. Therefore, it is worth observing the energy management characteristics linked with production process through the lens of Industry 4.0.

Nonetheless, energy management towards industrial energy efficiency has been widely discussed in academia, and several barriers are still persistent in the energy management domain. The identification of barriers is important because it hampers or slows down the adoption of energy efficiency measures [80]. Several studies have addressed the barriers which cover energy-intensive industries to SMEs and include regional, national, and transnational perspectives [15,26,27,81–84]. However, most comprehensive studies focusing on energy management have been discussed without really looking at the integration of energy management into production and operation management. An imperative avenue, therefore, lies to be further explored in future within this research domain.

5. Concluding Remarks and Future Research Avenues

The paper attributes a review of research works on the energy management model for energy management practices in the industries. Multiple models have been compiled and structured, maintaining the narrations. Moreover, the energy management frameworks were synthesized emanate from the findings in order to facilitate energy management in the industries by offering necessary benchmarks to the industrial experts. The review findings show that the narrated models can support an organization to assess energy management and incorporate insightful contribution to energy efficiency initiatives. Nonetheless, some of the studies did not comply with a full methodological description and exhibited shorter model validation. In addition, a gap exists between the theoretical concept and practical implementation of energy management and its practices. Precisely, majority approaches remain unsuccessful in replicating and scope of actions distinct in energy management due to the certain barriers [27,66,85].

Moreover, most of the models have looked the energy management as a single unit function, whilst it is a combination of multi-dimensional approaches with the involvement of several functional units in the industries. Let us not forget about the multi-dimensional operational activities in the industries which are conjugal part with energy management. Notably, multi-dimensional approaches are critical to support the process and operational oriented program [86]. Therefore, a comprehensive operational approach should be considered by integrating all the relevant energy flows. It infers to all forms of energy, including externally supplied energy sources as well as internal energy flows. Interestingly, relating the energy management into the operational framework integrates the resource efficiency also at the manufacturing level. The raw or auxiliary material consumption might be of interest, considering the direct and indirect impact on energy and resource efficiency in the manufacturing process. Moreover, keeping mind about the non-static nature of energy consumption, the dynamic consumption feature might unveil manifold resource optimization aspects [87].

Unfortunately, the integration of energy management into operational activities have been little explored. It becomes even more imperative while we look to adopt Industry 4.0 keeping in mind about the manifold complex technical features consist of Internet of Things (IoT), big data, cloud computing, and so on in the industrial plants. Many scholars predict that the exponential progress in the promises of manifold technical features offered by Industry 4.0 might affect the production activities in the industries inclusively. In addition, there are high chances of modification in the traditional industrial actions that cover the processing of elements and material, grinding, and assemble/ dismantle. This is emphasized in Industry 4.0 concept and its implementation, where we pursue to pool the common features with the enormous potential of digital technology [88]. However, it is understood as a necessary incremental approach aimed at further optimizing the energy system without seeking to disrupt it in principle. In the energy efficiency domain, the energy management and its practices have already influenced the production scenario in a broader aspect, and this inclination should remain as long as we allow the nexus between Industry 4.0 and energy efficiency. On the other hand, energy productivity investments in present as well as the recent technologies must be conveyed through the implementation of energy management and its practices [89]. Energy management practices and energy services are acknowledged as fundamental solutions; the diminutive effort is being paid in characterizing them [24]. Notably, assessment models for supporting the industrial decision-makers emphasizing detailed activities for better energy management is deficient. Therefore, it is imperative to consider the energy management in multiple aspects keeping mind about the complex nature of industrial energy system [31].

Interestingly, energy management has implications on asset maintenance, e.g., on maintenance procedures. As energy management includes the control of energy-consuming devices to optimize energy consumption, manual toggling on and off devices depending on requirement is a rudimentary custom of energy management. The initiation of mechanical and electrical equipment (e.g., timers for programmed toggling, bimetallic strip thermostat, pneumatic and electrical transmission system, and so on) provided means for early energy management schemes in the form of automatic temperature control. Nowadays, the inclusion of direct digital control in energy management has retrofit benefits that allow device monitoring linked to maintenance procedure, thanks to energy management and its practices. The comprehensive data recommend that while energy management does improve the accuracy and response of a system in the industries, the energy management routines facilitate partially asset maintenance [90]. It infers to monitoring or log building equipment performance while consuming the energy resulting increasing magnitude of all benefits covering maintenance and cost avoidance benefit. Unfortunately, much of the energy management studies have bypassed this retrofit fact while focusing on the energy management framework. So far, the integration of energy management with asset management has not been widely explored, and several questions remain unanswered at present. Therefore, more research needs to be undertaken to fit the asset maintenance into energy management framework in a comprehensive way.

In addition, the narrated models have little explored the sustainability feature integrated with energy efficiency, pointing to the optimization of resource utilization [91]. We must consider the paradigm that allows industrial energy management effective for the companies. In this context, it would be certainly interesting to visualize the energy management through Industry 4.0 technologies and solutions, may contribute to improved sustainability performances of the companies. If Industry 4.0 is expected to unveil enormous directions not only to energy management but also the sustainability field, the challenge definitely lies on the integrational aspects with energy–industry–sustainability nexus. Therefore, the future research avenues should reflect the energy management framework complying the diverse directions and encompassing the operational management, Industry 4.0 along with sustainability features.

Author Contributions: Conceptualization, A.S.M.M.H. and A.T.; methodology, A.S.M.M.H. and A.T.; formal analysis, A.S.M.M.H. and A.T.; investigation, A.S.M.M.H. and A.T.; writing—original draft preparation, A.S.M.M.H. and A.T.; writing—review and editing, A.S.M.M.H. and A.T.; visualization, A.T.; supervision, A.T. All authors have read and agreed to the published version of the manuscript.

Funding: This research received no external funding.

Conflicts of Interest: The authors declare no conflict of interest.

Nomenclature

EMP	Energy Management Practice
EEM	Energy Efficiency Measure
ESM	Energy Saving Measure
SME	Small- and Medium-sized Enterprise
GHG	Greenhouse Gases
CMM	Capability Maturity Model
PDCA	Plan-Do-Check-Act
CMMI	Capability Maturity Model Integration
EMMM50001	ISO 50001-based Energy Management Maturity Model
ISO	International Organization for Standardization
EnMS	Energy Management System
EUMMM	Energy and Utility Management Maturity Model

References

1. Faheem, M.; Gungor, V.C. Energy efficient and QoS-aware routing protocol for wireless sensor network-based smart grid applications in the context of industry 4.0. *Appl. Soft Comput. J.* **2018**, *68*, 910–922. [CrossRef]
2. Tesch da Silva, F.S.; da Costa, C.A.; Paredes Crovato, C.D.; da Rosa Righi, R. Looking at energy through the lens of Industry 4.0: A systematic literature review of concerns and challenges. *Comput. Ind. Eng.* **2020**, *143*, 106426. [CrossRef]
3. König, W. Energy efficiency in industrial organizations—A cultural-institutional framework of decision making. *Energy Res. Soc. Sci.* **2020**, *60*, 101314. [CrossRef]
4. Cagno, E.; Neri, A.; Trianni, A. Broadening to sustainability the perspective of industrial decision-makers on the energy efficiency measures adoption: Some empirical evidence. *Energy Effic.* **2018**, *11*, 1193–1210. [CrossRef]
5. Pye, M.; McKane, A. Making a stronger case for industrial energy efficiency by quantifying non-energy benefits. *Resour. Conserv. Recycl.* **2000**, *28*, 171–183. [CrossRef]
6. Bunse, K.; Vodicka, M.; Schönsleben, P.; Brülhart, M.; Ernst, F.O. Integrating energy efficiency performance in production management—Gap analysis between industrial needs and scientific literature. *J. Clean. Prod.* **2011**, *19*, 667–679. [CrossRef]
7. Sola, A.V.H.; Mota, C.M.M. Influencing factors on energy management in industries. *J. Clean. Prod.* **2020**, *248*, 119263. [CrossRef]
8. Introna, V.; Cesarotti, V.; Benedetti, M.; Biagiotti, S.; Rotunno, R. Energy Management Maturity Model: An organizational tool to foster the continuous reduction of energy consumption in companies. *J. Clean. Prod.* **2014**, *83*, 108–117. [CrossRef]
9. Hirst, E.; Brown, M. Closing the efficiency gap: Barriers to the efficient use of energy. *Resour. Conserv. Recycl.* **1990**, *3*, 267–281. [CrossRef]
10. Cagno, E.; Worrell, E.; Trianni, A.; Pugliese, G. A novel approach for barriers to industrial energy efficiency. *Renew. Sustain. Energy Rev.* **2013**, *19*, 290–308. [CrossRef]
11. Anderson, S.T.; Newell, R.G. Information programs for technology adoption: The case of energy-efficiency audits. *Resour. Energy Econ.* **2004**, *26*, 27–50.
12. Cagno, E.; Trianni, A. Analysis of the Most Effective Energy Efficiency Opportunities in Manufacturing Primary Metals, Plastics, and Textiles Small- and Medium-Sized Enterprises. *J. Energy Resour. Technol.* **2012**, *134*. [CrossRef]
13. Jaffe, A.B.; Stavins, R.N. The energy-efficiency gap What does it mean? *Energy Policy* **1994**, *22*, 804–810.
14. Johansson, M.T. Improved energy efficiency within the Swedish steel industry—The importance of energy management and networking. *Energy Effic.* **2015**, *8*, 713–744.
15. Brunke, J.C.; Johansson, M.; Thollander, P. Empirical investigation of barriers and drivers to the adoption of energy conservation measures, energy management practices and energy services in the Swedish iron and steel industry. *J. Clean. Prod.* **2014**, *84*, 509–525. [CrossRef]
16. Hasan, A.S.M.M.; Rokonuzzaman, M.; Tuhin, R.A.; Salimullah, S.M.; Ullah, M.; Sakib, T.H.; Thollander, P. Drivers and Barriers to Industrial Energy Efficiency in Textile Industries of Bangladesh. *Energies* **2019**, *12*, 1775. [CrossRef]
17. Soepardi, A.; Thollander, P. Analysis of Relationships among Organizational Barriers to Energy Efficiency Improvement: A Case Study in Indonesia's Steel Industry. *Sustainability* **2018**, *10*, 216.
18. Trianni, A.; Cagno, E.; Thollander, P.; Backlund, S. Barriers to industrial energy efficiency in foundries: A European comparison. *J. Clean. Prod.* **2013**, *40*, 161–176.
19. Trianni, A.; Cagno, E.; Farné, S. Barriers, drivers and decision-making process for industrial energy efficiency: A broad study among manufacturing small and medium-sized enterprises. *Appl. Energy* **2016**, *162*, 1537–1551.
20. Cagno, E.; Trianni, A.; Spallina, G.; Marchesani, F. Erratum to: Drivers for energy efficiency and their effect on barriers: Empirical evidence from Italian manufacturing enterprises. *Energy Effici.* **2017**, *10*, 855–869, 10.1007/s12053-016-9488-x.
21. Trianni, A.; Cagno, E.; Marchesani, F.; Spallina, G. Classification of drivers for industrial energy efficiency and their effect on the barriers affecting the investment decision-making process. *Energy Effic.* **2017**, *10*, 199–215. [CrossRef]
22. Apeaning, R.W.; Thollander, P. Barriers to and driving forces for industrial energy efficiency improvements in African industries—A case study of Ghana's largest industrial area. *J. Clean. Prod.* **2013**, *53*, 204–213. [CrossRef]

23. Gangolells, M.; Casals, M.; Forcada, N.; Macarulla, M.; Giretti, A. Environmental impacts related to the commissioning and usage phase of an intelligent energy management system. *Appl. Energy* **2015**, *138*, 216–223.
24. Sa, A.; Paramonova, S.; Thollander, P.; Cagno, E. Classification of Industrial Energy Management Practices: A Case Study of a Swedish Foundry. *Energy Procedia* **2015**, *75*, 2581–2588.
25. Trianni, A.; Cagno, E.; Bertolotti, M.; Thollander, P.; Andersson, E. Energy management: A practice-based assessment model. *Appl. Energy* **2019**, *235*, 1614–1636.
26. Hasan, A.S.M.M.; Hoq, M.T.; Thollander, P. Energy management practices in Bangladesh's iron and steel industries. *Energy Strateg. Rev.* **2018**, *22*, 230–236.
27. Thollander, P.; Ottosson, M. Energy management practices in Swedish energy-intensive industries. *J. Clean. Prod.* **2010**, *18*, 1125–1133.
28. Hasan, A.S.M.M.; Hossain, R.; Tuhin, R.A.; Sakib, T.H.; Thollander, P. Empirical Investigation of Barriers and Driving Forces for Efficient Energy Management Practices in Non-Energy-Intensive Manufacturing Industries of Bangladesh. *Sustainability* **2019**, *11*, 2671.
29. Backlund, S.; Thollander, P. Thollander the energy-service gap. What does it mean? In Proceedings of the ECEEE 2011 Summer Study, Hyères, France, 6–11 June 2011; pp. 649–656.
30. Good, N.; Martínez Ceseña, E.A.; Zhang, L.; Mancarella, P. Techno-economic and business case assessment of low carbon technologies in distributed multi-energy systems. *Appl. Energy* **2016**, *167*, 158–172.
31. Kannan, R.; Boie, W. Energy management practices in SME—Case study of a bakery in Germany. *Energy Convers. Manag.* **2003**, *44*, 945–959. [CrossRef]
32. Schulze, M.; Nehler, H.; Ottosson, M.; Thollander, P. Energy management in industry—A systematic review of previous findings and an integrative conceptual framework. *J. Clean. Prod.* **2016**, *112*, 3692–3708. [CrossRef]
33. Denyer, D.; Tranfield, D. Producing a Systematic Review. In *The SAGE Handbook of Organizational Research Methods*; Sage Publications Ltd.: Thousand Oaks, CA, USA, 2009; pp. 671–689. ISBN 9781412931182.
34. Higgins, J.P.; Green, S. *Cochrane Handbook for Systematic Reviews of Interventions: Cochrane Book Series*; John Wiley and Sons: Hoboken, NJ, USA, 2008; ISBN 9780470699515.
35. Martín-Martín, A.; Orduna-Malea, E.; Thelwall, M.; Delgado López-Cózar, E. Google Scholar, Web of Science, and Scopus: A systematic comparison of citations in 252 subject categories. *J. Informetr.* **2018**, *12*, 1160–1177. [CrossRef]
36. Podsakoff, P.M.; MacKenzie, S.B.; Bachrach, D.G.; Podsakoff, N.P. The influence of management journals in the 1980s and 1990s. *Strateg. Manag. J.* **2005**, *26*, 473–488. [CrossRef]
37. Kahlenborn, W.; Kabisch, S.; Klein, J.; Ina Richter, S.S. *Energy Management–Systems in Practice: A Guide for Companies and Organisations*; Federal Ministry for the Environment, Nature Conservation and Nuclear Safety: Bonn, Germany, 2010; p. 12.
38. Capehart, B.L.; Turner, W.C.; Kennedy, W.J. *Guide to Energy Management*, 5th ed.; The Fairmont Press: Lilburn, GA, USA, 2006; ISBN 0881734772.
39. O'Callaghan, P.W.; Probert, S.D. Energy management. *Appl. Energy* **1977**, *3*, 127–138. [CrossRef]
40. Ates, S.A.; Durakbasa, N.M. Evaluation of corporate energy management practices of energy intensive industries in Turkey. *Energy* **2012**, *45*, 81–91. [CrossRef]
41. Abdelaziz, E.A.; Saidur, R.; Mekhilef, S. A review on energy saving strategies in industrial sector. *Renew. Sustain. Energy Rev.* **2011**, *15*, 150–168. [CrossRef]
42. Rasmussen, J. The additional benefits of energy efficiency investments—A systematic literature review and a framework for categorisation. *Energy Effic.* **2017**, *10*, 1401–1418. [CrossRef]
43. Mckane, A.; Williams, R.; Perry, W. Setting the Standard for Industrial Energy Efficiency Permalink. In Proceedings of the Conference on Energy Efficiency in Motor Driven Systems, Beijing, China, 10–13 June 2007.
44. Christoffersen, L.B.; Larsen, A.; Togeby, M. Empirical analysis of energy management in Danish industry. *J. Clean. Prod.* **2006**, *14*, 516–526. [CrossRef]
45. ISO 50001:2018(en), Energy Management Systems—Requirements with Guidance for Use. Available online: https://www.iso.org/obp/ui/#iso:std:iso:50001:ed-2:v1:en (accessed on 18 October 2020).
46. Sa, A.; Thollander, P.; Cagno, E. Assessing the driving factors for energy management program adoption. *Renew. Sustain. Energy Rev.* **2017**, *74*, 538–547. [CrossRef]
47. Weber, C.V.; Curtis, B.; Chrissis, M.B. Capability Maturity Model, Version 1.1. *IEEE Softw.* **1993**, *10*, 18–27.
48. Becker, J.; Knackstedt, R.; Pöppelbuß, J. Developing Maturity Models for IT Management. *Bus. Inf. Syst. Eng.* **2009**, *1*, 213–222. [CrossRef]

49. Scott, J.E. Mobility, business process management, software sourcing, and maturity model trends: Propositions for the IS organization of the future. *Inf. Syst. Manag.* **2007**, *24*, 139–145. [CrossRef]

50. Lahrmann, G.; Marx, F.; Mettler, T.; Winter, R.; Wortmann, F. *Inductive Design of Maturity Models: Applying the Rasch Algorithm for Design Science Research*; Springer: Berlin/Heidelberg, Germany, 2011; Volume 6629, ISBN 9783642206320.

51. CMMI Product Team. *CMMI for Development, Version 1.3*; Software Engineering Institute, Carnegie Mellon University: Pittsburgh, PA, USA, 2010.

52. Proença, D.; Borbinha, J. Maturity Models for Information Systems—A State of the Art. In *Proceedings of the Procedia Computer Science*; Elsevier: Amsterdam, The Netherlands, 2016; Volume 100, pp. 1042–1049.

53. Jovanović, B.; Filipović, J. ISO 50001 standard-based energy management maturity model—Proposal and validation in industry. *J. Clean. Prod.* **2016**, *112*, 2744–2755. [CrossRef]

54. Ngai, E.W.T.; Chau, D.C.K.; Poon, J.K.L.; To, C.K.M. Energy and utility management maturity model for sustainable manufacturing process. *Int. J. Prod. Econ.* **2013**, *146*, 453–464. [CrossRef]

55. Gordić, D.; Babić, M.; Jovičić, N.; Šušteršič, V.; Končalović, D.; Jelić, D. Development of energy management system-Case study of Serbian car manufacturer. *Energy Convers. Manag.* **2010**, *51*, 2783–2790. [CrossRef]

56. Carbon Trust. *Energy Management—A Comprehensive Guide to Controlling Energy Use*; Carbon Trust: London, UK, 2011.

57. ENERGY STAR. *Guidelines for Energy Management*; ENERGY STAR: Oakland, CA, USA, 2012.

58. Fleiter, T.; Hirzel, S.; Worrell, E. The characteristics of energy-efficiency measures—A neglected dimension. *Energy Policy* **2012**, *51*, 502–513. [CrossRef]

59. Trianni, A.; Cagno, E.; De Donatis, A. A framework to characterize energy efficiency measures. *Appl. Energy* **2014**, *118*, 207–220. [CrossRef]

60. Worrell, E.; Laitner, J.A.; Ruth, M.; Finman, H. Productivity benefits of industrial energy efficiency measures. *Energy* **2003**, *28*, 1081–1098. [CrossRef]

61. Lung, R.B.; McKane, A.; Leach, R.; Marsh, D. Ancillary Savings and Production Benefits in the Evaluation of Industrial Energy Efficiency Measures. In *Proceedings of the 2005 American Council for an Energy-Efficient Economy Summer Study on Energy Efficiency in Industry*; ACEEE: Washington, DC, USA, 2005; pp. 103–114.

62. Sorrell, S. The economics of energy service contracts. *Energy Policy* **2007**, *35*, 507–521. [CrossRef]

63. Benedetti, M.; Cesarotti, V.; Holgado, M.; Introna, V.; Macchi, M. A proposal for energy services' classification including a product service systems perspective. In *Proceedings of the Procedia CIRP*; Elsevier: Amsterdam, The Netherlands, 2015; Volume 30, pp. 251–256.

64. Antunes, P.; Carreira, P.; Mira da Silva, M. Towards an energy management maturity model. *Energy Policy* **2014**, *73*, 803–814. [CrossRef]

65. O'Sullivan, J. Energy Management Maturity Model (EM3)—A Strategy to Maximize the Potential for Energy Savings through EnMS. 2012.

66. Finnerty, N.; Sterling, R.; Coakley, D.; Keane, M.M. An energy management maturity model for multi-site industrial organisations with a global presence. *J. Clean. Prod.* **2017**, *167*, 1232–1250. [CrossRef]

67. Prashar, A. Energy efficiency maturity (EEM) assessment framework for energy-intensive SMEs: Proposal and evaluation. *J. Clean. Prod.* **2017**, *166*, 1187–1201. [CrossRef]

68. Ashford, C.J. Energy Efficiency: A Managed Resource. *Facilities* **1993**, *11*, 24–27. [CrossRef]

69. ISO 50001:2011(en), Energy Management Systems—Requirements with Guidance for Use. Available online: https://www.iso.org/obp/ui/#iso:std:iso:50001:ed-1:v1:en (accessed on 11 April 2020).

70. Mao, M.; Jin, P.; Hatziargyriou, N.D.; Chang, L. Multiagent-based hybrid energy management system for microgrids. *IEEE Trans. Sustain. Energy* **2014**, *5*, 938–946. [CrossRef]

71. Blaauwbroek, N.; Nguyen, P.H.; Konsman, M.J.; Shi, H.; Kamphuis, R.I.G.; Kling, W.L. Decentralized Resource Allocation and Load Scheduling for Multicommodity Smart Energy Systems. *IEEE Trans. Sustain. Energy* **2015**, *6*, 1506–1514. [CrossRef]

72. Jokela, T.; Siponen, M.; Hirasawa, N.; Earthy, J. A survey of usability capability maturity models: Implications for practice and research. *Behav. Inf. Technol.* **2006**, *25*, 263–282. [CrossRef]

73. Krishnan, M.S.; Kriebel, C.H.; Kekre, S.; Mukhopadhyay, T. An Empirical Analysis of Productivity and Quality in Software Products. *Manag. Sci.* **2000**, *46*, 745–759. [CrossRef]

74. Fawcett, T.; Hampton, S. Why & how energy efficiency policy should address SMEs. *Energy Policy* **2020**, *140*, 111337.

75. Skumatz, L.A.; Dickerson, C.A. Extra! Extra! Non-Energy Benefits Swamp Load Impacts For PG&E Program! In Proceedings of the 1998 ACEEE Summer Study, Panel 8, Pacific Grove, CA, USA, 25–31 August 1998; pp. 301–312.

76. Skumatz, L.A.; Gardner, J. Methods and results for measuring non-energy benefits in the commercial and industrial sectors. In Proceedings of the Proceedings ACEEE Summer Study on Energy Efficiency in Industry, Washington, DC, USA, 2005; pp. 163–176.

77. Mills, E.; Rosenfeld, A. Consumer non-energy benefits as a motivation for making energy-efficiency improvements. *Energy* **1996**, *21*, 707–720. [CrossRef]

78. Gilchrist, A. *Industrial Internet Use-Cases*; Apress: Berkeley, CA, USA, 2016.

79. Tortorella, G.L.; Fettermann, D. Implementation of industry 4.0 and lean production in brazilian manufacturing companies. *Int. J. Prod. Res.* **2018**, *56*, 2975–2987. [CrossRef]

80. Fleiter, T.; Worrell, E.; Eichhammer, W. Barriers to energy efficiency in industrial bottom-up energy demand models—A review. *Renew. Sustain. Energy Rev.* **2011**, *15*, 3099–3111. [CrossRef]

81. Walsh, C.; Thornley, P. Barriers to improving energy efficiency within the process industries with a focus on low grade heat utilisation. *J. Clean. Prod.* **2012**, *23*, 138–146. [CrossRef]

82. De Groot, H.L.F.; Verhoef, E.T.; Nijkamp, P. Energy saving by firms: Decision-making, barriers and policies. *Energy Econ.* **2001**, *23*, 717–740. [CrossRef]

83. Harris, J.; Anderson, J.; Shafron, W. Investment in energy efficiency: A survey of Australian firms. *Energy Policy* **2000**, *28*, 867–876. [CrossRef]

84. Hossain, S.R.; Ahmed, I.; Azad, F.S.; Monjurul Hasan, A.S.M. Empirical investigation of energy management practices in cement industries of Bangladesh. *Energy* **2020**, *212*, 118741. [CrossRef]

85. Palm, J.; Thollander, P. An interdisciplinary perspective on industrial energy efficiency. *Appl. Energy* **2010**, *87*, 3255–3261. [CrossRef]

86. Nesticò, A.; Somma, P. Comparative Analysis of Multi-Criteria Methods for the Enhancement of Historical Buildings. *Sustainability* **2019**, *11*, 4526. [CrossRef]

87. Thiede, S. *Energy Efficiency in Manufacturing Systems*; Sustainable Production, Life Cycle Engineering and Management; Springer: Berlin/Heidelberg, Germany, 2012; ISBN 978-3-642-25913-5.

88. André, J.-C. *Industry 4.0: Paradoxes and Conflicts*; Wiley-ISTE: Hoboken, NJ, USA, 2019; ISBN 978-1-786-30482-7.

89. Backlund, S.; Thollander, P.; Palm, J.; Ottosson, M. Extending the energy efficiency gap. *Energy Policy* **2012**, *51*, 392–396. [CrossRef]

90. Stephan, A.; Roosa, W.C.T.; Doty, S. *Energy Management Handbook*, 9th ed.; Fairmont Press: Lilburn, GA, USA, 2018; ISBN 9781138666979.

91. Ghobakhloo, M. Industry 4.0, digitization, and opportunities for sustainability. *J. Clean. Prod.* **2020**, *252*, 119869. [CrossRef]

Publisher's Note: MDPI stays neutral with regard to jurisdictional claims in published maps and institutional affiliations.

MDPI

St. Alban-Anlage 66

4052 Basel

Switzerland

Tel. +41 61 683 77 34

Fax +41 61 302 89 18

www.mdpi.com

Energies Editorial Office

E-mail: energies@mdpi.com

www.mdpi.com/journal/energies